KU-158-883

OIL SPILL RESPONSE
in the
MARINE ENVIRONMENT

OIL SPILL RESPONSE
in the
MARINE ENVIRONMENT

J. W. DOERFFER
Ship Research Institute,
Technical University of Gdansk, Poland.

PERGAMON PRESS
OXFORD · NEW YORK · SEOUL · TOKYO

U.K. Pergamon Press Ltd, Headington Hill Hall, Oxford OX3 0BW, England

U.S.A. Pergamon Press, Inc., 660 White Plains Road, Tarrytown, New York 10591-5153, U.S.A.

KOREA Pergamon Press Korea, KPO Box 315, Seoul 110-603, Korea

JAPAN Pergamon Press Japan, Tsunashima Building Annex, 3-20-12 Yushima, Bunkyo-ku, Tokyo 113, Japan

First edition 1992

ISBN 0 08 041000 6

Cover illustration: Sinking oil tanker HAVEN on fire off Genoa, April 1991.

Frontispiece: Showing an example of the ecological disaster caused by the sinking of the HAVEN.

Printed in Great Britain by BPCC Wheatons Ltd, Exeter

Contents

Introduction

The rapid economic development of many countries since World War II has caused a considerable increase in marine transportation of raw materials, especially of crude oils and in offshore activities. However, a significant amount of oil comes into the sea from operational discharges of ships (ballast and bilge water) as well as from incidents such as collisions, groundings and contacts. Offshore exploration and exploitation of oil and gas is connected with the danger of blow-outs and major spills.

The first large oil spills were caused:

- in 1967 by grounding of the tanker "Torrey Canyon" (117,000 tons);
- in 1969 by a blow-out of the offshore platform "Santa Barbara" (13,600 tons).

These incidents have been the pivotal points in the approach of maritime nations to the protection of the marine environment through international legislation and implementation of rigorous requirements concerning the construction and exploitation of ships and offshore platforms, limiting the possibility and extent of oil spills. Regional and bilateral international agreements between neighbouring countries have been concluded, containing contingency plans, describing the means and methods of cooperation in case of a major spill.

A new danger for marine environments has ensued recently through the increasing maritime transport of chemicals, some of which are more persistent and toxic to the marine biota than oil. These problems are now being researched and new international requirements are being formulated.

General knowledge of marine environments, of oil and its behaviour when released onto the water surface, and of the methods and means of response to an oil spill is rather limited. The aim of this book is to introduce the reader to these problems, which have been presented in numerous publications issued up to 1988 and reflect the level and trends of development in this field. This book could be read by anyone interested in marine pollution response and especially by those who intend to become directly involved in combat teams.

As different standards and units are still used throughout the world, this book uses them in accordance with the original publications in order to make the reader fully conversant with them. It is hoped that in this way the book will better serve the purpose it is intended for.

The author wishes to thank all the friends who helped in making numerous publications available, for their kind advice and ecouragement to have this book written and published. Protection of the environment is at last becoming one of the main topics for future economic development of many regions in the world. Too much damage has been done until now and preservation of nature is the principal task of the present generation.

CHAPTER 1

Types and Characteristics of Oil

1.1 Oil Characteristics

1.1.1 Petroleum Hydrocarbons

The International Convention for the Prevention of Pollution from Ships, 1973 (MARPOL 73/78) defines oil in Annex 1 as follows: "Oil means petroleum in any form including crude oil, fuel oil, sludge, oil refuse and refined products (other than petrochemicals, which are subject to the provisions of Annex II of the present Convention) and, without limiting the generality of the foregoing, includes the substances listed in Appendix I to Annex I of the 1973 Convention".

Of special interest for the purpose of combating oil pollution at sea are crude oils and their refined products. Crude oils are complex mixtures of hydrocarbons of varying molecular weights and structures ranging from a light gas (methane) to heavy solids. Hydrogen and carbon are the most important and prevalent elements, comprising up to 98% of some crude oils and 100% of many refined products. Petroleum hydrocarbons are grouped into four basic classes according to the structural arrangement of the hydrogen and carbon atoms:

- alkanes, characterised by branched or unbranched chains of carbon atoms with attached hydrogen atoms (fig.1.1a). Alkanes have a general formula C_nH_{2n+2} and contain no carbon-carbon double bonds, i.e. they are saturated. Alkanes are also called paraffins and they are the major constituents of natural gas and petroleum. Alkanes containing less than

5 carbon atoms per molecule (n is less than 5) are usually gases at room temperature (methane) and those having between 5 and 15 carbon atoms are usually liquids, and straight chain alkanes having more than 15 carbon atoms are solids. Low carbon number alkanes produce anaesthesia and narcosis (stupor, reduced activity) at low concentrations and at high

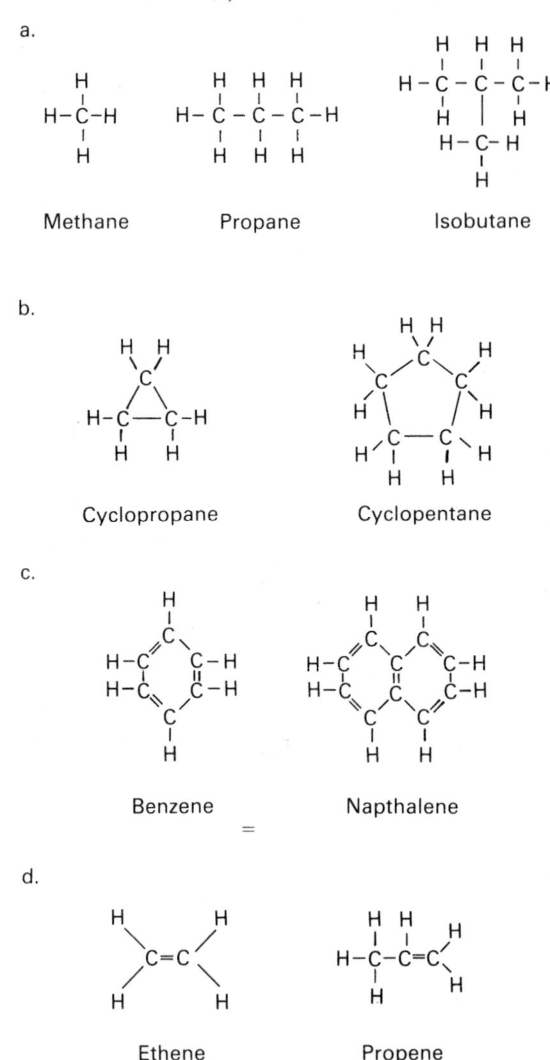

FIG.1.1 Structures of typical hydrocarbons: (a) alkanes; (b) naphthenes; (c) aromatics; (d) alkenes.

concentrations can cause cell damage and death in a variety of organisms. Higher carbon alkanes are not generally toxic,but may interfere with normal metabolic processes and communication in some species;

- naphthenes, characterised by the presence of simple closed rings of carbon atoms (fig.1.1b) have the general formula C_nH_{2n} and contain no carbon-carbon double bonds, i.e. they are saturated. They are insoluble in water and generally boil at temperatures 10 to 20 °C higher than the corresponding carbon number alkanes;
- aromatics, characterised by rings containing 6 carbon atoms (fig.1.1c). Benzene is the simplest aromatic and most aromatics are derived from this compound. They are the most toxic of the hydrocarbons found in oil and they are present in virtually all crude oils and petroleum products. Many aromatics are soluble in water to some extent, thereby increasing their danger to aquatic organisms. Certain aromatics are considered long-term poisons and often produce carcinogenic effects;
- alkenes, characterised by the presence of carbon atoms united by double bonds (fig.1.1d). Alkenes are also called olefins and have the general formula C_nH_{2n}. Alkenes containing 2 to 4 carbon atoms are gases at room temperature, while these containing 5 or more carbon atoms are usually liquids. Alkenes are not found in crude oils, but are often formed in large quantities during the cracking of crude oils and are common in many refined products such as cracking. They are generally more toxic than alkanes, but less toxic than aromatics.

The above classes of hydrocarbons occur in different proportions in various crude oils and refined products and the relative proportions present determine their physical and chemical properties. Physical-chemical properties of some hydrocarbon groups are given in table 1.1.

1.1.2 Crude Oils

In addition to hydrocarbons, crude oils also contain other com-

TABLE 1.1 Typical physical-chemical properties for hydrocarbon groups

Hydrocarbon	M Molecular weight (g/mol)	S Solubility (g/m^3)	P Vapoour pressure (Pa)	Density (kg/m^3)	K Oil-water partition coeff.	So Hydrocar. solubil. (g/m^3)
Lower alkanes (C_3 to C_7)	72	40	70,000	800	20,000	6
Higher alkanes (>C_8)	120	0.8	2,000	800	1,000,000	0.08
Benzenes	100	200	1,500	800	4,000	20
Naphthenes	160	20	5	800	40,000	0.4
Higher PNAs	200	0.1	.003	800	8,000,000	0.002
Residue	—	0	0	800	∞	0
Dispersant	300	100,000	0	1000	10	—

ponents, both organic and inorganic, which differ from one crude oil to another. Elements such as nitrogen, oxygen, sulphur, vanadium, sodium, nickel, iron, etc. have been detected. The number and form of the individual components in the crude oil give the oil different physico-chemical properties (table 1.2).

With respect to combating oil spills, the classification of crude oils based on expected spreading behaviour and evaporation is of special interest. Crude oils with a pour point above 5 to 10 °C will probably solidify very rapidly under ambient conditions with a seawater temperature of about 10 °C. These oils are therefore listed in Group 1 (Table 1.3). For the majority of oils with pour points below 5 to 10 °C the spreading behaviour will be comparable shortly after spillage and therefore further classification is based on predicted evaporative loss. Further classification into a group of oils which are expected to solidify during weathering (pour point of 200 °C+ fraction above 5 to 10 °C) and those which are not is very useful. This classification assists selection of the best clean-up methods and equipment to be used.

Variations in the basic characteristics of crude oil assist in tracing the source of an oil spillage, but oil spilled at sea spreads quickly and alters its properties due to "weathering", which is the combined effect of evaporation, oxidation and possible biodegradation.

The change in physical properties (notably density) will be observed as a shift to a heavier and more viscous oil and an in-

TABLE 1.2 Specification of crude oils

Category	Country	Type	Specific Gravity kg/l	Viscosity cS at 38 °C	Pour point °C	Pour point for residue 200 °C+ °C
1 High wax	Gabon	Gamba	0.872	28.5	30	not relevant
content	Libya	Es sider	0.841	5.7	9	not relevant
	Libya	Libyan h. pour	0.846	12.7	21	not relevant
	Libya	Sarir	0.847	11.9	24	not relevant
	Nigeria	Nigerian light	0.844	3.59	21	not relevant
	Egypt	El Morgan	0.847	13.0	13	not relevant
	UK	Argyll	0.833	3.2	9	36[b]
	UK	Auk	0.837	5.7	12	35[b]
	UK	Beatrice	0.835	8.15	27	42[b]
	UK	Dunlin	0.850	5.3	3	30[b]
2 Moderate	Qatar	Qatar	0.814	2.55	-18	5/10
wax content	Qatar	Qatar mar.	0.839	4.1	-12	5
	USSR	Muhanovo	0.835	4.18	0	13
	USSR	Romashkin.	0.859	6.9	- 4	5/7
	Algeria	Zarzaitine	0.816	4.56	-15	5
	Libya	Brega	0.824	3.6	-18	7
	Libya	Zueitina	0.808	2.9	-12	10
	Iran	Iranian light	0.854	6.6	- 4	10
	Iran	Iranian heavy	0.869	10.2	- 7	7
	Iraq	Northern Iraq	0.845	4.61	-15	10
	Abu Dh.	Abu Dhabi	0.830	3.42	-18	5/7
	Abu Dh.	Abu Dhabi-Z.	0.825	2.9	-15	7
	Abu Dh.	Abu Dhabi-U.	0.840	3.8	-15	5/10
	Norway	Ekofisk	0.808	2.4	-12	30[a]
	UK	Andrew	0.827	3.3	-12	33[b]
	UK	Brent	0.833	4.6	- 6	36[b]
	UK	Magnus	0.828	3.1	- 3	30[b]
	UK	Forties	0.842	4.43	-13	33[b]
3 Low wax	Algeria	Hassi Messaoud	0.802	1.95	<-30	<5
content	Algeria	Arzew	0.809	2.4	<-30	<5
	Nigeria	Nigerian med.	0.907	14.1	<-30	<5
	Nigeria	Nigerian export	0.872	5.8	<-30	<5
	Kuwait	Kuwait	0.869	10.6	-17	<5
	Saudi Ar.	Arabian light	0.851	5.45	<-30	<5
	Saudi Ar.	Arabian med.	0.874	9.7	-15	<5
	Saudi Ar.	Arabian heavy	0.887	19.1	<-30	<5
	Neutr. Z.	Kafji	0.888	18.1	<-30	<5
	Iraq	Southern Iraq	0.847	5.76	-13	<5
	Oman	Oman	0.861	8.7	-27	<5
	Venezuela	Tia Juana med.	0.900	16.8	<-30	<5
4 Very low	Venezuela	Bacchaquero	0.978	1280	-15/7	–
wax highly	Venezuela	Tia Juana Pos.	0.980	2983	-3	–
viscous						

(a) Sbm=Single buoy mooring; (b) Residue 342 °C+(648 °F+).

creased tendency to form stable emulsions. Higher concentrations of paraffin wax and asphaltenes contributes to higher emulsion stability.

TABLE 1.3 Classification of oils based on spreading behaviour and evaporation

Grup 1 Pour point>5-10 °C			
Agryll Auk Brega Cabinda	Cormorant Dunlin Gamba Lucina	Nigerian Light Ninian Sarir Schoonebeek	Soyo Suez Mix Thistle Zueitina
Group 2 Evaporative loss 0-20%(vol)	**Group 3** Evaporative loss 20-40%(vol)	**Group 4** Evaporative loss 40-50%(vol)	**Group 5** Evaporative loss >50%(vol)
PP 200+>5-10 °C	PP 200+<5-10 °C	PP 200+<5-10 °C	PP 200+<5- 10 °C
Lagunillas Tia Juana Pesado West Nederlands	Arabian Heavy Champion Export Khafji Nigerian Medium Santa Maria PP 200+>5-10 °C Maya	Arabian Light Arabian Medium Basrah Light Danmark Dubai Iranian Light Kuwait Nigerian Export Blend Oman Ural PP 200+>5-10 °C Buchan Es Sider Flotta Forties Iranian Heavy Nigerian Light Gulf	Zakum PP 200+>5-10 °C Abu Dhabi Berri Beryl Brass River Brent Spar Ekofisk Kirkuk Kole Marine Montrose Murban Murchison Nigerian Light Mobil Qatar Marine Saharan Blend Sirtica Statfjord

The most common analytical techniques used in spillage source identification are:

- high temperature gas or gas/liquid chromatography. This gives the exact hydrocarbon distribution, both in weathered and unweathered crude oil;
- atomic absorption (emission spectroscopy), mainly used to determine heavy metals like vanadium, nickel, etc.
- mass spectrometry, usually in tandem with gas chromatography;
- infrared spectroscopy for identifying the C-C (or other atoms)

bonds present;

- X-ray fluorescence. The sample is excited with X-rays and the intensity of emitted radiation is measured. Sulphur and heavy metals show a distinct response to this radiation.

1.1.3 Refined Oil Products

Refined products are obtained from crude oils using processes such as catalytic cracking and fractional distillation. Their physical and chemical characteristics differ depending on the source of the crude oil and the various processes to which they have been subjected. A graphical illustration of types of refined products which may be obtained from a whole crude oil at different temperatures is shown in fig.1.2. The light distillates and products do not include the nondistillable residuum, which often contains wax and asphaltic and polar materials (Table 1.3).

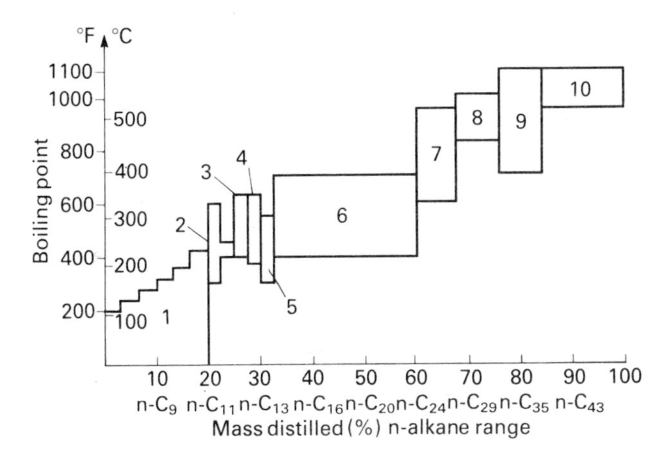

FIG.1.2 Distilled mass distribution (%) and alkane range versus boiling point for most crude oils: (1) gasoline fraction (2) kerosine; (3) heating oil; (4) diesel oil; (5) jet fuel; (6) light gas oil; (7) light lubricating oils; (8) heavy lubricating oils; (9) gas oils; (10) residuum.

1.1.4 Physical and Chemical Characteristics of Oils

The physical and chemical characteristics of oil which affect its behaviour on water and the efficiency of clean-up operations include:

TABLE 1.4 Physical characteristics of some crude oils and refined products

	Specific Gravity (15 °C)	API Gravity (15 °C)	Viscosity cS (38 °C)	Pour Point °C	Flash Point °C	Initial Boiling Point °C
Crude Oils	0.8 to 0.95	5 to 40	20 to 1,000	-35 to 10	Variable	30 to 500
Gasolines	0.65 to 0.75	60	4 to 10	n/a	-40	30 to 200
Kerosene	0.8	50	1.5	n/a	55	160 to 290
Jet Fuel	0.8	48	1.5	-40	55	160 to 290
No.2 Fuel Oil (diesel,stove)	0.85	30	15	-20	55	180 to 360
No.4 Fuel Oil (Plant Heat.)	0.9	25	50	-10	60	180 to 360
No.5 Fuel Oil (Bunker B)	0.95	12	100	-5	65	180 to 360
No.6 Fuel Oil (Bunker C)	0.98	10	300 to 3,000	+2	80	180 to 500

– specific gravity;
– surface tension;
– viscosity;
– pour point;
– flash point;
– solubility in water;
– changes in the above parameters with time.

The above oil characteristics are measured at a standard or constant temperature and atmospheric pressure. However, the physical properties of an oil will vary greatly depending on local (ambient) environmental conditions and may deviate from values reported for standard conditions.

1.1.5 Specific gravity

Specific gravity is the ratio of the mass of a substance to the mass of an equivalent volume of water. The specific gravity of an oil is a measure of its density. All oils have a specific gravity less than 1.00 with the exception of some heavy crudes and residual fuel oils (fig.1.3). After considerable "weathering", i.e. after evaporating volatile fractions some oils reach specific gravities greater than 1.00 and such oils will sink below the water surface. The API gravity scale has been developed by the American Petroleum Institute. It expresses the ratio of weights of equal volumes of oil and pure water at an ambient temperature of 16

°C and one atmosphere pressure. Water with a specific gravity of

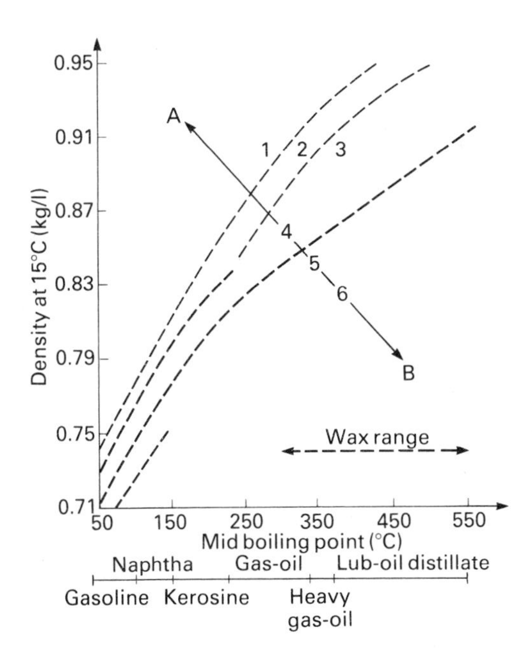

FIG.1.3 Comparison of crude oils on density/midboiling point basis; (A) naphthenic and low wax; (B) paraffinic and high wax; (1) heavy naphthenic - low wax (Venezuela); (2)medium naphthenic - low wax; (3) intermediate (Middle East); (4) naphthenic and waxy (Nigeria); (5) intermediate and waxy (U.K.); (6) highly waxy and paraffinic (Libya and Brazil).

1.00 has an API gravity of 10 °. Oils of progressively lower specific gravity have higher values of API gravity. Crude oils have an API gravity of approximately 50 ° to 300 °, whereas gasolines have an API gravity of approximately 60 °. The API gravity can be calculated from the specific gravity using the formula:

$$API^{\circ} = \frac{141.5}{spec.grav.at 16\ ^{\circ}\mathrm{C}} - 131.5 \qquad (1.1)$$

Thus oils with low specific gravities have high API gravities and low viscosities, low adhesion properties and high emulsification tendencies, whereas oils with high specific gravities have low API values, high viscosities, high adhesion properties and low emulsification tendencies. The API value is used commercially.

High API degree oils contain a larger proportion of the more profitable and easily extractable gasoline hydrocarbons and they fetch higher price.

1.1.6 Surface Tension

Surface tension is the force of attraction between the surface molecules of a liquid. This force and the viscosity determine the rate of spread over the surface of water or land, or into the ground. Oils with low specific gravities such as light crudes and lighter fuel oils generally have a greater potential spreading rate. Surface tension decreases with increasing temperature and increases the rate of spreading of a spill.

1.1.7 Viscosity

Viscosity is the property of a fluid (gas or liquid), by which it resists a change in shape or movement. The lower the viscosity, the more easy it flows. Viscosity changes with temperature and the lower the temperature, the higher the viscosity (fig.1.4). The viscosity of crude oil depends on its content of light fractions and the ambient temperature. As oil weathers, its viscosity increases due to the progressive loss of low molecular weight, volatile fractions (Table 1.5).

The viscosity of spilled oil influences the rate of spreading of the spill, the adhesion abilities of the oil, its penetration into the soil and beach sediments as well as the ability of the pumps used in a clean-up operation to remove oil from the surface.

TABLE 1.5 Change of viscosity with loss of light fractions.

Evaporative loss (% weight)	residual oil viscosity (cS at 10 °C)	Approximate time scale
0	23	zero
15	86	1 hour
20	197	2 hours
27	1023	4 hours
33	2650	1 week

1.1.8 Pour Point

The pour point of an oil is the temperature at which it becomes semi-solid or plastic and will not flow. This is the result of

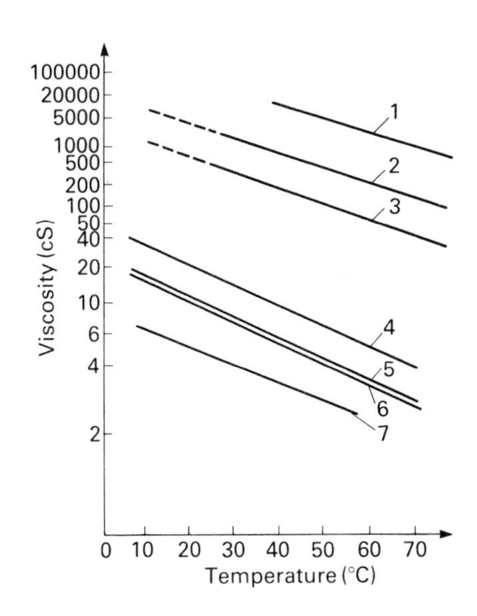

FIG.1.4 Temperature/viscosity chart: (1) Boscan; (2) HFO no.2; (3) HFO no.1; (4) Kuwait; (5) Arabian light; (6) gas oil (ASTM no.2); (7) Libya Brega.

the formation of an internal microcrystalline structure, which overrides the effects of viscosity and surface tension. The pour point of crude oils varies from −57 °C to +32 °C. Lighter oils with low viscosities have lower pour points. Gasoline spilled on seawater will remain a liquid, but a heavy bunker oil may become a solid when its temperature is lowered to the temperature of seawater. This characteristic is very important with respect to combating spills and the damage done to beaches. Free flowing oils rapidly penetrate beach substrates, while semi-solid oils are deposited on the surface and penetrate the beach substrate, if it is very coarse or during warmer weather conditions in the spring and summer.

1.1.9 Flash Point

The flash point of an oil is the temperature at which its vapours will ignite when exposed to an ignition source such as a naked flame. Light oil and many freshly spilled crude oils can ignite under most ambient conditions. When lighter components have dispersed or evaporated, the flash point drops and they become less dangerous in clean-up operations. Bunker and heavy fuel oils do not cause serious fire hazards when spilled on a water surface.

1.1.10 Solubility

Solubility is a process by which a substance (solute) will dissolve in another substance (solvent). The solubility of oil in water is extremely low (generally less than 5 p.p.m.). This process is very important in relation to the toxicity of hydrocarbons to aquatic organisms because certain "slightly" soluble hydrocarbons and various mineral salts present in oil are dissolved in the surrounding water.

1.2 Behaviour of Oil in Marine Environments

1.2.1 Marine Environments

For anyone involved in combating pollution in marine environments, it is essential to have some knowledge of these environments, of typical processes taking place in the oceans and of the interactions with shores. These processes will get more complicated when oil is spilled in large quantities over the surface and may be more seriously disturbed by inappropriate human action than if Nature was left to deal with the oil spill. Seawater is a complex solution of dissolved minerals, elements and salts. As water (H_2O) is a compound of hydrogen and oxygen, these two are the most abundant elements. Sodium chloride (NaCl) forms the majority of dissolved salts, with magnesium, calcium and potassium chlorides and carbonates forming the rest.
Seawater, being a mixture of many different types of salts, has virtually the same composition wherever and whenever readings are taken. In mid-oceans the ratio between these various salts is remarkably constant, whereas in inshore waters this ratio may vary considerably due to the input of water by rivers.
The density of pure water at 4 °C is 1,000 g cm^{-3} or 1,000 kg m^{-3}. If it is heated or cooled, the density of water decreases.

Saline water has a higher density and lower freezing point than pure water, depending on the content of salts. Wind action produces a mixed layer on the surface to a depth of 50 to 200 m which is almost isothermal in the vertical. A zone below this extends for another 500 to 1,000 m over which the temperature decreases rapidly. Below this zone there is a deep region of cold water, in which the temperature decreases very slowly.

The salinity distribution at the surface of the world's oceans is completely controlled by rainfall patterns. Regions of high rainfall and low evaporation (tropical regions) have low salinity due to rainfall dilution of the seawater. Regions of low rainfall and considerable evaporation (subtropical regions between 20° and 40° latitude) have high surface salinity. Vertical salinity distribution is related to the density of the water. Less dense water lies above more dense water. A salinity increase of 1‰ produces the same density change as a 4 °C decrease in temperature.

Breaking waves at the sea surface aerate the water and dissolve atmospheric gases (O_2 - 21%; N_2 - 78%; CO_2 - 0.03%). The solubility of oxygen (O_2), which supports life in aquatic systems depends both on the salinity and the temperature. The normal range is from 7×10^{-6} to 14×10^{-6} mg of O_2 per kg of water with hot water holding less oxygen. Water can be supersaturated due to vigorous stirring or as a result of plant photosynthesis, which converts carbon dioxide into oxygen. This process takes place in sunlight and its maximum in heavily vegetated waters is reached at noon. Low values of dissolved oxygen (below 4 p.p.m.) indicate a rapid multiplication of bacteria in salt water, which depletes the water of oxygen faster than it can be replaced by either the plants or the atmosphere. Such water is polluted and can not support aquatic life.

Dissolved nitrogen (N_2) in seawater is not altered by biological changes, except under unusual conditions. When saturation reaches 127%, fish appear to suffer from nitrogen-induced disease.

Carbon dioxide (CO_2) has a very complicated chemistry which favours the formation of bicarbonates. The standard measure of gaseous carbon in natural water is the pH. Pure water has a pH value of 7. Values of pH below 7 denote acids and the lower the value the stronger the acid. Values of pH above 7 correspond to alkaline solutions. The pH of surface water is very stable, usually ranging between 8.1 and 8.3 with a direct relation between salinity and pH. This implies a dominance of bicarbonate in seawater. Plants utilise carbon dioxide and raise the pH, while the respiration of organisms acts in the opposite direction.

Life depends on photosynthesis and the availability of carbon, oxygen, nitrogen and phosphorus (nutrients) as well as water. In the oceans these constituent elements are available in solution as dissolved bicarbonate, phosphate and nitrate. The relative proportions of nutrients used in photosynthesis are characterised by a chemical equation of the form:

$$106CO_2+90H_2O+16NO_3+PO_4+\text{energy} \rightarrow 154O_2+\text{protoplasm}$$

Thus incoming light energy of 5.40×10^9J produces 3.258 kg of protoplasm in the proportions 106 C, 180 H, 46 O, 16 N and 1 P. When this protoplasm is burnt, it releases 5.4×10^7J of heat energy.

Increased urbanisation and widespread overapplication of fertilizers has led to increased nutrient loads into the waterways and seas. This produces high nitrogen loads through urea based fertilizers and high phosphorus loads from superphosphate fertilizers.

The feeding relations of species in the aquatic community delineate an ecosystem, which is determined by the flow of energy and nutrient materials from the physical environment to plants. These form the first trophic level, they act as first food producers and from them to higher trophic levels as consumers (fig.1.5.).

The rate of primary supply depends on the supply of light and nutrients. The light energy necessary for primary production decreases with depth and is limited to surface waters. The nutrients are recycled from detritus and are most accessible near the bottom. The two regions overlap in high productivity areas. When sufficient nutrients are present, then the rate of phytoplankton photosynthesis is proportional to:

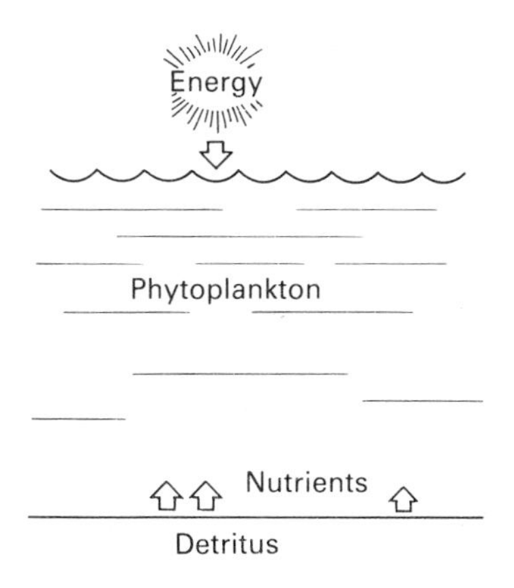

FIG.1.5 The trophic web.

- the relative photosynthesis rate;
- the concentration of phytoplankton present, expressed in terms of chlorophyll per m^3 of water in the water column;
- inversely proportional to the extinction coefficient of light.

In upwelling areas there is an abundance of nutrients for phytoplankton, so primary production is limited to availability of light and nitrogen. If there is sufficient nitrogen, then the growth is controlled by the availability of phosphorus. Thus the growth rate of an organism is limited by the availability of growth factors, that is the shortest in supply (Liebig's law of the minimum - fig.1.6.). In this example trace elements such as copper are in abundant supply, but the rate of primary production will follow the heavy line and be limited first by iron, then as more iron becomes available by nitrogen and finally phosphorus. Oil spilled on the water surface limits the supply of light and thus is very harmful to the production of phytoplankton.

1.2.2 Waves and mixing

The wind is responsible for generating waves. Wave motion is periodic i.e. it is repeated through fixed periods of time T. The horizontal distance between successive crests or troughs is called the wavelength.

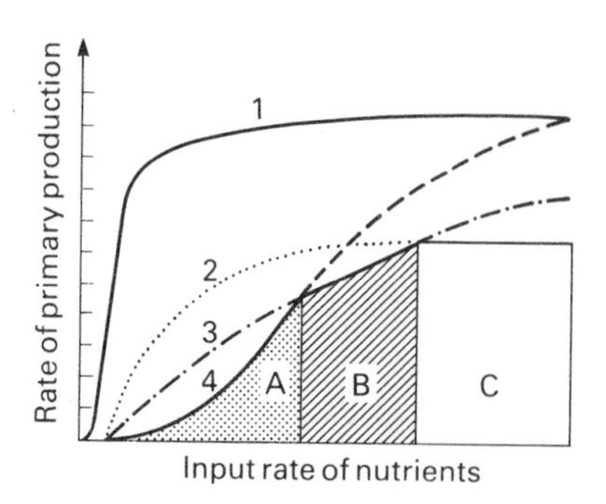

FIG.1.6 Potential hydrocarbon production areas: (1) copper; (2) phosphorus; (3) nitrgen; (4) iron; (A) zone of iron limitation; (B) zone of nitrogen limitation; (C) zone of phosphorus limitation.

As a wave travels through the water, the particles of water move up and down in a simple harmonic motion and they also move forwards and backwards in a simple harmonic motion. The energy in the wave is equally divided between potential energy, which is the energy arising from the displacement of water above its equilibrium position, and kinetic energy, which is the energy associated with particle movement. The wave carries its energy along until it is dissipated through friction - in which case the wave dies away and the water heats up very slightly - or through wave breaking. A breaking wave redistributes some of its energy into stirring up sediments, some into heat and some into sound. Turbulent flow is always present in rivers, oceans and coastal waters. It interchanges the fluid properties through their fluctuations about their mean values. Where the density of the water increases strongly with depth, the strong hydrostatic equilibrium introduces different eddy sizes in the horizontal and in the vertical. Fluids with marked density stratification, with light fluids on top of dense fluids, are very stable, but if there is relative motion between two layers, then the boundary between the two will deform. If the relative motion is strong, then the boundary may become unstable and turbulence ensues and mixing will occur. The process of vertical circulation being driven by vertical density differences is called convection, whereas horizontal circulation of water is called advection. Convection occurs when the air temperature drops below the temperature of the surface water. The surface water cools and sinks and is replaced by subsurface water, which also cools and sinks.

The wind acting on the surface of the sea stirs the upper layers of the ocean - the well-mixed layer. This layer entrains cold water from beneath the mixed layer, provided that the wind imparts sufficient energy. This is due to the phenomenon whereby a stratified fluid can become turbulent within a thin layer if there is a velocity shear across this layer. This phenomenon is called turbulent entrainment. If the wind is sufficiently energetic, it entrains some cold water and then mixes it throughout the well-mixed layer. This produces an upper layer of virtualy constant temperature, sharply separated from the lower regions in which the temperature decreases with depth. This separation region constitutes a thermocline. It progresses downwards till the energy from the wind is sufficient to entrain and to completely mix the lower waters.

As has been pointed out, many mixing processes depend on waves and wind. A universally adopted description of weather and sea conditions is the Beaufort Scale. Wind speeds can be measured accurately, whereas measuring the waves is very difficult. It should be also borne in mind that a lag effect exists between the wind rising and the sea increasing. The height of waves will depend on water depth, local currents, etc.

1.2.3 Weathering Processes

When crude oil or petroleum products are released into marine environments they are immediately subjected to a wide variety of weathering processes (fig.1.7) such as:

- spreading and drift;
- evaporation;
- dissolution and advection;
- dispersion of whole oil droplets into the water column;
- photochemical oxidation;
- water-in-oil emulsification;

- microbial degradation;
- adsorption onto suspended particulate material (SPM);
- ingestion by organisms;
- sinking and sedimentation.

The oil composition and the position of oil release greatly influence on the relative magnitudes and importance of each of the above named processes, as well as affecting the relative changes in oil composition and associated rheological properties due to physical/chemical weathering in the sea. Fundamental knowledge of these processes is therefore required to predict this behaviour. A number of these processes can be described quite satisfactorily, while others can only be described qualitatively as gaps in our knowledge still exist. A review of the various processes and identification of the relevant physico-chemical characteristics of crude oils and refined products which influence their behaviour when exposed to open sea conditions will be carried out. Of the refined products only the heavy fuel oils need to be considered in detail as most others will disappear very quickly. The boiling point and the nondistillable residuum (bp > 425 °C) are extremely important with respect to changes in rheological properties and the behaviour of oil during weathering at sea.

Light distillates and products do not contain a nondistillable residuum, which often contains wax and asphaltic and polar materials. Refined products do not form stable water-in-oil emulsions (or mousse) due to the absence of surface active materials, which are contained in the higher boiling residuum. Compounds with boiling points below that of n-C_{14} compounds (250 °C) are generally the only ones subject to partitioning into the atmosphere under normal, temperate marine conditions. Aromatic hydrocarbons in the same boiling point range (80 to 250 °C) are the only ones subject to significant dissolution. Thus certain refined products such as gasoline, light kerosines, aviation fuels, etc. may be subject to nearly complete evaporation (and limited dissolution) processes, whereas heavier distillates such as gas oils, lubricating oils and Bunker C residual fuel oil, etc. would not undergo significant weathering as a result of these processes.

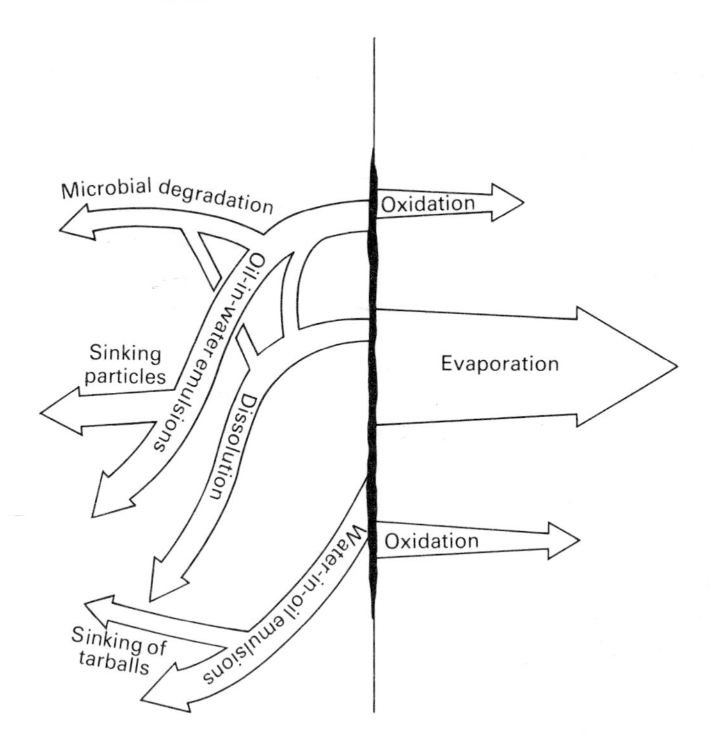

FIG.1.7 Gross weathering processes of oil spills on the seawater surface.

Practically all crude oils contain all or most of the components which constitute the refined products, but it should be noted that the aromatic fractions of crude and refined oils are generally characterised by a predominance of alkyl-substituted aromatics relative to the parent aromatic compounds and are therefore more toxic.

1.2.4 Spreading and Drifting

A major process, which affects the behaviour of crude oil and refined products during the first hours after release into the sea, provided that the pour point is lower than the ambient temperature, is the spreading. This process increases the overall surface area of the spill, thus enhancing mass transfer via evaporation and dissolution. In calm water this leads to a continuous decrease in the oil slick thickness in a circular pattern, which finally reaches a minimum value of $h_0 = 10^{-2}$ to $10^{-3} cm$.

To measure oil thickness, surface samples are collected with a 23 cm square sheet of polypropylene oil-sorbent pad, which is placed on the water surface within the slick. The pad sampling should be coordinated with IR surveillance from a remote sensing aircraft. The oil is immediately absorbed by the exposed area of the pad. After lifting from the surface, the pad is folded and placed in a precleaned 350 ml wide-necked glass bottle and transported immediately after the trials to a laboratory for extraction and analysis. The oil-coated sorbent pad is extracted with 300 ml of pentane and analysed photometrically. By comparing the response of oil on the pads with a standard calibration curve of the same oil, the surface oil thickness can be found.

Apart from evaporation and dissolution, surface active agents (surfactants) cause the formation of unstable or stable water-in-oil emulsions (chocolate mousse) with steadily increasing densities and viscosities and with a growing tendency to disperse. These processes are effectively accelerated by the influence of wind, waves and currents, which disrupt the oil slick into discontinuous stripes and patches. The principal forces influencing the lateral spreading of oil on a calm water surface are gravitational, inertial, frictional and those due to surface tension (fig.1.8). The spreading behaviour in open ocean situations is much more complicated due to competing sea state effects.

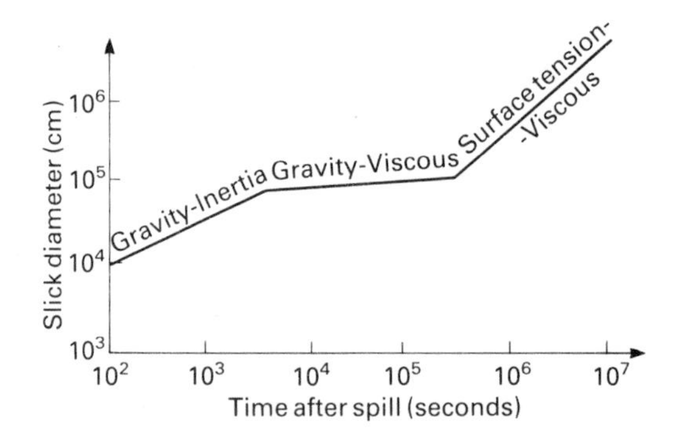

FIG.1.8 Three phases of oil slick spreading.

During the initial period of spreading, the volume of oil spilled has a direct impact on the rate of spreading, because the force of gravity plays a predominant role. Thus very large spills spread faster than small ones (fig.1.9), but the change from one spreading phase to another occurs after longer periods of time. When the major spreading force eventually becomes surface tension, the rate of spreading decreases, the volume of the spill has little effect on the spreading rate and the inner portion of the slick may be thicker than the edges.

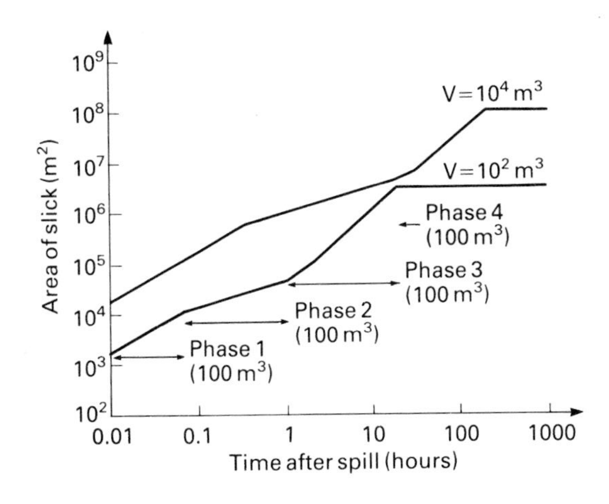

FIG.1.9 Influence of spill volume on spreading rate.

The surface tension or spreading coefficient is the difference between the air/water surface tension. Although density does have some effect on the rate of spreading, particularly shortly after spillage, many oils tend to spread on the water surface at about the same rate even though they may have different densities (fig.1.10) The initial difference between the area of a slick after spillage of a very light (0.700 kg/l) and heavy crude oil (0.980 kg/l) is at most a factor of 2, and as soon as 6 hours after a spill of that size the areas are equal, since from that point onwards spreading coefficient determines the rate of spreading. The spreading coefficient of different crude oils with spreading coefficients of 10 and 30 mN/m respectively and similar densities may differ by a factor of 3 (fig.1.11).

On clean water uninfluenced by wind or currents oil quickly spreads into a circular patch. 1 m^3 of Middle East crude oil can spread in 10 minutes to a circle of 48 m diameter with an average thickness of 0.5 mm and in 100 minutes this would grow to 100 m diameter with an average thickness of 100 μm. However, the spreading of oil does not always follow the above pattern due to formation of a much more viscous water-in-oil emulsion, which reduces the spreading rate of the oil spill. This happens when there is sufficient wave motion, when the oil is of a particular type and, most importantly, when the spill is large enough. In a concentrated oil spill the oil at the centre of the spill remains in a relatively thick layer long enough for agitation by wave motion to produce a water-in-oil emulsion. In this condition the oil is likely to float in large islands, which may be several centimetres thick, separated by clean water. Seawater temperature and salinity also influence spreading of the oil, but these are of minor importance except when they influence the pour point.

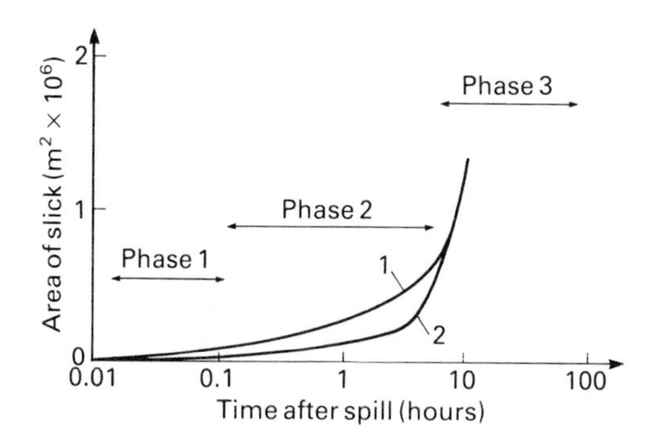

FIG.1.10 Influence of oil density on spreading rate for a spill of 10^3m^3 and spreading coefficient 25 mN: (1) density of oil 0.700 kg/l; (2) density of oil 0.980 kg/l.

A leakage from an oil pipe line on the sea floor causes oil to spread on the water surface. Oil may be released in two ways:

- oil is allowed to rise freely to the surface from the point source due to the difference in density between oil and water (fig.1.12);
- oil is forced out of the point source by positive pressure (fig.1.13).

From tests the following conclusions could be made:

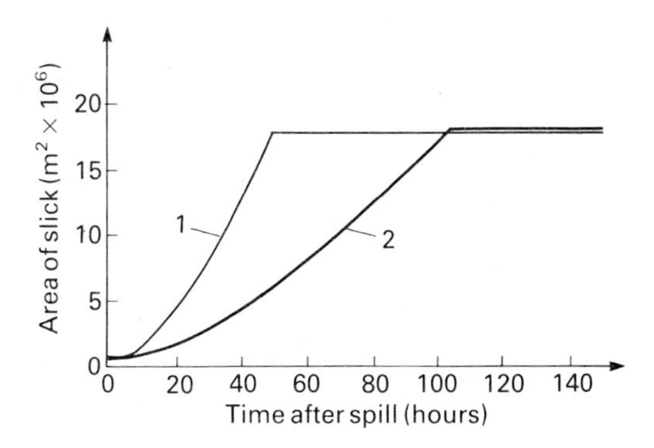

FIG.1.11 Influence of the spreading coefficient on the spreading of oils in a spill of $10^3 m^3$ and density 0.807 kg/l: (1) spreading coefficient 30 mN/m; (2) spreading coefficient 10 mN/m.

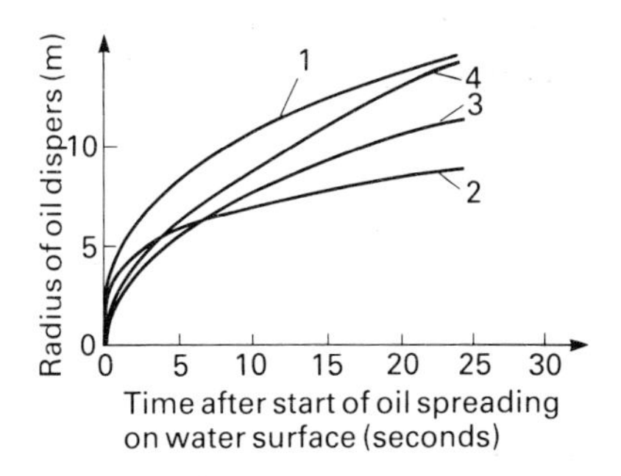

FIG.1.12 Comparison of dispersion of oil in still water and in waves (1) Kerosene in still water; (2) Kerosene in waves; (3) Equivalent C-oil in still water; (4) Equivalent C-oil in waves.

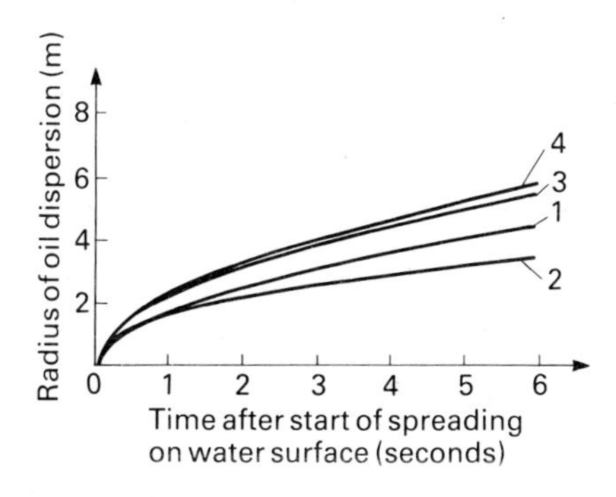

FIG.1.13 Comparison of oil dispersion with water surface conditions and oil leak pressure; (1) pressure 10 kg/cm^2, still water; (2) pressure 10 kg/cm^2, waves; (3) pressure 30 kg/cm^2, still water; (4) pressure 30 kg/cm^2, waves.

- early dispersion of a small volume of oil on the water surface can be expressed by the formula:

$$R = a \times t^b \tag{1.2}$$

where: R – the radius of the oil dispersion on the water surface;
t – the time after the start of dispersion on the water surface;
a and b – coefficients determined experimentally.

- the higher the oil pressure or the greater the volume of oil leak per unit time, the larger the spreading radius will become;
- the area of oil spreading is reduced in waves and the rate of reduction is larger for light oil;
- oil forced out of the leak outlet by high pressure is divided into small oil drops and when the pressure is increased the oil is emulsified.

Drift is characterised by the movement of the centre of mass of the spill and is independent of spreading and spill volume. The slick will move in the same direction and at the same speed as the surface water. Tidal currents move the surface water, but they are cyclic with only a small residual movement of surface water in any direction. The movement of a thin layer of oil is affected to a far greater extent by wind. The exact dependence is not known with high accuracy, but movement at a rate of approximately 3% of the wind speed in the direction of the wind is to be expected. As a very rough first approximation the velocity U_o of the centre of mass of the oil moves as a vector sum of the surface current U_c and a fraction ξ of the wind velocity W. Thus:

$$U_o = U_c + \xi \times W \tag{1.3}$$

The fraction is the wind factor, whereas the wind induced drift ($\xi \times W$) is called the leeway. Assume that the wind stress is continuous across the surface of the oil and that the drag coefficients on both media are identical. From the fluid dynamical stress equations:

$$\xi = \sqrt{\frac{\varrho_a}{\varrho_{oil}}} \tag{1.4}$$

where: ϱ_a = the density of air (1.3 kg/m^3)
ϱ_{oil} = the density of oil (from 700 to 980 kg/m^3) depending on its composition.

Thus the wind factor (about 4%) is high enough to dominate the motion of the slick.
Due to the earth's rotation the movement of the slick in the northern hemisphere is 20° to 30° to the right of the wind direction and in the southern hemisphere slightly to the left due to the Coriolis effect. When current and wind are significant, their speed component vectors should be added considering the direction of incurred forces. With very strong winds the slick may even move "up-current". With winds in excess of 16 km/h, causing rough waters, the oil may disperse or emulsify, thus making it harder to see and to remove. The slick breaks up into streaks and windrows. The presence of excessive surface debris will also affect the movement of an oil slick.
Generally there is a lack of information on the residual current, and local winds may differ considerably from those measured and predicted for the next forecast period. The best available data should be used to plot and estimate the movement of the slick. This information is of great importance for the entire combating strategy. Theoretical models based on actual, predicted or historical wind and current data never give accurate forecast of the real trajectory of spill movement or information as to if and when the spill will reach the coastline. Consequently it is extremely difficult to forecast patterns of slick movements and therefore there is no substitute for actual observation by aircraft, ship or satellite.

1.2.5 Evaporation

Evaporation is the most important weathering process during the first 24 to 48 hours of an oil spill with regard to mass transfer (molecular partitioning) and to removal of the more toxic, lower molecular weight components from the spilled oil. Evaporative loss is controlled by several factors, the most important of which are:

- the composition and physical properties of the oil;
- the slick surface area and its thickness;
- the wind velocity and sea state;
- air and sea temperature;
- intensity of solar radiation.

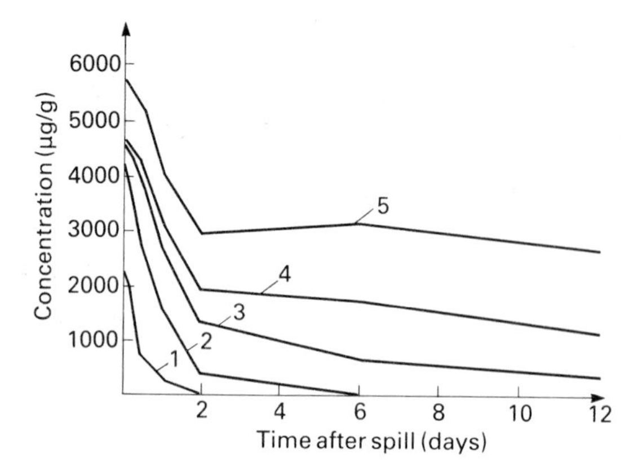

FIG.1.14 The effect of weathering time on individual compound concentrations in oil phase: (1) n-C_8; (2) n-C_9; (3) n-C_{10}; (4) n-C_{11}; (5) n-C_{12}.

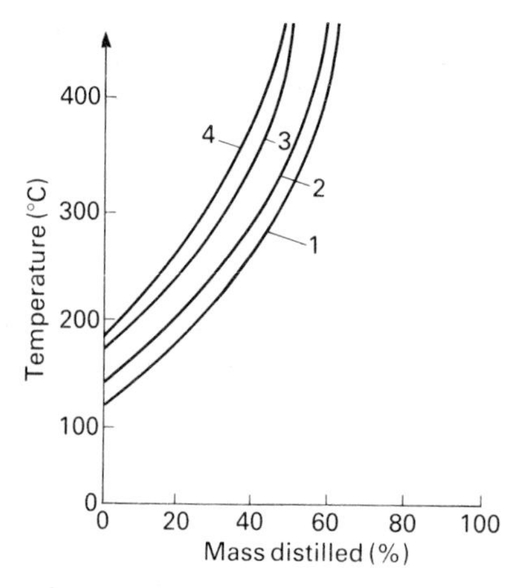

FIG.1.15 TBP destillate curves for various crude oils as effect of weathering: (1) 0 hours; (2) 12 hours (3) 144 hours; (4) 288 hours.

The rates and magnitudes of evaporative loss are extremely dependent on the composition of the oil and on the environmental conditions. Compound-specific losses are generally difficult to predict. It could be stated that compounds with vapour pressures greater than that of n-C_8 compounds will not persist in the slick, whereas compounds with vapour pressures more than those of n-C_{18} compounds do not evaporate appreciably under normal environmental conditions. Component-specific concentrations remaining in the slick (fig.1.14) show the loss of individual compounds due to evaporation and to a lesser extent due to dissolution. Compounds below n-C_{14} are readily removed, but these compound-specific data alone do not allow an overall accounting of the total mass balance of the slick. For this reason a distillation characterisation of weathered crude oil is required as shown on fig.1.15 The initial boiling point of weathered residual material increases with time and the overall loss due to evaporation processes of 15 to 20% from a slick over a 288 hour period can be derived. The relative nondistillable material has also increased from 34% to above 50%.

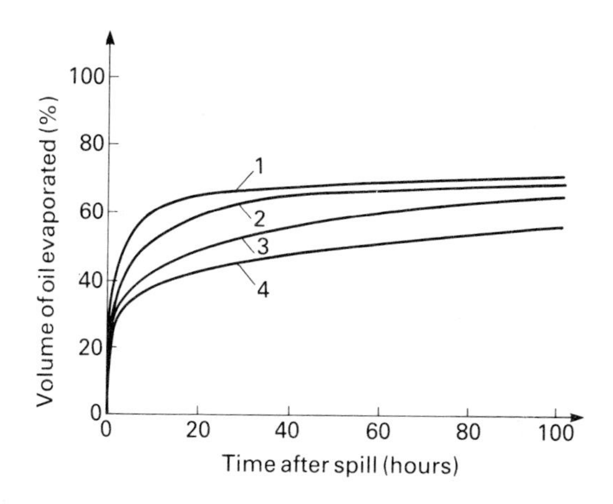

FIG.1.16 Relationship between time after spillage and percentage of oil evaporated for different spill volumes of Kuwait oil at wind speed 8 knots and a temperature of 10 °C: (1) 10 m^3; (2) 100 m^3; (3) 1000 m^3; (4) 10000 m^3.

The influence of spill volume on the relative rate of evaporation is shown in fig.1.16. The larger the spill volume, the smaller the evaporation rate, since the surface/volume ratio of the slick thickness increases. The effects of wind speed and ambient temperature on the rate of evaporation are shown in fig.1.17. The rate of evaporation increases with increasing wind speed and temperature.

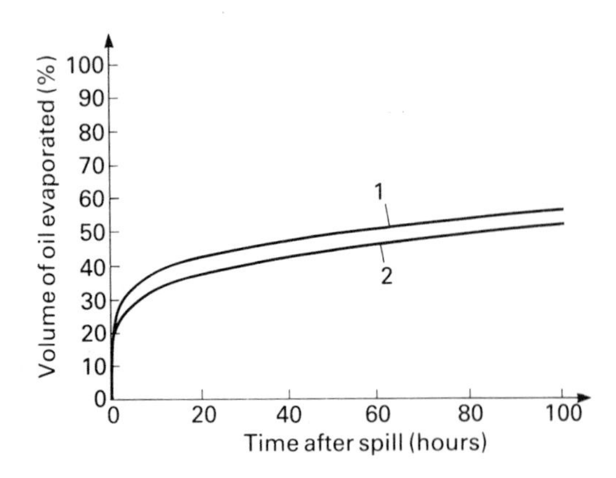

FIG.1.17 Relationship between time after spillage of 104 m^3 and percentage of oil evaporated at different wind speeds; Kuwait oil at temperature 10 °C: (1) wind speed 8 knots; (2) wind speed 2 knots.

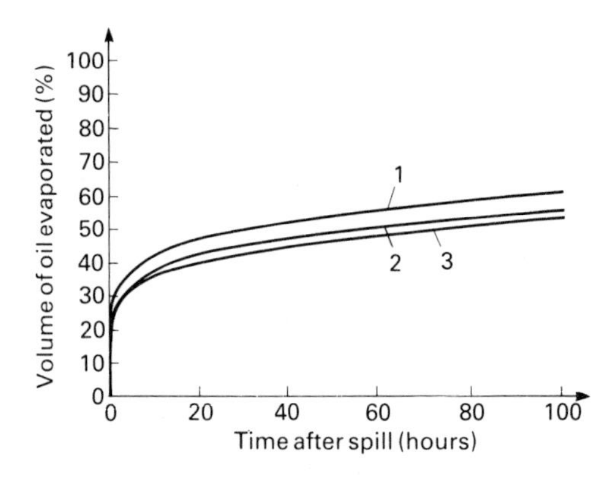

FIG.1.18 Relationship between time after spillage of 100 m^3 and percentage of oil evaporated at various temperatures of Kuwait oil at wind speed 8 knots: (1) temperature 25 °C; (2) temperature 10 °C; (3) temperature 3 °C.

The difference in rates of evaporation for some typical crude oils with different volatility characteristics are shown in fig.1.19. Thus it is advisable to classify different crude oils with respect to their evaporative behaviour after spillage at sea, based on the expected evaporative loss 24 hours after a spillage of 10,000 m^3 of crude oil under average conditions (wind speed 8 knots and temperature 10 °C - Table 1.6).

As a consequence of the removal of the lighter hydrocarbons through evaporation the remaining volume of oil decreases, but

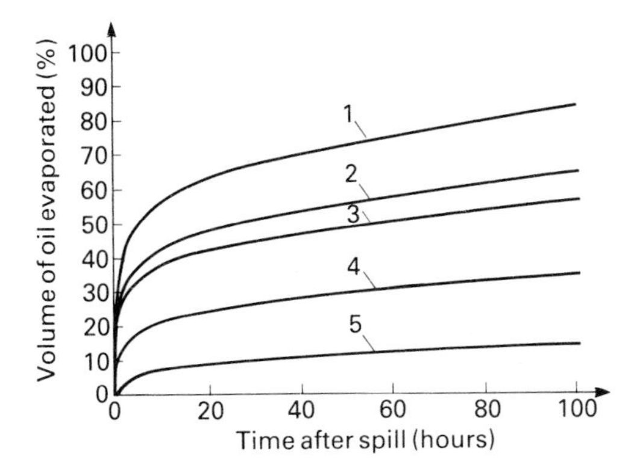

FIG.1.19 Relationship between time after spillage and percentage of oil evaporated for different crude oil sources under the following environmental conditions: volume 104 m^3; wind speed 8 knots; temperature 10 °C: (1) Ekofisk; (2) Forties; (3) Kuwait; (4) Gamba; (5) Tia Juana Pesado.

density and viscosity increase. Using the spreading and evaporation model to calculate the volume fraction evaporated under ambient conditions and to correlate evaporative loss with physico-chemical properties of various fractions obtained by distillation, it is possible to estimate the density and viscosity of the remaining oil after weathering at sea. However, in practice emulsification dramatically affects the viscosity.

TABLE 1.6 Classification of crude oils by expected evaporative loss under average meteorological conditions: wind speed 8 knots (4.17 m/s), ambient temperature 10 °C.

Group of crude oils	Evaporative loss in 24 hours (spill volume 10,000 m^3)
1	Pour point greater than sea water temperature
2	0 to 20%
3	20 to 40%
4	40 to 50%
5	> 50%

The relationship between the density of the oil remaining after distillation and the volume fraction distilled is shown in fig.1.20. It is a close to linear relationship, which shows in turn that the relationship between the density of oil remaining after evaporation at sea and the time after spillage is qualitatively comparable (fig.1.21). The densities of oils are temperature dependent and in most cases the reference temperature for the determination of density is 15 °C.

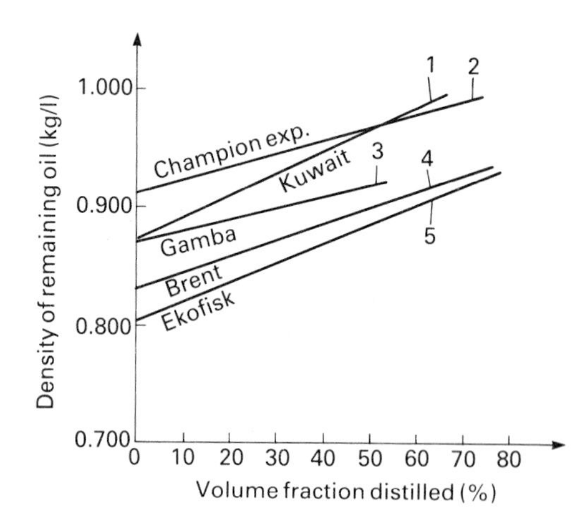

FIG.1.20 Relationship between volume fraction distilled and density of remaining oil for different crudes.

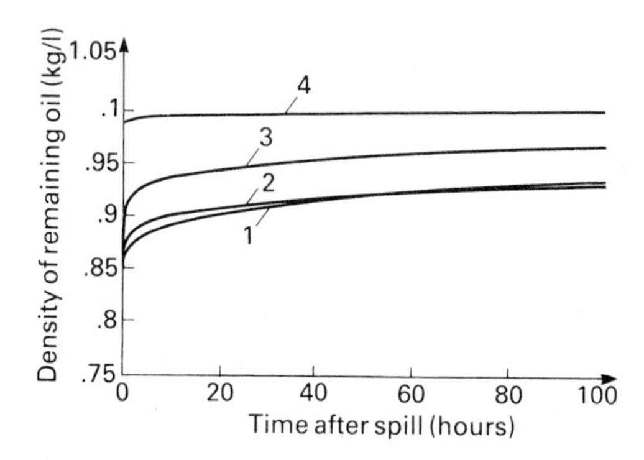

FIG.1.21 Relationship between time and density of remaining oil for different crudes: (1) Ekofisk; (2) Forties; (3) Kuwait; (4) Gamba.

The density at 3 °C is about 1% higher than that at 15 °C. Thus for evaluation of the behaviour of oil spills at sea the influence of the temperature on the density can initially be neglected.
A relationship exists between the viscosity of the oil remaining after distillation and the volume fraction, but the viscosities can not be calculated as simply as the densities and if the viscosities of the fractions have been measured at different temperatures, recalculation to a standard temperature will be necessary. The relationship between the viscosity of oil remaining after distillation (presented as $log.log(^{\circ}F)$) and the volume fraction distilled

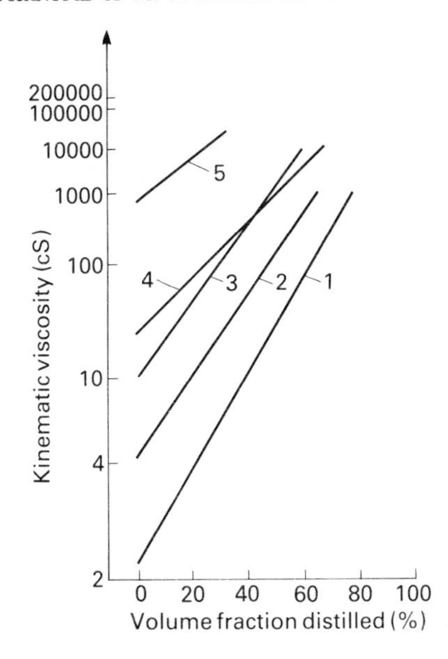

FIG.1.22 Kinematic viscosities of remaining oil, measured at 40 °C, versus volume fractions distilled; (1) Ekofisk; (2) Forties; (3) Kuwait; (4) Gamba; (5) Tia Juana Pesado.

can be described by a linear curve with a constant slope, although some exceptions do exist (fig.1.22). Using this linear relationship between the viscosity and the volume fraction distilled, the viscosity of the oil remaining with time after spillage at sea can be calculated (fig.1.23) This calculation should be considered with great care when the pour point of the weathered oil approaches the ambient temperature. It may be useful in choosing the clean-up technique, e.g. whether or not dispersants should be used.

1.2.6 Dissolution

Dissolution may have important biological consequences, although it is of less importance in terms of the overall mass balance of an oil spill. The extent of dissolution depends on the point of oil release. Subsurface releases of crude oil, which happen in offshore operations, enhance the dissolution of lower molecular weight aromatic components. As the oil floats upwards through the water column to the surface, the lower molec-

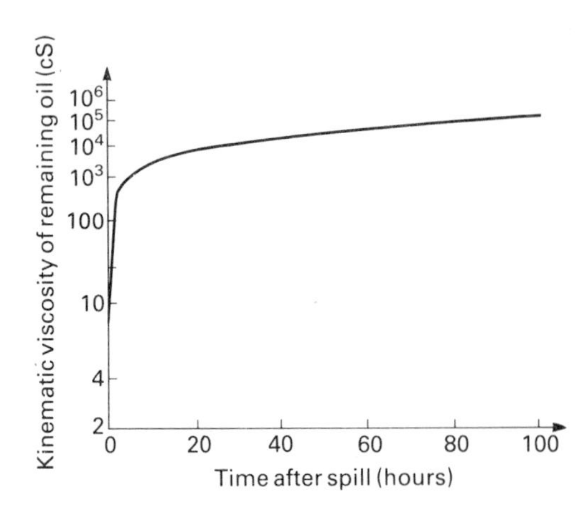

FIG.1.23 Kinematic viscosity of Kuwait oil remaining as a function of time after spillage (volume 10^4 m^3; wind speed 8 knots; temperature 10 °C).

ular weight aromatics, such as benzene partition into the water column and are almost completely removed from the oil mass by the time it reaches the surface, giving elevated (greater than 100 micrograms/liter) levels of benzene when measured directly beneath the water surface. Thus evaporation processes on the surface will be slightly delayed, although predominant over other simultaneous weathering processes by as much as a factor of 100 to 1.

Dissolution of individual molecular components from the slick is controlled by partition coefficients (oil/water concentration ratios), often referred to as M-values. The dissolution of individual components is controlled by the mole fraction of each component in the slick, the oil/water partition coefficient and interphase mass transfer coefficients, but not by pure component solubilities.

Time-series concentrations of selected lower molecular weight aromatics dissolved in a water column are shown in fig.1.24. These values were obtained from wave tank experiments. The peak concentrations of dissolved compounds occur approximately 8 to 12 hours after the initial release of the oil to the water surface. The concentration then drops off in an exponential manner, reflecting the loss of these compounds by evaporation

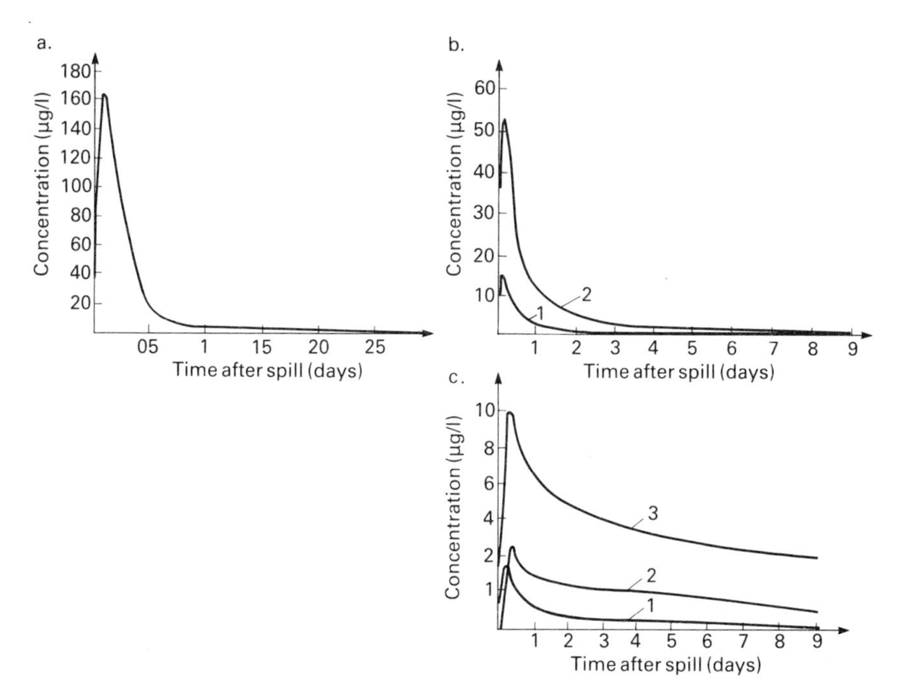

FIG.1.24 Dissolved aromatic component concentrations in a water column as a function of weathering time: (a) tuolene; (b) 1 ethylbenzene; 2 P-xylene; (c) 1 N-propylbenzene; 2 methylnaphthalene; 3-C_2-naphthalene.

from both the slick and the water column and the removal of dissolved components by advection (water mass transport). The early time-series concentrations of total aromatics are in the 10 to 100 parts per billion (p.p.b.) range. After a relatively short time these concentrations resulting from advection and diffusion quickly yield values in the low parts per trillion level in an open ocean spill. These predictions are in agreement with measured values in subsurface water column samples collected within 2 to 3 kilometres of the slick generated during the 1979 IXTOC I blow-out in the Bay of Campeche (Gulf of Mexico).

1.2.7 Dispersion

Dispersion of whole oil droplets is the most important process in the break-up and disappearance of a surface slick. The sea surface turbulence has a direct impact on droplet dispersion, although dispersion of oil droplets into the water column or processes of spontaneous emulsification from calm seas can also occur. The dispersion rates are given in table 1.7 as percentages per day for various sea states independent of individual crude oil characteristics and were derived from observations of real spills.

TABLE 1.7 Natural dispersion effects.

Sea state	Percent of oil lost per day		
	1 to 3 days	4 to 5 days	6 days and more
Low	10 to 30	5 to 15	0 to 5
Medium	20 to 40	10 to 20	0 to 7
High	30 to 50	20 to 30	0 to 10
Very high	40 to 60	25 to 35	0 to 10

Natural dispersion is in fact the net result of three separate processes:

- the initial process of globulation, i.e. the formation of oil droplets from a slick under the influence of breaking waves;
- the process of dispersion, i.e. the transport of oil droplets into the water column as a net result of the kinetic energy of the oil droplets supplied by the breaking waves and the rising forces;
- the process of coalescence of the oil droplets within the slick.

The various relevant physico-chemical parameters which influence these individual contributing processes are shown in Table 1.8. The thinner the slick, the more frequently wave breaking occurs, because the damping effect decreases with thinner slicks and the increased probability of the formation of small droplets (globulation increases), which become permanently dispersed (dispersion increases and coalescence decreases).

TABLE 1.8 Slick properties influencing natural dispersion.

Property	Slick thickness	Oil/water interfacial tension	Density	Viscosity
Process	(↓)	(↓)	(↓)	(↓)
Globulation	↑	↑	↓	↑
Dispersion	↑	—	↓	—
Coalescence	↓	↓	↑	—

decrease ↓ increase ↑

The oil/water interfacial tension affects globulation and coalescence, but does not affect dispersion (transport) of oil droplets into the water column. The smaller this oil/water interfacial tension, the more likely the formation of oil droplets (globulation increases) and the less the tendency of droplets to coalesce within the slick.

The density and viscosity of the spilled material also affects natural dispersion. The higher the density of oil and thus the smaller the density difference between oil and water, the more easily small oil droplets are formed from the slick. Globulation increases and the tendency of droplets to rise decreases, resulting in a greater depth of dispersion. The heavier the oil, the more important is natural dispersion.

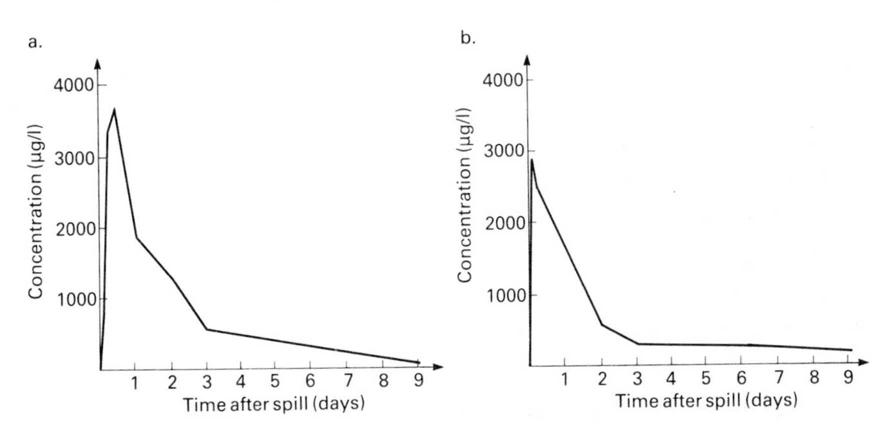

FIG.1.25 Whole dispersed oil and particulate bound oil concentrations as a function of weathering time and SPM type: (a) total dispersed/particle bound oil with SPM (air temperature 13 °C, water temperature 11 °C); (b) total dispersed/particle bound oil with no SPM (air temperature 13 °C, water temperature 11 °C).

Dispersion of oil droplets increases the surface area, allowing enhanced dissolution of lower molecular weight aromatics, and provides materials which may be ingested by organisms and ultimately deposited in sediments in the of faecal pellets. The larger droplets (greater than 0.1 mm) will return to the surface, when the turbulence is reduced, but the smaller droplets (less than 0.1 mm) will remain in the water column even at a great depth, as in the case of the tanker "Arrow", where particles less than 0.1 mm were found at a depth of 80 m. Dispersion of smaller droplets with changes in slick rheological properties (viscosity and interfacial surface tension) and as a result of water-in-oil emulsification becomes selflimiting within 2 or 3 days. Fig.1.25 presents time- series profiles of water-borne concentrations of dispersed/particulate bound oil droplets. After oil viscosities approached 2000 centipoise further dispersion was largely inhibited. The break-up of the slick is then limited to higher turbulence regimes, wherein 1 to 3 cm oil droplets can break away from the main slick and disperse into the water column. However, these droplets will return to the surface unless adsorbed onto suspended particulate material, ingested (whereby they could be deposited as faecal pellets) or ultimately used (as tar balls) by sessile organisms as a substrate to support growth.

1.2.8 Photochemical Oxidation

Several mechanisms are proposed for photochemical oxidation of oil:

- free radical oxidation in the presence of oxygen;
- singlet oxygen initiation of hydroperoxide formation;
- ground-state triplet oxygen combining with free radicals to form peroxides.

Rates of photo-oxidation are considered to be dependent on the wave length, but they are also affected by turbidity and SPM concentrations, particularly for higher molecular weight aromatics. The presence of inhibitors such as sulphur compounds or beta-carotenes can restrict the formation of radicals or inhibit singlet oxygen-mediated peroxide formation. Humic substances may reduce the photolysis rates of UV-sensitive compounds, but humic materials can also photosynthesise transformations of organic compounds through an intermediate transfer of energy to molecular oxygen. Some photo-oxidised compounds have an enhanced water solubility and consequently are quickly removed from surface slicks and diluted in underlying waters.

1.2.9 Emulsification

Emulsification of crude oils and/or refined products involves the dispersion of water droplets into the oil medium. As the percentage of water increases (up to maximum 75 to 80%) the resulting emulsion, called "chocolate mousse", undergoes a dramatic increase in viscosity, its effective volume increasing by a factor of up to 4, its density approaching that of the surrounding water and colour changes occurring, ranging through browns, oranges and reds. These properties of resulting emulsions make clean-up operations more difficult. The evaporation of components from an emulsion is also diminished.

Certain oils have a tendency to form stable water-in-oil emulsified mixtures. This tendency appears to depend on wax and asphaltic materials contained in the crude oil. Refined petroleum products do not contain waxes and asphalts and thus they do not form stable oil-in-water emulsions.In stabilising water-in-oil emulsions an important role is played by oxy-generated surface active materials, which are generated by photochemical and microbial degradation processes. Surface active compounds are believed to surround the water droplets, which usually range in size from less than 10 micrometres, preventing water/water droplet coalescence and ultimate oil/water phase separation.

TABLE 1.9 Consequences of emulsification of oil with time.

Oils	After 4 hours			After 24 hours			After 168 hour		
	Volume Change %	Water %	Visc. cP 20 °C	Volume change %	Water % 20 °C	Visc. cP %	Volume change %	Water 20 °C	Visc. cP
Ekofisk	-10	0	6.5	+360	85	-(1)	+220	86	1800
Kuwait	0	3	77	+100	70	400	+230	82	6100
Arabian Light	+360	88	1600	+260	83	2950	+150	78	6900
Iranian Light	+450	82	1400	+350	84	1770	+240	86	9000
Marine Diesel Oil	+515	84	-	+537	82	-(1)	+580	82	-

Significant changes in rheological properties of the slick or oil mass can be observed during the formation of water-in-oil emulsions (Table 1.9). Fig.1.26 presents time-series data on rheological properties obtained from wave tank experiments. Over 12 days the viscosity of the mixture increased to well over 2,800 centipoise, whereas the oil/air interfacial surface tension changed very little and the oil/water interfacial surface tension dropped from 27 to 13 dynes/cm. Subsequent changes in tension did not occur over a 12 month period. Formation of such stable water-in-oil emulsions affects the clean-up operations and combustability. It also severely limits additional weathering and removal of components from the emulsified mass. So additional evaporative and dissolution weathering can be significantly reduced. Thus additional degredation by either microbial or photochemical oxidation processes are limited to the exterior surface of tar balls or mousse.

As the result of weathering evaporation of the light ends (increase in density), dissolution of surface active components (decrease in spreading coefficient) and emulsification, the rate of spreading will gradually decrease under natural conditions (Table 1.9).

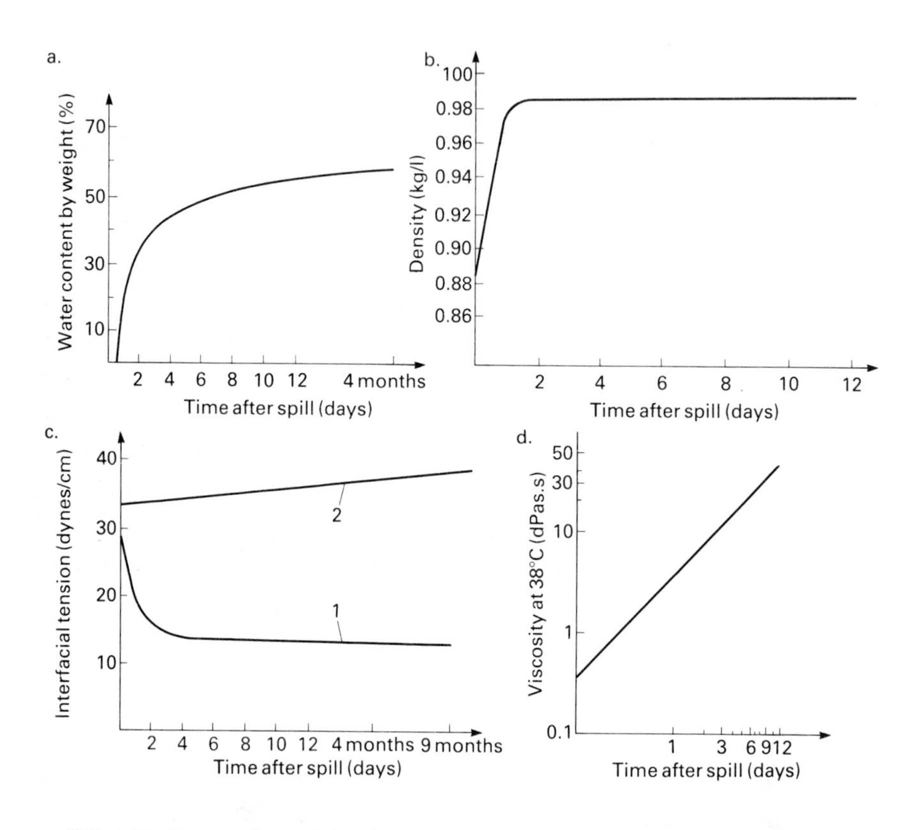

FIG. 1.26 Changes in various physical properties of crude oil as a function of weathering time: (a) water incorporation by oil; (b) density; (c) water/oil (1) and oil/air (2) interfacial tensions; (d) viscosity.

1.2.10 Adsorption onto Suspended Particulate Material (SPM)

Interactions between spilled oil and suspended particulate material represent an important mechanism for the rapid dispersal and removal of oil from surface waters. Oil spills in nearshore waters with high suspended particulate loads experience rapid dispersal and removal of the oil due to sorption onto SPM along frontal zones. At concentrations greater than 100 mg/litre massive oil transport may occur with the potential for significant adverse impacts on the benthos. Thus the adsorption of dispersed oil onto suspended particulates may provide an efficient mechanism for sedimenting fractions of oil mass (up to 15%).

1.2.11 Ingestion by organisms

Active biodegradation of oil droplets takes place due to the presence of microorganisms in seawater. This natural cleansing process is described in Chapter 2.1.

1.2.12 Sinking and sedimentation

Thick oils have specific gravities close to 1.00 at a temperature 15 °C and the thicker the oil the greater the risk of sedimentation, which increases at low temperatures. When the specific gravity of the oil due to weathering processes and to particles in the water increases towards 1.00, the oil will start to sink. There is also the possibility of tar lumps or balls forming, which are compact semi-solid masses of highly weathered oil aggregated with particles present in the water column. They can be transported by floating in the water column or along the sea bed. Tar balls may present serious shoreline problems on sandy beaches.

1.2.13 Shoreline Interactions

Fresh oil can penetrate coarse grained intertidal sediments quite rapidly. Ultimately oil can reach the depth of down to 0.5 m in gravel and coarse sand. With recurring high tides, dissolved components may be removed from the interstitial waters, but significant amounts of higher molecular weight (greater than n - C_{12}) aliphatic and aromatic components remain for several years. Finer grained sediments - silts and clays - limit oil penetration, which will be confined to holes from burrowing organisms. During "Amoco Cadiz" spill, fresh oils from the surface mud flat sediments were lifted off the substrate with repeating tides. Under these circumstances oil will continue to undergo increases in viscosity with evaporation/dissolution and mixing in the intertidal zone. With time, however, the weathered oil material will be encrusted with a greater amount of finer grained sedimentary material and after several days an asphalt-like material may be formed, which will be transported into upper or middle intertidal zones. With increased storm activities, this material can be lifted from the beach surface and deposited in offshore berms.

When water-in-oil emulsions containing up to 60% water are deposited on the beach, little penetration of the mousse into the intertidal regime takes place. The mousse is primarily subject to mixing with sand and reflotation with incoming tides. If a large oil mat covering several kilometres of shoreline is present, this continuing lifting and redepositing ultimately causes accumulation of sedimentary materials in the oil, resulting in an oil/sand, asphalt-like mat. Such a mousse may reach a thickness of 0.5 to 15 cm and may be coated with vegetation. It is resistant to further microbial or chemical weathering, although limited dissolution of lower molecular weight aromatics may occur, affecting nearshore waters for several months or even years. Oil/sediment mats may be covered by clean sand deposits, but they may be reexposed after some time by a storm and subsequently removed to subtidal berms in water depths down to 30 m, where they may support algal and bacterial growths.

Most whole crude oils will not undergo weathering to the point where their density will be greater than that of seawater. These densities can increase sufficiently to sink crude oils due to:

- incorporation of detrital material and sand during deposition on the shore;
- interaction with suspended particulate material in a water column;

- attachment of pelagic organisms such as barnacles.

1.2.14 Ice Conditions

During the last 15 years rich oil and gas deposits have been found in the cold ocean regions, where the presence of ice is a dominant factor. Thus the structure of ice, its behaviour and movements, its deformation and fracture mechanics are of special interest in designing and operating systems on polar seas. Polymorphic ice, formed under atmospheric pressure and temperature is lighter than its melt. This characteristic results in ice floating on water with the resulting ice growth downwards. The floating ice cover acts as an insulating layer, so the ice growth slows down as it gets thicker. Thus the Arctic ice in one year never gets thicker than 2.5 m.

Ordinary ice has hexagonal crystallographic symmetry. Ice crystals or grains can vary in shape from tubular to granular to columnar. Crystal dimensions are in the range of 1 mm to several centimetres. The axis of symmetry of an ice crystal is perpendicular to the so-called basal plane. Snow ice, including icebergs, is generally composed of granular crystals of random orientation, whereas ice which grows on the surface of water will be composed mostly of long continuous crystals (columnar). These crystals have generally basal planes, which are oriented vertically, but not necessarily parallel to each other (fig. 1.27) "Secondary" forms of ice such as refrozen ice rubble will generally contain columnar ice pieces, but randomly arranged so that they could be considered isotropic.

Impurities in the form of salt and gases (air) are expelled during the growth of ice to the platelet boundaries within the crystals. At normal sea ice temperatures the brine inclusions remain unfrozen and create additional weakness along the basal plane, which cause the sea ice to be weaker than fresh water ice at the same temperature. During the seasonal cycles of warming and cooling, the brine is expelled gradually, so that multiyear ice is stronger than annual or first year ice.

Wind drag acting over the ice causes the ice to move - if free, the

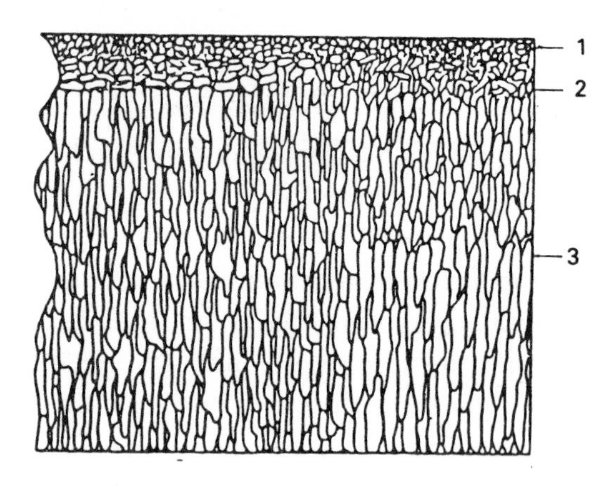

FIG.1.27 Typical crystalline structure of sea ice: (1) fine grain zone (d = 1 mm); (2) transition zone (d = 1 to 4 mm); (3) columnar zone (6 × 12 × 24).

ice will attain a speed of about 2% of the wind speed. But the ice usually can not move freely, and the wind drag creates ice pressure, which leads to ridge formation. Most ridges are formed in the thinner annual ice, which is always present between multiyear floes. First year ridges are initially a loose accumulation of ice blocks held together by gravity and buoyancy forces. They attain much larger thicknesses than the surrounding sheet ice. A typical first year ice ridge is shown in fig.1.28. Its total thickness is about 20 m, most of which is below the water line. Multiyear ridges are consolidated to a much greater depth. The average ridge heights in the Beaufort and Chukchi Seas tend to be of the order of 1.4 to 1.6 m. The average ridge spacing tends to be of order of 100 to 200 m. Extreme ridge heights of the order of 4 m high have been measured. The largest measured ice keel depth is 47 m, but evidence from sea floor ice gouging suggests keel depths in excess of 60 m.

Offshore pack ice, which is composed mostly of multiyear ice, rotates in the Arctic in a clockwise direction at a rate of several kilometres per day. One complete rotation of the Beaufort Gyre takes about 10 years. Close to the shore the ice is composed of first year ice (about 2 m thick). This inshore ice becomes landfast and is stationary to water depth about 20 to 25 m by

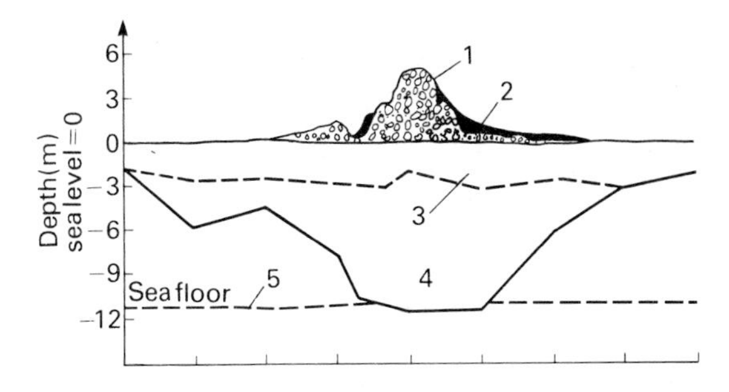

FIG.1.28 Typical profile of a first year ridge in the Beaufort Sea: (1) ridge; (2) snow cover; (3) consolidated zone; (4) porous zone; (5) sea floor.

early January. Prior to that time the ice is moved to and fro by winds and extensive first-year pressure ridges can be formed, which may become grounded (fig.1.29), creating a landfast zone.

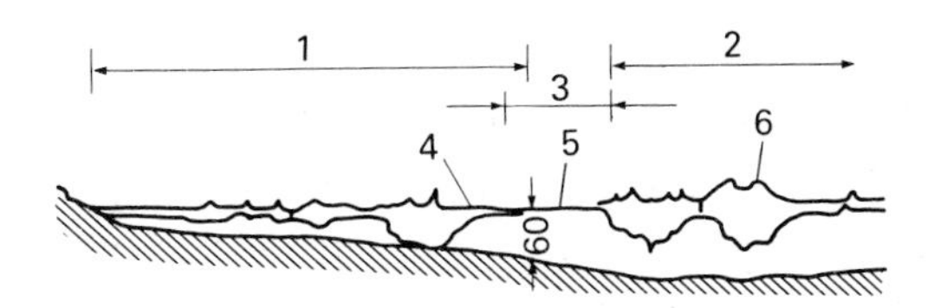

FIG.1.29 Nearshore winter ice conditions (Beaufort Sea); (1) land fast ice; (2) mobile pack ice; (3) shear zone; (4) grounded ridge; (5) open lead; (6) multiyear floes.

Between the landfast zone and the pack ice is a shear zone or transition zone where the stresses created between the mobile pack and the stationary landfast ice create extensive pressure ridging. Some of these ridges may become grounded and even scour the sea floor extensively.

Ice under loading behaves differently to conventional materials such as metals, rocks and soils. At typical Arctic temperatures ice exists close to its melting point and thus it creeps significantly under load. (fig.1.30). This is a typical creep curve. The time required to reach tertiary creep and failure is a function of the stress level, the ice temperature and salinity. For low stress levels this time could be many hours, whereas for higher stress levels the time to failure might be only minutes. For rapid loading the ice will fail in an elastic-brittle fracture.

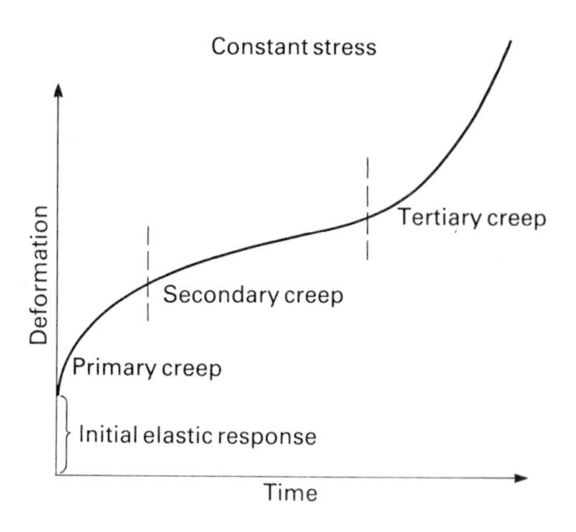

FIG.1.30 Typical ice deformation versus time.

Ice is a very brittle material. The fracture toughness of sea ice is in the range of 50 to 100 kPa m. This is less than glass and about one thousandth that of steel. The cracks propagate quite easily and if the loading or strain rate is high enough, the fracture of ice will be dominated by crack propagation or fracture.

1.2.15 Arctic Environments

The weathering of petroleum in the presence of sea ice is dependent on the point of release and the extent of ice cover. If oil is spilled in open or broken sea ice, it will tend to pool in the open leads. The structure of the ice will have an effect on the subsequent weathering and partitioning of the oil. Oil weathering processes are spreading on ice, under ice and within open leads and encapsulation (Fig.1.31). Oil spilled during winter in the presence of level ice may be deposited on the ice, underneath the ice, or in both environmental pools. A subsea pipeline rupture or blow-out could result in extensive under-ice coverage. The thickness of an under- ice slick is primarily controlled by the depth of under-ice depressions. An ice lip will form in winter conditions within 24 to 48 hours around the pools, preventing horizontal movement of oil. Some dissolution of lower molecular weight aromatics may occur. In winter conditions this ice lip will continue to grow around and underneath the oil, eventually encapsulating the spill and stopping any alterations to its physical properties until the following spring, when brine channels will form in the first year ice. After the brine channels have penetrated to the depth of the encapsuled oil, the oil may migrate slowly upwards through the ice and eventually spread on the ice surface, further enhancing localised melting. Finally open leads form and the weathering processes are taking place similar to open ocean processes.

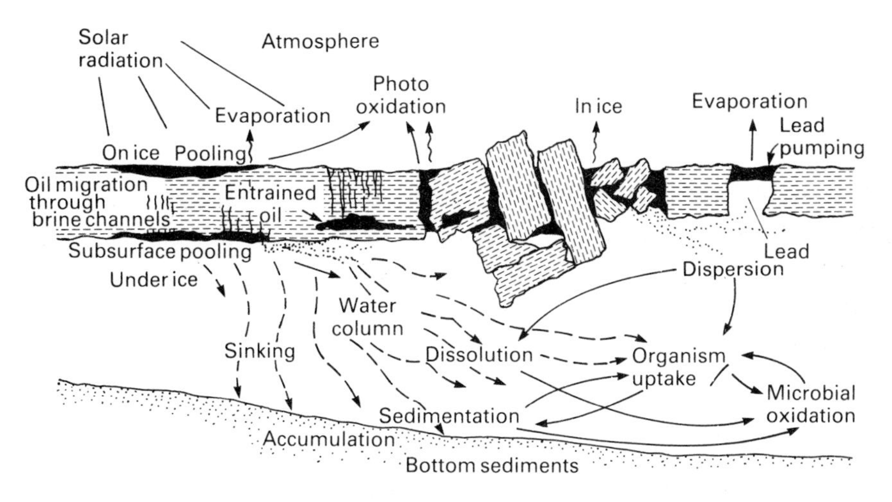

FIG.1.31 Environmental reservoirs and dispersion mechanisms for oil spilled in an Arctic marine environment with ice.

Oil released on the surface of broken ice is subjected to a variety of weathering processes including spreading between leads, evaporation, dissolution and emulsification. Evaporation is the dominant weathering process and is extremely dependent on the kind of oil and the temperature. Emulsification is of particular importance because it leads to an increase in viscosity and to a decrease in pour and flash points of the resultant product. The oil on the ice surface is trying to enter the leads and cracks between the floes, spreading to various equilibrium thicknesses, depending on the ice concentration and the physical properties of the oil. Accelerated mousse formation and sinking of the water-in-oil emulsion beneath the grease ice may result in a considerable

increase in dispersed oil micelles in the water column, further enhancing immediate dissolution of aromatic hydrocarbons. Snow cover and ice ridges would confine oil and thus limit the extent of spreading. Weathering of spilled oil during winter is relatively slow due to the low temperature, reduced solar radiation and diffusion-controlled restriction of component evaporation due to limited area and greater oil layer thickness.
If the oil persists until the following year's freeze-up, the oil rises to the surface of the forming grease ice or is pumped laterally due to combined lead/matrix effects onto the surface of pancake ice. As the pack ice or multiyear ice are in constant motion with associated high lateral under-ice currents, the resulting area of contamination could be very large. Most oil under the multiyear ice would most likely emerge at its surface by late summer, because pack ice melts on the top and grows from beneath, but the oil could remain entrapped for periods up to 10 years.

1.3 Impact of Oil on Marine Environment

1.3.1 Physical Contamination

When oil is spilled on a calm water surface, only soluble components in the oil affect organisms in the underlying water. Most waters however, are not calm and waves and currents mix oil into the underlying water. The growth of marine organisms depends basically on the quantity and quality of the primary production of phytoplankton (algae). Apart from the toxic effects of oil, marine microfauna can experience indirect food effects since algal production can be changed after an oil spill. Type and concentration of oil compounds dissolved in water determine the effects of oil on phytoplankton. After a relatively large spill (>0.1 l/m^2), rapid mechanical removal of oil from the water surface is necessary. Treatment of the oil with a dispersant will generally aggravate the effects caused by high dissolved oil concentration in the water as a direct result of dispersion. Alternative combat methods, such as mechanical methods are needed. To prevent long-term effects, transportation of oil to the sediments must be prevented. Depending upon the oil and the mixing energy, natural break up of the slick occurs with droplets of oil remaining in the water column. Water soluble fractions (WSF) cause immediate toxic effects on marine and freshwater organisms. The WSF has also been used for bioassay testing, which is difficult with oil that adheres to vessel walls and organisms, particularly with flow-through systems. Field exposures are increasingly diluted with time. The generally used bioassay expression (96 h LC_{50}), i.e. concentration of oil, producing 50% mortality of the test organisms, does not permit a direct comparison of bioassays with different exposure times. Changing the bioassay values and field dilution concentrations of organisms exposures (doses) permits a better comparison of results. This concept multiplies the time of exposure and the mean concentration to give an organism's exposure expressed as p.p.m.-days or p.p.m.-h. A premise is made here that all the organisms will respond to a toxicant in the same manner, if exposed to 20 p.p.m. for 1 hour or to 1 p.p.m. for 20 hours. There are obvious limits to this concept because some organisms may be killed instantly when concentrations are high. When concentrations are low and the time is long, many organisms can metabolize hydrocarbons and live without apparent adverse effects. Some bioassays show the greatest mortality during the first 24 hours of a 4 day exposure, whereas other bioassays show little mortality during the first day, but increased mortality after that.

Experiments based on 96 hours exposure have shown that hydrocarbon toxicity is unrelated to hydrocarbon type, but toxicity occurs when a given molar concentration is attained in an organism's tissues. This is controlled by the pure hydrocarbon solubility and the partitioning of the hydrocarbon between water and the organic phases of the organism. In general, the lower the solubility, the higher the partitioning from water to organisms. Surprisingly, saturated hydrocarbons are more toxic than aromatic hydrocarbons, although the latter have generally been considered to be the most toxic soluble hydrocarbons, i.e. hexane and cyclohexane are more toxic than benzene, octane and decane than all the monoaromatic hydrocarbons.

The above transfer path is relatively long compared to an oil droplet or oil film that attaches to an organism or egg, or is ingested by the organism. The diffusion path then is short and there may be almost direct exchange of hydrocarbons between the oil droplet and the organism's organic phases.

Mortality for more than 50 species ranged from 9 p.p.m.-h to >1900 p.p.m.-h. Larval stages of organisms were no more than twice as sensitive to the WSF of crude oil as were adults. The kelp shrimp larva was the most sensitive at 14 p.p.m.-h. Limited studies have indicated that chemically dispersed crude oil is about twice as toxic as the WSF of crude oils based upon total C_1 to C_{10} hydrocarbon exposures using chum salmon fry.

The measured field exposures to C_1 to C_{10} hydrocarbons from untreated and chemically dispersed crude oils are much lower than those observed to kill a wide range of organisms in laboratory bioassays (from 150 times to over a million times lower). Thus oil spills and chemical dispersion of crude oil slicks are unlikely to have an adverse effect on larval, juvenile or adult organisms in the water column. The spill response implications are that the oil slick should be chemically dispersed in the water column to protect biologically sensitive shoreline habitats and water areas occupied by sensitive organisms such as birds and sea otters.

1.3.2 Tainting and Recovery of Fish and Crustaceans

Fish, crustaceans and molluscs under prolonged exposure to concentrations of oil may acquire an objectionable, oily odour or flavour, thus loosing their market value; in extreme cases they are inedible. After "Amoco Cadiz" spill, oysters in several farms had to be destroyed because of unacceptably high levels of contamination.

The degree of hydrocarbon contamination of intertidal zones, as well as the rate of natural depuration was assessed using brown mussels (*Perna perna*) as an indicator. Mussels were collected from the shore area where spilled oil was most visible: two days after the major oil spill, two weeks and two months later. Mussels from a neighbouring area were used as controls. After two weeks, there were still indications of the presence of oil spill components, but after two months there was little evidence.

The concentrations of aromatic species in samples collected are shown in table 1.10. The results indicate that the accumulated hydrocarbons are readily lost and that after two months the levels approach those of control mussels.

TABLE 1.10 Hydrocarbons found in brown mussels *Perna perna* (expressed as μg hydrocarbons/g wet weight of mussels).

Aromatic fraction	At time of spill	2 weeks later	2 weeks later	2 months later	Control site
Low	181.0	10.90	12.30	9.42	11.20
Medium	87.5	5.18	5.39	4.27	3.36
Heavy	33.9	4.86	0.38	0.95	0.47
Overall	82,6	5.22	4.02	3.62	3.33

The ratio of lower molecular weight species to heavy components is much higher in the mussel sample than in fuel oil. The lighter components have been preferentially accumulated, presumably because of their higher solubility in the water column. Brown mussels readily lose their accumulated hydrocarbon burden and there is no definete indication that any components are strongly retained by the mussels.

After initial destruction of intertidal or benthic organisms, repopulation by the original species may be low, but some resistant or opportunistic species may undergo dramatic population increases and fluctuations in the affected area.

Recovery in water column is generally very rapid and the oil effects in nearshore systems are local and transitory, with full recovery in a few weeks.

Direct deaths of adult fish have been observed in only a few oil spills, but modest oil contamination of muscle tissue was found more frequently. Chronic exposure to low concentrations of high-molecular-weight polynuclear aromatic and heterocyclic compounds may provide the potential for carcinogenic effects in marine organisms or for accumulation of carcinogens in marine organisms used as human food. The formation of tumours has also been detected in field populations of fish and molluscs.

Fish from coastal areas that receive repeated oil spills along with chronic discharges of oily wastes from ships and from land-based sources, however, may pose higher risks to human consumers. Changes in future catches can not be attributable to spill effects due to the great mobility of fish with the exception of extremely localised or confined areas.

1.3.3 Marine Mammals

The vulnerability of mammals to contamination depends upon their behavioural responses to oil. It is unclear to what extent marine mammals in their natural habitat detect and attempt to avoid floating oil and plumes of dispersed oil. There are still many uncertainties about the impact of oil on marine mammals. Occasional reports of stranded oil-fouled animals do not give sufficient evidence to pinpoint the cause of death.

Heavy, tar-like oil, made viscous by cold, may foul swimming appendages, adhere to fur in lumps and clog body openings. Heavy oiling, which can result in exhaustion of the animal, appears rare. Coating of the body with lighter oils causes few mechanical effects. Volatile oils may severely irritate sensitive membranes such as those around the eyes. This is reversible after exposures of up to 24 hours, but permanent eye damage is a possible consequence of longer exposure.

Animals depending on fur or hair for thermal insulation (sea otters, fur seals and polar bears) suffer loss of body temperature when their hair is fouled with oil. Most other sea mammals (whales, porpoises, manatees, phocid seals) rely on blubber and vascular constriction for controlling their body temperatures and they are more resistant to the thermal effects of oiling. Hair seals rely on a combination of a blubber layer and a coat of insulating hair that traps a thin layer of stagnant, warm air or water next to the skin. Their ability to retain heat is little affected by surface oiling, although dark staining of the fur may increase solar absorption in "hauled-out" animals, causing overheating. New-born young have little fat, but they rely on a dense layer of fur (lanugo), which is sensitive to oiling. As they grow, the proportion of fat increases and the fur is lost. The northern fur seal also has a waterproof fur that traps a thin layer of stagnant air next to the skin. Loss of waterproofing reduces the insulation and can result in chilling, hypothermia and death.

Marine mammals may ingest petroleum hydrocarbons by inhaling volatile fractions, grooming oiled fur or feeding on contaminated material. Marine mammals confined close to the source of a spill or surfacing repeatedly in fresh oil slicks will inhale petroleum vapours. Short term inhalation is likely to produce only mild irritation of mucous tissues, while prolonged inhalation of high vapour levels would cause death or damage to the nervous system. Animals further from the immediate area of a spill are not likely to suffer serious consequences of inhalation.

Sea otters and polar bears are likely to ingest quantities of oil by frequent grooming. It is also possible that they might feed on contaminated carcasses. Baleen whales may ingest large quantities of floating or dispersed oil in a water column, but they are more likely to consume oil that was ingested first by their zooplanktonic prey.

Unless the animals are quickly and properly cleaned, fouling of sea otter fur is likely to cause death either through hypothermia or as a result of oil ingestion or aspiration of oil into the lungs during grooming. Although ingestion of oil for short periods usually does not result in serious effects, prolonged consumption may lead to organ damage or hormonal imbalances.

1.3.4 Effects of Oiling on Sea Birds

The effects of oiling on birds may be twofold:

- external effects associated with oiling of plumage;
- internal effects associated with the pathological effects of ingested oil.

The external effects of oil are the most noticeable and most immediately debilitating. Oil destroys the waterproofing and insulating properties of the plumage. The bird will suffer from chilling and it is often unable to fly or remain afloat in the water. The bird has difficulty in obtaining food or escaping predators. In addition to the decreased foraging abilities of the bird, the presence of oil in the environment usually results in loss of food. Irritation or ulceration of the eyes and clogging of the body openings and mouth often accompany oil contamination. The weakened bird becomes susceptible to secondary infections, both bacterial and fungal.

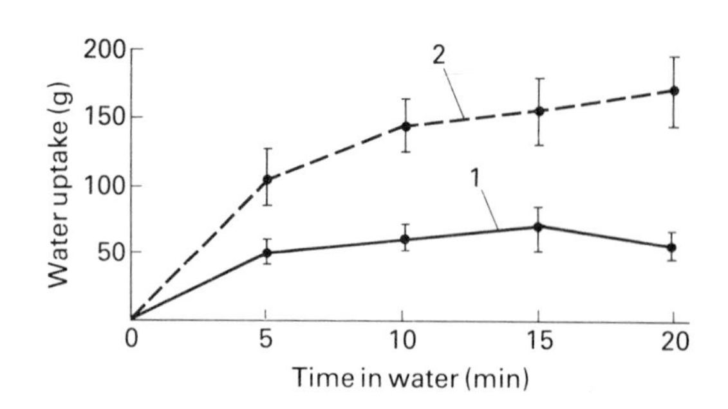

FIG.1.32 Water uptake of penguins relation to time spent in water: (1) clean; (2) oiled.

The insulating properties of plumage provide thermoregulation, especially in an aquatic environment. By trapping a layer of air between the feathers and the skin, the interlocking microstructure of the feathers provides a strongly water-repellent surface. The bird becomes waterlogged, water penetrates to the skin and the bird is no longer insulated. This wetting of the plumage can be measured by weighing the birds before and after immersion; the difference in weight will be a measure of water uptake onto the feathers. Investigations were conducted in South Africa on penguins, as these birds spend much time on the water surface and are therefore particularly likely to encounter floating oil. Like the auks of the northern hemisphere, penguins are more vulnerable to oil pollution than flying birds. Fig.1.32 shows the rates of water uptake onto the feathers of clean (waterproof) and oiled (non-waterproof) penguins, at 5 minute intervals for 20 minutes. Oiled penguins took up significantly more water in the first 5 minutes than clean birds and for the remaining 15 minutes the oiled birds continued to be wetted faster (4.3 g/min) than the clean birds (0.6 g/min).

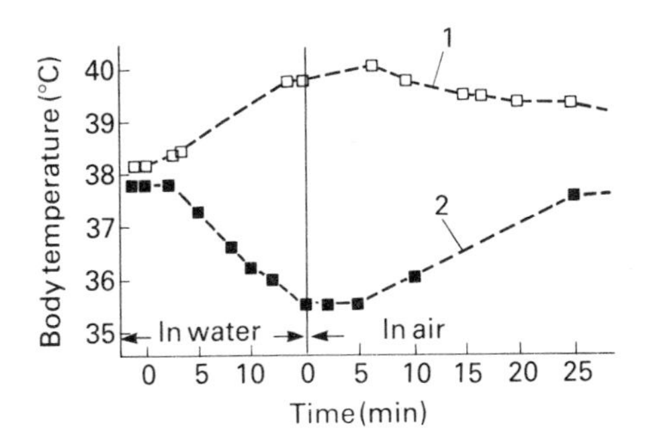

FIG.1.33 Changes in body temperature of penguins induced by 15 min. of swimming and following 40 min. on land; (1) clean; (2) oiled.

The waterlogging of feathers results in penetration of water through the plumage to the skin, with subsequent loss of insulation by the plumage. The changes in body temperature (T_b) of oiled and clean jackass penguins subjected to 15 minutes swimming followed by 40 minutes of recovery on land are presented in fig.1.33. Water temperatures varied from 19.5 to 20.5 °C when the air temperatures varied from 18.0 to 25.0 °C. The temperature T_b of the clean, waterproof birds increased in water and stayed above resting T_b for the remainder of the experiment. This increase in T_b is the result of increased activity while swimming and due to heat dissipation while active in the water at these temperatures. The oiled bird's resting temperature T_b was similar to that of clean birds, indicating that it was able to thermoregulate in air at these temperatures. Despite the increased activity of swimming, the oiled birds rapidly became hypothermic when placed in water and if not removed from the water would probably develop fatal hypothermia at these relatively warm water temperatures. This result illustrates the importance of intact plumage for thermoregulation in these penguins.

The heat production of oiled and clean jackass penguins, expressed as metabolic rate (kJ/kg/day), were determined from oxygen consumption at a range of temperatures, both in air and in water (fig.1.34). The heat of clean unoiled birds in air at 5 °C was particularly low and was achieved behaviourally by fluf-

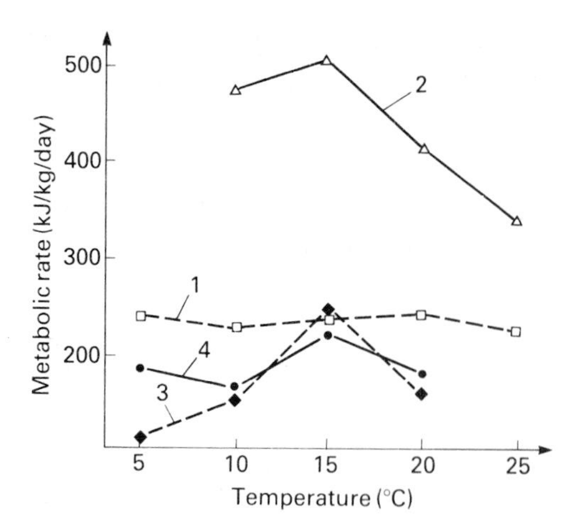

FIG.1.34 Inactive metabolic rates at various temperatures for penguins: (1) oiled in air; (2) oiled in water; (3) clean in air; (4) clean in water.

fing their feathers and keeping still at this low temperature. At other temperatures tested, however, the heat production of clean penguins in air and water did not differ significantly ($p<0.05$) while the birds were inactive despite the heat sink effect of the water. This again emphasizes the importance of unimpaired insulation by plumage. The mean heat production of oiled birds in air increased by only 22% over that of clean birds, indicating that oil-covered plumage still provided insulation in air, although these birds were not capable of reducing their heat production at lower temperatures by piloerection of the oil-caked feathers. Oiling therefore deprives these birds of the behavioural mechanism for decreasing their heat production at low air temperatures.
When these oiled birds were placed in water they showed a strong increase in heat production while inactive. When swimming they experienced heat production rates more than four times the inactive heat production of an unoiled penguin.
Internal effects of oil, while not as apparent as external ones, are equally life threatening. Direct toxic effects on the gastrointestinal tract, pancreas and liver have been documented. These effects frequently result in ulceration and haemorrhaging in the gastrointestinal tract with severe loss of digestive and absorptive ability. Oil aspiration pneumonia is not uncommon in oiled

birds. Visceral gout due to kidney damage as a direct toxic effect of oil or due to dehydration has been documented.
Extensive investigations carried out on jackass penguins have indicated differences in the microorganism population of oiled and control birds. Thus a greater diversity of potentially pathogenic bacteria may be the reason for secondary infection in these already stressed birds, playing an important role in their mortality. After oiling, penguins have shown an initial loss of appetite lasting on average 5 days, although two birds (22%, n = 9) refused food and eventually died. Lost weight and temporary nitrogen deficits were associated with this loss of appetite by oiled penguins. After recovering their appetite the birds recovered body weight and stabilized their nitrogen balance. The weight loss of a starved oiled penguin (zero voluntary food intake) was higher (101 g/day) than that of an unoiled starved penguin (56 g/day), reflecting the higher energy demands of oiled birds (fig.1.34). Penguins accidently oiled at sea rapidly recovered body weight, when fed in captivity. The main cause of death of oiled penguins would be starvation resulting from inability to forage, although secondary infections may play a role once the birds are heavily stressed. It is thought that aerial foraging birds, which also experience hypothermia in water when oiled, may still be capable of foraging by restricting the time spent in water.

1.3.5 Rehabilitation of Oil Contaminated Birds

Rehabilitation of oiled birds begins with stabilizing of the bird, which should be kept warm and quiet, away from people and other stressors. The mortality of oiled birds is affected by the stress they experience in rehabilitation. On receipt, an oiled bird should be banded with a temporary, numbered plastic leg band and its record should be scrupulously kept. The amount and distribution of oil on the body, and the weight and cloacal temperature should be recorded. Oil is removed from the mouths and the body openings. The vent is checked for oil and matted feathers, causing impaction. The eyes are flushed with a sterile saline solution or ophthalmic irrigation. If possible, blood should be taken at this time to check packed cell volume and total solids. A number of oiled birds are dehydrated as a result of enteritis. Tube feeding the bird with a warm electrolyte solution serves to rehydrate the bird, while flushing the oil from the gut. An enteric coating agent is also administered orally by tube. Nutrients are later added to the solution and tube feeding is repeated every 4 to 6 hours until the bird is permitted free access to food and water after cleaning. Excessive preening of oil-covered feathers is prevented (especially after contamination with highly refined oils) by making a poncho for the bird with a disposable diaper or a pillow case loosely taped around the bird. Holes should be cut in the pillow case for the head and legs, and a hole left for the cloaca.

Some birds benefit from being placed with the conspecific, but overcrowding can be stressful. Curtains are hung to limit visual contact with humans. People should speak in low voices when in the bird holding area. The bird should not be washed until it is responsive, alert and restored to normal fluid balance. Effort should be made to stabilize the bird and wash it within 8 to 24 hours. The longer the oil is left on the feathers, the more difficult it is to remove and the more damage may be done to the skin. Highly refined oil can be absorbed through the skin, intensifying internal problems.

There is a strong correlation between body weight and body temperature for survival in washed birds. The average weight and temperature of the affected birds was determined. Those birds that were tested at or above the norm in weight and temperature for their sex had a 100% release rate. Those birds under the norm in one or both areas evidenced reduced rates (temperature - 66%; weight - 75%; both - 38%).

Oil must be removed from the feathers without damaging the delicate feather structure. Most effective is a warm detergent bath. The water must be above 102 °F in order to lift the oil. The detergent must suspend the oil and hold it in suspension. It must be minimally irritating to the body, must not damage the structure of feathers and must easily rinse off the bird. Any detergent residue left on the bird will impede waterproofing. The cleaning procedure should be done in a warm, quiet area, free from drafts.

A single bird is placed in a 10 gallon tub containing 4 to 15% detergent in 103 to 105 °F water. A 10 gallon tub is adequate for birds ranging in size from small ducks to a small Canada goose. Swans and large geese must have larger containers such as human bath tubs holding at least 15 to 20 gallons of water. One worker ladles the detergent solution over the bird's body and wings, while a second person gently strokes the bird in the direction of feather growth. A third worker may be needed to hold a large bird. It is important that the feathers should never be scrubbed or rubbed, since this could damage the delicate feather structure, which is vital to waterproofing. A container of clean water or normal saline should be kept at hand and the bird's eyes should be flushed with eyechoppers frequently to prevent detergent irritation. The bird's head must be gently restrained during washing. Many birds attempt to duck their heads under the water. The birds should be removed from the tub when the water gets dirty and the entire washing process is then repeated three or more times. Weathered or heavy crude oils are particularly difficult to hold in suspension. If the oil does not lift easily, then the wet bird will have to be removed from the tub, wrapped in a large dry towel and placed in a dry box for 4 to 6 minutes. This allows the bird to calm down slightly and also provides time for the detergent solution to soak into the oiled feathers. Very stubborn oils may require softening with warm mineral oil. This should only be used as a last resort, since mineral oil itself is a contaminant, which will have to be washed out from the plumage.

The feathers must be completely rinsed if the bird is to be rehabilitated. Any detergent residue can impede waterproofing. Rinsing is carried out with a combination of spray rinses and tubs of clean water at temperatures of 102 to 105 °F. Special attention should be given to the coverts under the tail, under the wings and the neck of the bird. The bird is not sufficiently rinsed until diamond-like beads of water roll freely from the feathers. Failure to rinse the bird adequately is probably the most common cause of unsuccessful rehabilitation efforts. The bird's feathers should be blotted dry with a clean towel and the eyes should be flushed one last time.

If all the materials and tubs of water are assembled in advance, the entire cleaning time should take from 15 to 30 minutes. After rinsing, the bird should be put in a pen separate from clean birds to dry. After stabilization (12 or more hours) the bird will need to be rewashed.

The pen should be lined with sheets or towels, curtained to minimise human intrusion and provided with heat lamps to allow the bird to find a comfortable ambient temperature. The heat lamps should be at two heights from the bottom of the pen (approx. 24" and 36" for a mallard-sized bird) with an unheated area also available. The birds generally do not respond well to forced hot air from driers and prefer the ambient warm air of heat lamps, immediately beginning to preen their feathers back into alignment.

Diving ducks require as much as 3" of foam padding under the sheeting to prevent breast abrasions and open sores. For birds which are totally unable to walk on dry land (i.e. common loons), small plastic bags stuffed with shredded foam should be provided to make slightly flat pillows under the sheeting. The birds soon learn to position these pillows under their keels in a comfortable position. Scrupulous care must be given to the pens of birds which are not mobile on land. The birds can incur serious feather damage if they became contaminated with faeces and old food.

Free access to water and a variety of foods can now be provided. The birds are checked to see which ones are selffeeding. The droppings are monitored for blood and oil and these birds should be treated for entritis with Pepto-Bismol and tubings of easily absorbed nutrients.

After 24 hours the birds are allowed to swim. They are provided with sufficient water to swim and preen. For birds that are smaller than Canada geese, pens 4 ft x 8 ft with a sloped ramp

should be provided. These ramps allow the birds to enter the water and to leave it at will. The pens are curtained and the elevated platform has a heat lamp on one side to prevent chilling. Children's pools with access ramps could be used, but the water must be changed 3 to 4 times a day or more. Tepid water is used for the first swim, but cool to cold water thereafter.
Waterbirds usually take to water readily and when they begin to get wet, they leave the water to preen. As the bird continues its efforts to swim, then preen, it realigns its feathers and restores the original feather structure. This alignment of feathers insures the bird's waterproofing. The feather structure does not require, but is further enhanced by, application of oil from the bird's own uropygial gland. This natural oil seems to assist in maintaining the feather structure. Birds which are waterproof will show the diamond-beading of water on their feathers and they will be able to remain in water, depending on the species, for from 10 to 50 minutes without getting wet. Certain diving birds (such as loons, scoters, grebe sp.) may have slightly wet outer contour feathers, but the down should remain absolutely dry. A rehabilitated bird should be of average weight for its species and sex. It should be adequately muscled so that it can forage normally in the wild and it should show no signs of any disease.
A bird which is waterproof should slowly be exposed to temperatures comparable to outside weather, which might be critical in cold winter months. Seabirds must be prepared for return to the ocean by being tube fed on a solution of normal saline (0.9%) for three days before release. This stimulates the function of the nasal salt gland. After marking with appropriate bands, the birds could be released early in the day into their habitat.

1.3.6 Coral Reefs

Coral reefs are important in supporting coastal fisheries, protecting tropical coast lines from wave action and erosion and providing a basis for tourism and recreation.
Platforms of fringing reefs form extensive shallow habitats covered with algae, seagrasses and invertebrates. After a few days of

exposure to an oil spill at low tides a band of substratum 1 to 3 m wide at the reef edge becomes nearly barren of the normal assemblage of sessile invertebrates and algae, which contribute to the lower abundances of organisms. Oil spills cause substantial mortality among fish and invertebrates (including lobsters, crabs, gastropods, bivalves, octopus, sea urchins, sea stars and sea cucumbers) in intertidal areas, on the surfaces and margins of coastal fringing reef platforms, and in adjacent shallow subtidal areas.

In the three months following a spill a reef edge was colonized by a thin transparent mat of algae such as *Cladophora*, which in a short time become thicker, overgrowing both the vacated substrata and the sessile organisms that survived the oil spill. This algal mat covered more than 54% of hard substratum - more than 4 times the average abundance and more than twice the maximum abundance measured in prespill surveys.

The two common species of zoanthids on a reef flat, *Palythoa caribaeorum* and *Zoanthus sociatus* were less abundant after the oil spill. The reef flat population of *Palythoa caribaeorum* was concentrated in the area where oil accumulated at low tide. Before the spill, its coverage ranged between 10 and 12% in a 2 m wide band at the seaward fringe of the reef flat. Three months after the spill the overall spatial covarage was less than one tenth the average prespill abundance and in 6 months was still less than at any time before the spill.

Most of the *Porites spp.* (scleractinian corals) were found in the same habitats as *Polythoa.* After the spill, the area covered by *Porites* was less than 1% of what it had been before, but the corals were always present. The abundance of crustose coraline algae, the main reef builders at the reef crest, also decreased in the aftermath of the oil spill to 10%. The percent coverage of the calcaerous green alga *Halimeda opuntia* was reduced slowly after the spill due to delayed mortality.

Numerous colonies of shallow water corals were dead or dying at a depth of 1 to 2 m in heavily oiled areas. The proportion of dead or dying colonies averaged between 17 and 30% on oiled reefs.

Many corals generate large quantities of mucous when exposed to oil and this may protect them from more serious damage. After the initial toxicity has been dissipated, the recruitment of plankton larvae and adult organisms can begin from nearby unaffected areas. Recovery of impacted reef communities occurs in most cases within a few years.

1.3.7 Beaches

Selfcleaning is faster on more exposed, sandy beaches than on sheltered, vegetated and muddier shores. The oil concentration on sandy shores with low wave energy levels is reduced by mixing, dilution with new sediment and possibly by *in-situ* water washing, rather than by rapid oiled sediment erosion.

Mousse treatment results in more patchy surface oil distribution. Mousse infiltration is less deep and more variable, especially on muddy shores. Visible contamination by mousse is longer lasting than that by crude oil.

Surface oil cover remains almost complete until coverage by the tide occurs, when some of the oil is translocated laterally. The oil is often buried by new sediment deposits (transported by wind, wave or tide) and thus is invisible from the surface. Aerial surveys even a few days after a real spill would not reveal the full extent of oil contamination.

Some oil treatments increase the firmness of the sediments so that even on fine or medium grained sands the formation of an "asphalt pavement" may take place. The increased shear strength of oil sediments is thought to be a major factor in increasing the residence time of oil spills on sediment shores.

From relatively high-energy, poorly drained muddy sand flat, most of the oil is removed by the first tidal cover due to hydrological protection of the surface sediments and subsequent tidal flushing. Clean-up may be completed by sediment current ripple migration under high-energy conditions.

High water tables on moderately exposed tidal flats contribute to their low vulnerability to oil pollution. The slightly higher vulnerability of sheltered tidal flats and salt marshes is also contingent on surface sediment drainage.

1.3.8 Salt Marshes

Salt marshes are extremely valuable components of many estuaries and coastlines. Salt marsh environments are extremely productive and provide nursery grounds for fisheries and are major energy sources in many coastal estuaries. Marshes control tidal erosion, buffer the impact of coastal storms and act as an interface between the land and sea. This last role prompts concern for the vulnerability of marshes when inundated with crude oil or refined petroleum products.

The most damaging effects occur when marsh plants are completely covered with oil. Even on these plants, the effects are limited primarily to the death of leaf structures above the ground, with normal growth resuming in as little as two weeks to several months. The potential for recovery of the vegetation damaged by oil is difficult to predict. Past studies have shown that the most important factors that appear to control whether oil will have a negative effect on the vegetation are:

- the degree to which oil comes into contact with the aerial portions of the plant;
- the extent of oil penetration in the sediment;
- the toxicity of oil.

There are three regions of vegetation (fig.1.35.):

a - creek edge;
b - midmarsh;
c - high marsh.

Plant stress symptoms are generally similar in all three vegetation zones but the symptoms progress at different rates. Symptoms progress fastest in the midmarsh zone and slowest in the high marsh zone. In oil-only treatments in all vegetation zones, oiled portions of leaves became chlorotic (yellow) within 3 days. Chlorosis was initially restricted to oiled portions only and unoiled tissues remained apparently healthy. Seven days after oil-

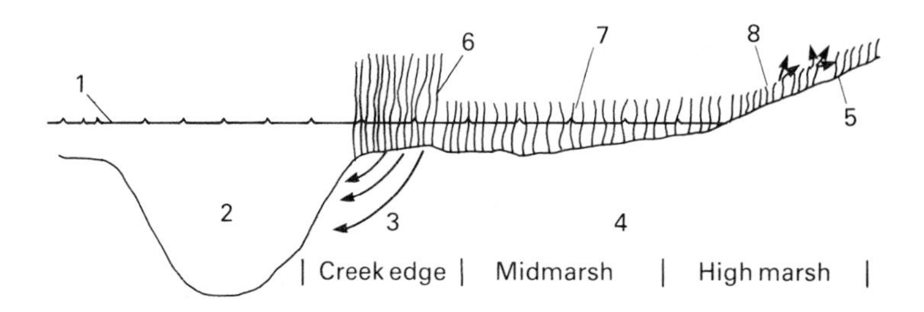

FIG.1.35 Schematic of major vegetation zones and dominant macrophytes in an Atlantic coast salt marsh: (1) high neap tide; (2) tidal creek; (3) drainage on most tides, good exchange of aerobic soils; (4) region of poor exchange, always waterlogged, anaerobic sulphate-reducing soils; (5) region of soils usually aerobic; (6) tall *S. alterniflora*; (7) short *S. alterniflora*; (8) short patens and other species.

ing, plants in the midmarsh became encrusted with salt, suggesting disruption of osmoregulatory functions. This sympton was restricted to the midmarsh samples. With time, chlorosis and eventual death progressed up the stem and plant mortality continued to increase with time. 20 days after oiling, 47% of plants in the midmarsh were judged to be dead, while in the creek edge and high marsh zones only 5% of the plants died. After 68 days, mortality had increased to 79% in the midmarsh, 34% in the high marsh and 23% at the creek edge.

Dispersant-treated plants became slightly chlorotic over the entire plant within 3 days of application. By day 5, the leaf tips had become severely chlorotic (white) and this chlorosis proceeded down the leaves over the following days, moving fastest in the midmarsh samples. Chlorotic areas died within a few days and by day 20 over 87% of the plants died in the midmarsh, 20% at the creek edge and 8% in the high marsh.

Two kinds of deformities of plant morphology were observed in high marsh *S. patens*, treated with either oil or oil/dispersant mixture. The first was the occurence of wavy ridges on the adaxial leaf surfaces. The second abnormality was more obvious and found in approximately 10% of the plants. In these, a U-shaped crook or loop developed in the stems, usually occuring immediately above the oiled portions of the plants. Microscopic comparison of deformed and normal stems revealed that the inner layers of leaves (young grass shoots having concentric layers of leaves in various stages of development) were almost fully developed in the deformed stems. All stems in deformed plants appeared to be developmentally mature, while in normal stems the inner layers were much less developed than the outer layers.

Reproductive biomass was severely affected by all treatments. The oil-only treatment had least effect, whereas the oil with dispersant had the greatest impact. The oil treatment had little long-term effect on plant height. The oil with dispersant treatment reduced the average plant height more than the oil treatment, but less than the dispersant-only treatment. High marsh showed less impact than other communities.

The largest reductions in stem density were associated with treatment with an oil/dispersant mixture in creek edge and midmarsh communities. The oil-only treatments generally resulted in the lowest reductions in stem density. The midmarsh community appeared to be the most sensitive to any of the treatments, while the creek edge community appeared to incur the lowest reductions in stem density. The highest reductions were of the order of 90%. Dispersant alone and in combination with oil had the largest impacts on density of flowering shoots, virtually eliminating reproduction in most instances.

Differences in plant response are probably linked to a large number of factors, such as type of oil or dispersant, amount applied, method of application, extent of oil weathering, time of year, species of plant and environmental stress experienced by the plant before treatment.

Mousse does not penetrate deeper sediment layers, whereas crude oil penetrates algal and litter layers and deeper sands. Mousse alone has little effect on plant cover with many plants flowering and setting seeds. Fewer plants were found in the dispersed mousse and crude-only plots, while none occured in the dispersed-crude plots. The larger plants, particularly *Spartina*, will recover quickly during the growing season by new growth from perennial root stock beneath the bare marsh surface. The growth may continue next season and help to stabilize the marsh.

If the authorities responsible for the clean-up can resist any pressure to clean the visually offensive remainder for 3 to 5 months, tidal flats will self-clean. Amenity and ornithological interests are likely to be the major sources of such pressure. The following clean-up methodology is reiterated:

- emphasis should be on protection rather than clean-up and dispersants may be favoured over floating oil (i.e. short term impact on fauna as opposed to long-term impacts on their habitat);
- the need for a clean-up should be ecologically, rather than aesthetically determined;
- clean-up techniques should be appropriate and not harsh;
- an ecologist should be present at large spill clean-up of marshes to conduct baseline surveys and monitor the clean-up.

1.3.9 Mangroves

Of the four mangrove species the red mangrove is the most seaward in occurence and makes up nearly all of the fringing forest. The labyrinthine prop roots of red mangroves serve as a firm substrate that support a number of plants and animals.

After a spill, the mangrove trees in oiled areas become defoliated and the leaves yellow. This is particularly so in areas where the trees are rooted in a berm of intertidal sediments. In most places, the berm apparently intercepts and partially absorbs the oil, blocking further movement of oil into the forest. The trees of the inner fringe that are rooted in the subtidal sediments suffer less defoliation. The trees that are rooted in supralittoral sediments of the oiled region suffer less defoliation than the trees rooted at lower levels. If mortality follows stress and defoliation of the red mangroves, the impact may be much wider than loss of the trees themselves. The thickets of prop roots of this species serve as breeding and nursing areas for many marine species and as a substrate and shelter for a diverse group of others, including economically important species of fish, molluscs and crustaceans. Sediments retained by the root masses could be released, if the roots decompose, which is a potential threat to nearby corals and other organisms that are intolerant to siltation.

1.4 Input of Oil to the Marine Environment

Oil enters the marine environment from many sources such as natural sources, offshore production, marine transportation, atmospheric deposition, waste waters, run-off and ocean dumping. No exact figures are available due to scarcity of data and only rough estimates could be made.

1.4.1 Natural Oil Input

Oil enters the marine environment naturally by means of two processes:

- submarine seepage especially due to earthquakes;
- the erosion of sedimentary rocks by rivers and underwater currents.

There are considerable difficulties in observing natural seepages in marine environments and residual evidence in the form of tar and asphaltic stains is hidden by water. Many thousands of seawater samples have been taken from the surface, near-surface and bottom, as well as bottom sediment samples for chemical analysis. Bottom and subbottom geological characteristics were also recorded with acoustic devices. Chemical analyses of gas bubbles migrating towards the surface indicate that they are of petrogenic and not biogenic origin. Tar samples have been collected frequently from water surfaces and beaches. Marine organisms are occasionally found on some of the tar samples and on the beaches tall grasses have been recorded growing through tar lumps.

Marine hydrocarbon seepages are frequently found in sedimentary rocks, although seepages are also found in different geological formations all over the world. There is a strong relationship between areas of high seepage and deformed areas of the earth's crust.

Recent scientific studies provide evidence for significant, naturally occurring oil and gas seepages that have existed in the marine environment for thousands of years. Many samples of asphalt have been found, which was used by prehistoric tribes to decorate pottery and to make it waterproof. There are many references in the literature, especially in the literature of Spanish explorers (XVI and XVII century), to caulking of ships with tar found on the beaches of Texas and Louisiana.

Investigations conducted in modern times clearly indicate that hydrocarbons in the marine environment are not necessarily detrimental to marine life. Large amounts of bacteria developing in waters enriched by hydrocarbons may serve as food sources for oysters and clams, which grow prolifically in these waters. Pictures taken from submersibles show fish swimming through streams of naturally occurring gas rising from beneath the sea floor into the water column.

The results of much research on the specific effects of naturally occurring hydrocarbons upon marine environments have indicated that low intensity, persistent introduction of hydrocarbons over thousands of years into the ecosystem has not been deleterious to the marine environment. An ecosystem influenced in this manner can continue to be biologically active and should not be considered to be irreparably harmed.

1.4.2 Oil Input Due to Human Activities at Sea

Offshore production introduces oil to the marine environment due to discharges of production waters after processing to minimise the content of entrained oil. According to international regulations the oil content should not exceed 50 mg/l. Minor spills (below 7 tons or 50 barrels) happen fairly frequently although significant technological progress has been made in drilling and monitoring systems.

Offshore exploration and exploitation of oil and gas is connected to the danger of blow-outs and considerable spills. Several major offshore blow-outs have been registered so far (table 1.11).

Such a major spill due to a blow-out might happen at any time

TABLE 1.11 Major offshore blow-outs.

Location	Year	size of spill [m^3]
Santa Barbara	1969	13,600
Breton Sound	1970	8,300
Bay Marchand	1970	14,500
Ekofisk Bravo	1977	21,800
Campeche Ixtoc	1979	>65,000

and in any location and its effects might be catastrophic. Spills from underwater pipe lines are caused mainly by anchor dragging and corrosion.

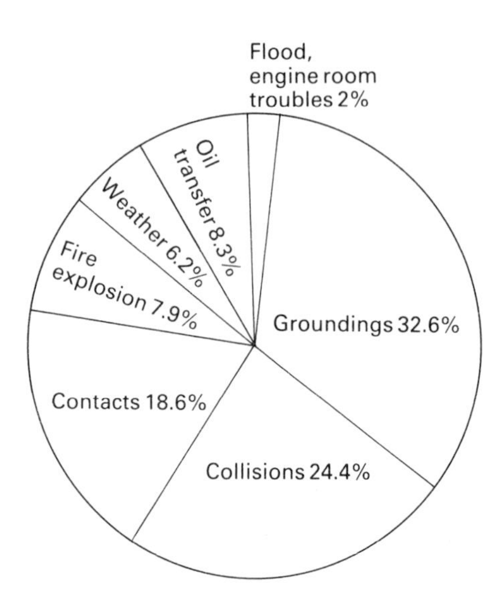

FIG.1.36 Significant Spill Causes.

Marine transportation has increased enormously in recent years. Over 1.5 billion tons of crude oil and oil products are transported each year by more than 7000 tankers. A significant amount of oil comes into the sea from operational discharges (ballast and bilge waters). The remainder is distributed among terminals, drydockings, fuel oil spillages from all ships while refuelling and accidents involving tankers and other ships. Analysis of significant spills shows that about 75% were caused by groundings, collisions and contacts and about 8% by oil transfer operations (fig.1.36).

The International Maritime Organisation (IMO) has undertaken the very difficult task of limiting operational and accidental discharges by introducing several very strict requirements concerning ships as well as seafaring specified in such conventions as MARPOL 73/78, SOLAS 1974, COLREG 72, STCW 78 and Annexes to these conventions.

1.4.3 Landbased Oil Input

The primary pathway for atmospheric deposition appears to be the removal of particulate material from the air by rain. Secondary pathways involve dry deposition of atmospheric particulate matter, scavenging of trace gases by precipitation and direct gas exchange with the oceans.

Municipal waste water appears to be the largest contributor, followed by industrial discharges and urban run-offs. Industrial waste from refineries includes only those refineries that discharge petroleum from their own waste water facilities. A sizable fraction of nonrefinery industrial waste is discharged into municipal waste water systems, but there is a quantity that directly enters the marine environment through coastal nonrefinery effluent discharges.

River discharges contain oil coming from untreated and treated waste waters, from urban and rural run-offs and spills.

Ocean dumping of wet sludge from municipal treatment plants and of dredge spoils adds quite a lot of oil to the marine environment.

1.4.4 Quantitative Damage Estimate of Oil Input

The range of total input of oil from all sources is very difficult to estimate. There are many approaches to this problem and many heated discussions occur over the estimates of probable ranges. Table 1.12 gives the range to be between 1.7 to 8.8 mln tons per year . The best single number estimate of total input is 3.2 mln tons per year.

Although the amount of oil and oil products transported by sea and production of oil offshore has increased in the last few years, the oil input did not increase over that period of time because of

TABLE 1.12 Input of oil into marine environment.

Source	Probable Ranges [mln tons]	Best Estimate [mln tons]
Natural Sources		
Marine Seepage	0.02-2.0	0.2
Sediment Erosion	0.005-0.5	0.5
Offshore Production	0.04-0.06	0.05
Transportation		
Tanker Operations	0.4-1.5	0.7
Drydocking	0.2-0.05	0.03
Marine Terminals	0.01-0.03	0.02
Bilge and Fuel Oil	0.2-0.6	0.3
Tanker Accidents	0.3-0.4	0.4
Non-tanker Accidents	0.02-0.04	0.02
Atmospheric Deposition	0.05-0.5	0.3
Wastewaters, Run-off and Ocean Dumping		
Municipal Wastes	0.4-1.5	0.7
Refineries	0.06-0.6	0.1
Nonrefining Industrial Wastes	0.1-0.3	0.2
Urban Run-off	0.01-0.2	0.12
River Run-off	0.01-0.5	0.04
Ocean Dumping	0.005-0.02	0.02
TOTAL	1.7-8.8	3.2

strict international requirements for ships, shipping and offshore operations. This action considerably reduced the operational and accidental release of oil into the sea. The "Torrey Canyon" oil spill (117.000 tons) in England (1967) and the "Santa Barbara" blow-out spill (13,600 tons) in the USA (1969) were pivotal points, because for the first time in the history of seafaring large spills occurred where the losses incurred due to damage to the environment by far exceeded the value of the ship and her oil cargo. During the "Torrey Canyon" clean-up operation highly toxic dispersants were directly applied to shorelines, causing greater ecological damage than the spill itself.

At the "Santa Barbara" blow-out efforts to contain the oil spill failed and oil had to be removed from the beaches in a very primitive and labour intensive manner. A cleaning station for oiled birds was set up, but survival rates were very low. Other disasters were to follow.

A wide range of research programmes was started to devise new clean-up methods and to identify their ecological impacts. Enormous progress has been made during the last 20 years in specialised spill control equipment and techniques.

Apart from major spills there are still a number of smaller incidents each year which cause shoreline damage. There are a small number of ships which discharge dirty ballast during bad weather or at night, when they feel they are safe from being detected. So, many nations are now introducing aerial control using the most modern equipment (aerial surveillance).

The damage resulting from oil spills caused by ships or platforms is usually local and of limited duration, but exceptionally large spills such as the "Torrey Canyon" (117.000 tons) and the "Amoco Cadiz" (230.000 tons) can persist for several years, especially when they take place in the Arctic or subarctic regions, like the "Exxon Valdez" spill of 11 million gallons of crude oil in Prince William Sound (Alaska) on 24 March 1989. Although oil spills are often dramatic, land based sources of oil pollution are much more significant, because they take place constantly and thus cause major changes in affected areas (Table 1.13).

TABLE 1.13 Summary of occurrence rates for accidental oil spills.

Source	Number of spills per billion bbl	
	>1,000 bbl	>10,000 bbl
platforms	1.00	0.44
pipeline	1.60	0.67
tankers - total	1.30	0.65
at sea	0.90	0.50
in port	0.40	0.15

There are various levels of biological effects of oil. These include:

- human hazards through eating contaminated seafood;
- decrease of fisheries resources and damage to wildlife such as sea birds and mammals;
- decrease of aesthetic values due to unsightly slicks and oiled beaches;
- modification of marine ecosystems by elimination of certain species with an initial decrease in diversity and productivity;
- modification of habitats, delaying or preventing recolonisation.

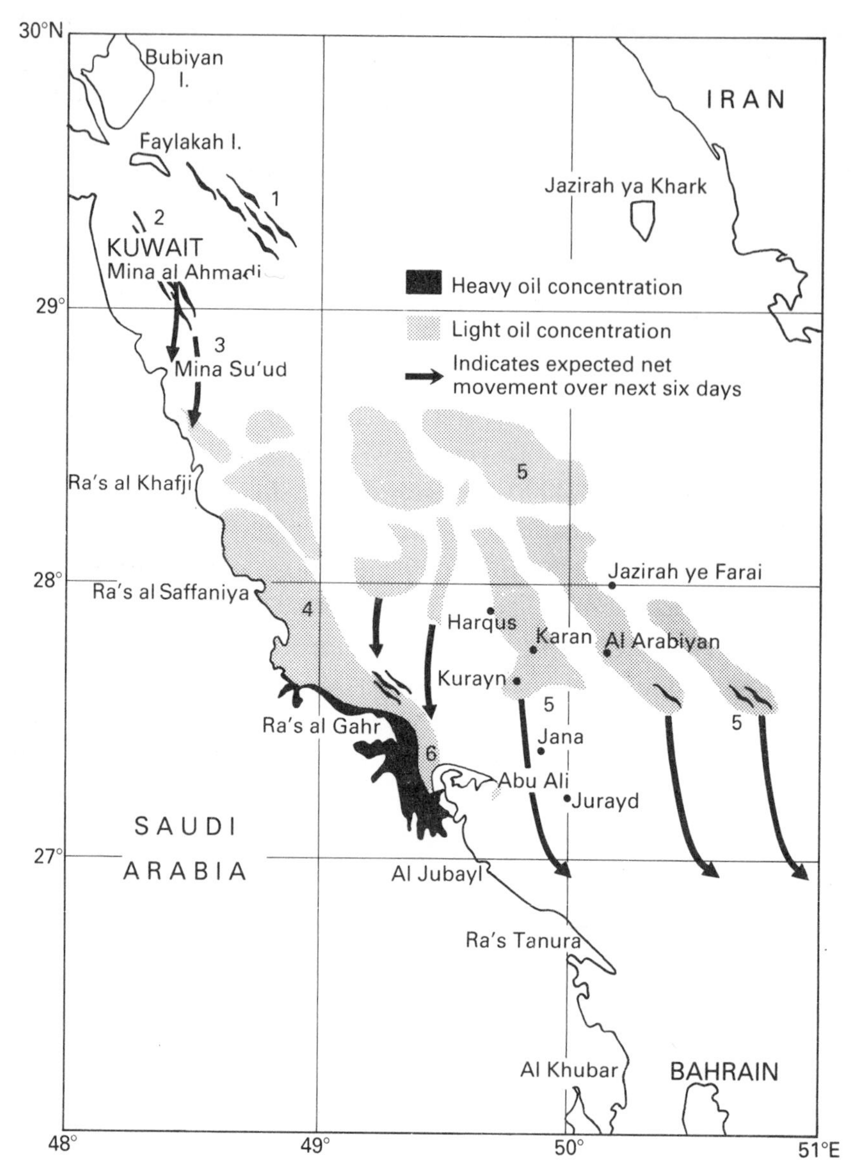

FIG.1.37 The extent of the Kuwaiti oil spill as per March 4, 1991 at 15.00 hours est.; numbers denote oily patches named in the text.

Risk assessment of an oil spill should be made and quantified in probabilistic terms. Contingency planning and preparations for the immediate combating of any size of oil spill should be carried out well in advance. Once the spill occurs, then accurate forecasts of the speed and direction of the slick, as well as of its spreading rate, are of great importance. Based on that information the best method of clean-up has to be determined and implemented.

As oil moves onto a beach or into a bay, its spreading is reversed and it tends to thicken. The onshore winds and waves will drive the oil onshore, so it continues to float in a trapped pool, part of which is deposited on the beach, when the tide ebbs or the storm drops.

The Gulf War produced some new experiences of the reasons and effects of oil spills. A very heavy oil spill was caused by Iraqui forces in January 1991 by pumping about 2 million barrels of crude oil into the sea, mainly through the loading terminal at Al Bakar. This oil quickly spread unhindered due to hostilities, over the surface, covering large coastal and nearshore areas. The extent of the spill is shown on the enclosed chart (fig.1.37) as per March 4, 1991 at 15.00 hours based on observations by an SLAR aircraft. Overflights on this date revealed several areas of floating oil in Kuwaiti waters. The first area (1) is a long streak of sheen and mousse that appears to be heading in a SSW direction from the channel south of Bubiyan Island. A second area (2) consists of a series of small sheen and mousse patches that appear to be drifting in a southerly direction from around the terminal at Al Bakar. The third region (3) is a relatively large area of scattered sheen and mousse at Mina Su'ud.

South of the Kuwait-Saudi Arabia border, floating oil appears as scattered patches of sheen with some lines or bands of heavier water-in-oil emulsion. One of these patches is centred about 20 miles east of Ras al Saffaniya (4). A second and larger patch (5) is further to the east and appears in SLAR imagery as a series of widely scattered sheen patches extending from about 80 miles east of Ras al Khafji to about 25 miles east of Abu Ali. The heaviest concentrations of oil (6) are seen along the nearshore region between Abu Ali Island and Ras al Tanajib.

According to the weather forecast, the winds for the next few days would be from N and NNW. A pronounced sea breeze was expected to move the oil closer to the shore each afternoon. Offshore patches of oil would move generally south, but were likely to shift to the west, closer to the coast. Of particular interest were oil patches currently north and east of Abu Ali (5). These would move south of the point, then toward the nearshore zone, threatening coastal impacts near Al Jubayl and along the coast to the south.

The oil concentrations along the Kuwaiti coast line would be expected to move south and toward the shore. Thus oiling was expected along the Kuwaiti coast, south of Mina al Ahmadi.

Just before the ground offensive of the Gulf War started and during the retreat from Kuwait, the Iraqui forces blew-up over 600 oil wells. Each day about 2 million barrels of crude oil went up in flames. The smoke from burning wells was so dense in some places that cars were forced to drive during the day with their lights on and the temperature dropped by 10 °C, lowering the desert region to a European spring chill. It is impossible to predict the full impact, but climatic effects of oil burn-off would be felt 2,000 km away from Kuwait and black rain contaminating public water supply and damaging agricultural economy would be expected at that distance. Hydrocarbons in the black rain could cause cancer and are likely to be consumed by human beings, livestock and fish in affected regions. If acid rain pollutes underground water sources, some areas could be rendered uninhabitable.

CHAPTER 2

Chemical Response Technology to an Oil Spill

2.1 Natural Cleansing

2.1.1 Natural Biodegradation

Natural cleansing, allowing oil to be degraded and removed by natural means, could for some spills be the most ecologically sound method, although it is not an active method and it takes a long time to be fully effective. On the open ocean, where there is no threat to any sensitive onshore or offshore habitats or species, the spills will evaporate, weather, disperse and degrade naturally. But such a spill should be monitored in order to intervene if the necessity to combat the spill should arise. Pollution of seawater by oil results in the formation of a continuous film or slick of oil, which tends to spread continuously. In open seas, this oily film is undesirable because it constitutes a barrier to the transfer to seawater of air and light, which are indispensable to the support of marine life. In coastal waters, oily slicks damage crustacea beds and beaches.

Active biodegradation of oil droplets suspended in the water column requires the presence of a high concentration of microorganisms at the oil/water interface, but they are present in limited quantities in seawater. Thus it is necessary to speed up the proliferation of these microorganisms in order to stimulate biodegradation. The organisms need not only the oxygen and carbon contained in seawater and oil respectively, but also nitrogen and phosphorus, which in natural seawater are present in small quantities, preventing the required rapid rate of development of these bacteria. Thus natural biodegradation of oil is a slow process, requiring months and sometimes years to complete this process.

On shorelines with high energy environments such as exposed rocky intertidal habitats the natural processes result in more rapid cleansing. In some low energy habitats such as marshes, if oiling is light, natural cleansing might be preferred, because almost all the clean-up methods involving pedestrian or vehicular traffic are potentially damaging, in some cases even more so than the spill itself. This traffic causes disruption of natural water circulation patterns, damages plant root system and works oil into the sediments. Burning and cropping the vegetation often delay recovery compared with natural cleansing. So in most cases the natural cleansing process should be considered first.

2.1.2 Enhanced Biodegradation

When oil is spilled onto the sea surface, natural processes such as evaporation, dissolution and biodegradation play a large role in the fate of the oil spill. These processes are usually safe, but they are slow with the exception of evaporation. Such organisms as bacterial yeasts and fungi are ubiquitous and can metabolise most of the oil components, but the process is very slow. The time necessary in natural conditions for degradation varies from two to six months and sometimes even longer.

Hydrocarbon biodegradation is a biological oxidation and therefore most microorganisms involved are aerobic species. Biodegradation is limited by several factors, of which the most important are:

- oxygen content of the water;
- temperature, as the biological activity is related to temperature: oxygen content increases, when temperature decreases;
- nutrient availability - in the vicinity of an oil slick carbon is abundantly available, but usually there is a lack of nitrogen and phosphorus, which are essential for microbiological activity.

The highest level of biodegradation will be reached when the limiting factors are at the optimal level. The oxygen content and the sea temperature can not be changed, but the nitrogen and phosphorus concentrations at the oil/water interface can be increased by the addition of nutritive products, which contain nitrogen and phosphorus at the correct ratio, are insoluble in seawater or have very low solubility, are stable when stored, are biodegradable, are nontoxic to marine fauna and are as cheap as possible. It should also contain carbon, that would be easily assimilated by microorganisms. Such bacteria, capable of metabolizing oil, are found in every sea of the world in sufficient quantities and there is no need whatever to seed them in the region of an actual spill.

A suitable nutrient should:

- be of the surfactant type, i.e. be both lipophilic and hydrophilic in order to remain at the surface of oil droplets and not to be washed away in water. The adverse influence of inverse emulsions on biodegradation rate is shown in fig.2.1, where much of nutrient is entrapped at the oil/water interphase inside the oil globule and not available to the bacteria;
- contain all the nutritive compounds, the concentration of which can be adapted to any specific problem;
- be of an oleophilic formulation, as the external phase is oleic acid, which when introduced into the sea will mix easily with an oil slick;
- be reasonably inexpensive;
- induce a high rate of biodegradation.

Fig.2.2. shows the variation of effectiveness in function of hydrophilic/lipophilic balance (HLB) of the main emulgators of the dispersant without and with nutrient. If the nutrient is added to the optimal dispersant (HLB 8.5) without correction, the effectiveness drops from 24% to 15% (WSL test). But with the correction of HLB of the main emulgators, the effectiveness can be fully recovered and even surpassed. Fig.2.3. shows how different chain lengths have a different influence on effectiveness. Short chain nutrients are too soluble in water and are washed

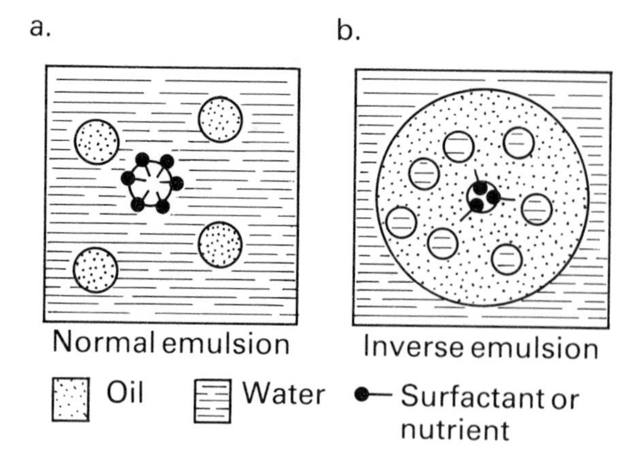

FIG. 2.1 Schematic drawing of oil-in-water (normal) emulsion and water-in-oil (inverse) emulsion.

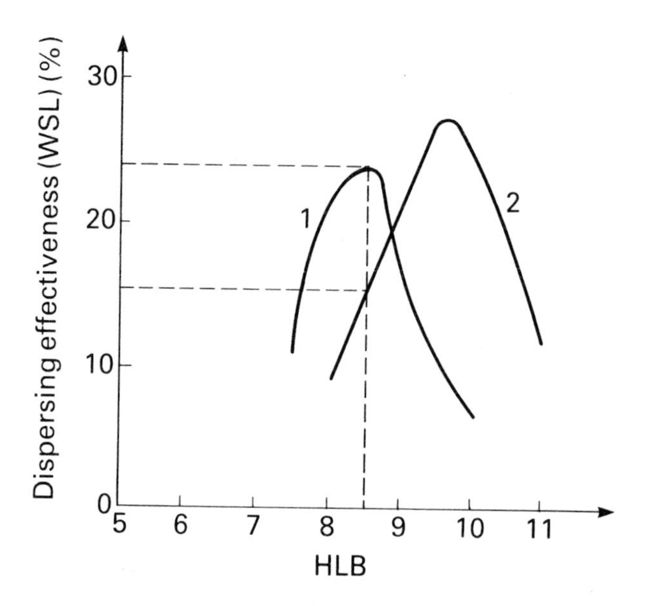

FIG.2.2 Influence of addition of nutrient to dispersant on dispersing effectiveness; (1) dispersant without nutrient; (2) dispersant with nutrient.

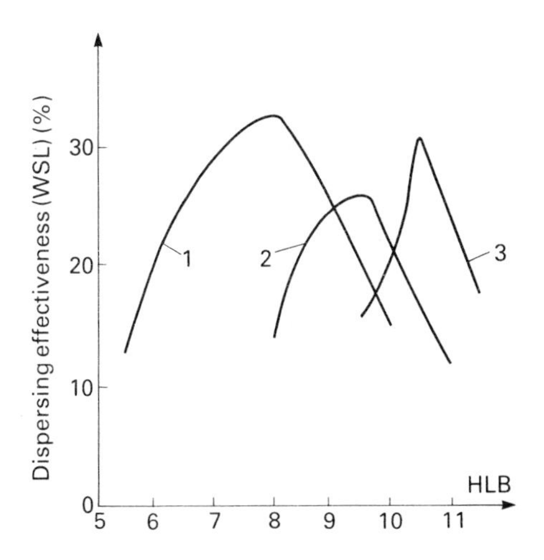

FIG.2.3 Influence of chain length of nutrients on the effectiveness of dispersants; (1) octyl; (2) dodecyl; (3) oleyl.

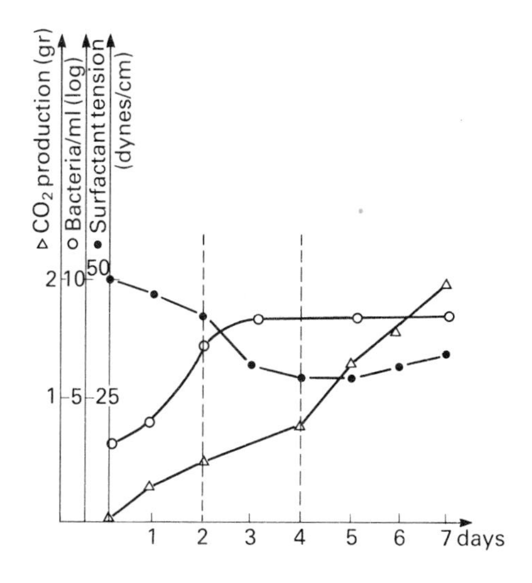

FIG.2.4 Evolution of carbon dioxide production, of bacteria concentration and of surfactant tension.

into the water before biodegradation starts. The result becomes better when the chain length increases. When they become too lipophilic, they induce the formation of inverse emulsions, which limit biodegradation.

Fig.2.4 shows the evolution of carbon dioxide and biomass production and the evolution of surface tension during a test. The first phase comes after a very short period of latency. Biomass increases exponentially with the assimilation of the carbon, giving rise to CO_2 production. The surface tension changes a little. The second phase is a rather slow step. The biomass does not increase very much and the surface tension decreases as production of biosurfactants increases.

In the third phase high production of CO_2 corresponds to a high level of biodegradation of the oil. At the end of this phase, the surface tension increases, when not enough hydrocarbons are available for further bacterial action to continue.

Experiments carried out in 18 m^3 tanks with an oil layer of 500 microns have shown that:

- addition of the nutrient increases the rate of elimination of oil from 14 to 24% to 62 to 70% and even 79%; in sea tests the rate of elimination of oil increased from 10% to more than 90%;
- temperature plays a role in the evaporation rate, but does not affect the total amount of oil evaporated. This is probably because evaporation and biodegradation affect the same components of crude oil;
- the type of crude oil used makes no significant difference;
- a nutrient/oil ratio of 0.1 seems sufficient; there is usually no advantage in increasing this rate;
- different methods of introducing the oil make no significant difference.

Enhanced biodegradation can be considered to be a new tool in fighting an oil spill. However, it is not and never will be the solution to any single problem and it should be used in conjunction with other treatments. The best areas of application are:

- spills occurring from an offshore blow-out;
- spills in remote and inaccessible coastal areas, where clean-up operations are technically and/or economically impossible;
- spills in sensitive areas, where many treatments would be more harmful than the oil itself;

– spills in ice-infested waters or under ice.

2.2 Chemical Barriers, Oil Sinking Agents and Sorbents

2.2.1 Chemical Barriers

Certain chemicals act as surface tension modifiers and inhibit the spread of oil. When placed on the water surface next to the floating oil, the oil is pushed away as the result of the stronger spreading force of the chemical (lower surface tension). Thus the surface collecting agents are used to decrease the area of an oil spill and thereby enhance the potential for mechanical removal techniques. In 1939, Garrett observed that surface films with spreading forces of 40 dynes/cm or greater will operate effectively on oil slicks with thicknesses of 1.0 cm or less. Garrett also indicated that such collectants would not concentrate films effectively against winds greater than 2 m/s. Practical sea trials have shown that they could keep a rapidly spreading slick in check for at least 5 hours, allowing mechanical equipment to be deployed and limiting the area of spill to a minimum.

Collectants are of the same family as dispersing agents. They both are or contain surface active agents. A practical distinction can be made on the basis of solubility. Collectants by the virtue of their lower solubility will:

- remain effective for a longer period of time (persistence), because the film remains intact longer;
- have a lesser effect on biological oxygen demand and a lower toxicity to the underlying water column.

A surface collectant spreads rapidly over the water surface. Initial spreading rates are at least 30 cm/s, falling with time to about 5 cm/s. During the first 6 minutes the average rate is about 10 cm/s. If the initial application covers a 1 m wide strip of water, then after 6 minutes the covered strip will have a width of 35 m. At an application rate of 8 litres per linear kilometre the initial coverage would be about 5,000 mg/m^2. After spreading to 35 m width the coverage would be 140 mg/m^2. If the collectant

was dissolved completely in the top 1 m layer of the underlying water immediately after application, the concentration resulting from the 5,000 mg/m^2 layer would be 5 p.p.m., but in 6 minutes spreading time this concentration would be 0.14 p.p.m. Owing to the relative insolubility of the collectant and the great speed of spreading, actual concentrations of 1 p.p.m. can be expected at a rate of 8 litre/linear kilometre.
Technical data for the collectants used in experiments are given in Table 2.1:

TABLE 2.1 Technical data of collectants

Collectant	Viscosity (cS (°C))	Flash point (°C)	Density lb/gall.	Spec.grav. at °C	Pour point (°C)
Corexit OC-5	15 (37.7)	82	7.64	0.917 (15.6)	- 35.5
Nalco 3WP-086	60 (37.7)	84	8.22	0.985 (20.0)	- 3.9
Shell Oil Herder	8 (25.0)	78	6.52	0.926 (15.6)	- 2.0

Experiments have revealed no significant differences in the ability of the three collectants to concentrate and thereby to thicken the oils tested. The reduction in oil slick areas varied, depending upon the average thickness of the initial (untreated) oil layer. Thin films (0.001 to 0.01 mm thick) were reduced 85 to 95 % in area. Such thin films were thereafter as much as 20 times thicker after collectant application. Oil layers at a near cold-water equilibrium thicknesses of 1 to 3 mm experienced as much as 50 % reductions in area, therefore doubling in thickness. Thick layers (4 to 6 mm) responded to the collectants with minor reductions in overall coverage and correspondingly smaller increases in thickness. The data suggest that it would be impractical to treat oil layers of 1 cm or more in thickness. None of the tests suggested that mechanical recovery techniques (such as sorbents or skimmers) would suffer due to the presence of surface collecting agents. The resulting thicker layer would enhance the efficiency of most recovery systems.
The use of collectants will enhance the chances of igniting and sustaining combustion of an oil slick. Thin layer (<0.001mm) oil films, however, will probably remain below the critical thickness for sustained combustion (approximately 1 mm) even after

collectants are applied. Such films will also have experienced evaporative losses, which make ignition difficult.

An effective herding agent can push oil against a current of about 0.25 knots. At higher currents the system collectant-plus-compressed oil will be carried with the current. The collectant can push oil against the drag exerted by a wind of about 2 to 5 knots. At higher wind speeds the system will be driven downwind. Collectants are less effective with viscous oils and thus they should be used immediately after the spill.

Application from a helicopter is more effective than application from a boat, because:

- a given area could be encircled very much faster by helicopter;
- the high angle of view makes it easier to see the thin film's outer boundary, than from the low angle of a boat.

Collectants should be applied in accordance with the manufacturer's instructions. According to international regulations their use requires approval of an appropriate governmental agency. Practical application rates will not exceed one litre per linear kilometre per 6 hour period with up to 3 applications in 24 hours. This mode of application has several advantages:

- low application rates per linear kilometre will not cause any adverse effects to the marine environment;
- re-applying the collectant every 6 hours on an as needed basis maintains its necessary persistence;
- three applications 6 hours apart present an opportunity to limit the collectant to the minimum amount to achieve an oil clean-up.

2.2.2 Oil Sinking Agents

Removing oil from the water surface could be achieved by sinking it to the bottom by applying sinking agents such as sand, clay, chalk, fly ash or cement. Applied to an oil slick they adsorb oil, the density of the resulting oil/agent mass above 1 kg/l and the mass sinks to the sea floor, remaining there for a long time. To prevent long-term effects such transportation of oil to the sediment should be prevented for the following reasons:

- bottom flora and fauna are likely to be adversely affected, because of the smothering effect exerted by the oil/agent mass;
- the oil passes through a broad spectrum of marine life between the surface and the bottom, possibly clogging the gills of organisms;
- the sunken oil/agent mass will be migrating over wider areas of sea floor than originally affected and therefore spreading the adverse effects of the spill;
- the oil/agent mass can foul fish bottom nets and gear and otherwise taint commercially valuable species;
- under certain turbulence and temperature conditions, the entrapped oil can be released, resulting in recurrence of surface slicks and possible shoreline contamination.

For the above reasons this method of removal of oil from the water surface is not recommended and in some countries it is prohibited.

2.2.3 Sorbents

Sorbents are commonly used for the final clean-up of trace amounts of oil or to remove oil from areas inaccessible to skimmers. Sorbents are very effective as "polishing" methods, i.e. in removing a thin layer from the water surface or a small amount of oil from beaches. Ecological effects of sorbent use are minimal except when deployed and recovered by hand in habitats sensitive to disturbance (marshes) or when they are not properly recovered.

Absorbents function by capillary action and the more porous the substance, the greater the opportunity for uptake of oil into its capillaries. But this capability will depend upon the specific gravity and viscosity of the spilled oil, because lighter oil will draw further up into the capillaries. The absorbtive capacity will depend on the surface area to which the oil can adhere. A good sorbent picks up oil strongly and picks up water weakly. The oil is taken up not only by surface adsorbtion, but also by entrapment in voids such as pores and/or capillaries. Thus oil sorbents work best on heavy, highly viscous oils and their performance on lighter oils or more freely flowing water-insoluble organic liquids can be poor. The uptake time of a good sorbent is fast - less than one minute - but it must retain oil when it is lifted off the water.

In order to improve the ability of the sorbent to attract oil instead of water and to stay afloat on the water surface, sorbents are treated with oleophilic compounds, which attract oil, and hydrophobic agents, which repel water. Some materials which attract oil rapidly allow it to drain out easily again.

There are three groups of sorbents (Table 2.2.):

- natural organic;
- mineral based;
- synthetic organic.

Natural sorbents absorb three to six times their weight in oil, are nontoxic and relatively nonpersistent in the environment as they are biodegradable. The oil is trapped in natural crosslinked strands, rather than by strand capillarity. Peat moss and chicken feathers, treated to increase their oil sorption potential are often used, but they absorb water as well as oil and sink rapidly when saturated with water, resulting in severe clean-up problems. Recovery of large volumes of natural sorbents requires considerable manpower and extensive disposal of oil-contaminated sorbents by burning or burial. Natural sorbents release low viscosity oils, while being lifted out of the water. To minimise this problem some natural sorbents are available in compressed sheets.Sawdust, pine bark and wood chips are available worldwide. Peat dried to a water content of less than 30% is an excellent sorbent, but it will sink unless pretreated.

Mineral-based sorbents have slightly better recovery efficiency than natural sorbents (four to eight times their own weight). They show several disadvantages such as:

- distribution difficulties in windy weather because of their very low specific weight;
- potential respiratory irritations due to inhalation of their dust;
- persistence in the environment because of nonbiodegradation;
- abrasiveness of sorbents causing sometimes damage to reco-

very devices.

Synthetic organic sorbents show extremely high recovery efficiency quite independent of oil viscosity. They usually are in the form of plastic foams or plastic fibres, which are quite easy to spread and to recover. Some synthetic foams can be reused after the oil has been squeezed out, but they readily absorb water. Synthetic foams are available in the form of sheets, booms and rolls. Polyurethane foam can be produced at the spill site by mixing two component liquids, which immediately expand to collate the sorbent foam. However these materials are relatively expensive and must be completely recovered to avoid persistence in the environment.

TABLE 2.2 Maximum oil sorption capacity (grammes oil/grammes sorbent)

Sorbent		Test oils			
		Bunker C	Heavy crude	Light crude	No.2 Fuel
	Viscos.at 77 °F (cS)	2,800	2,600	7.8	3.1
	Spec.grav.at 77 °F	0.942	0.977	0.854	0.856
Natural sorbents					
Corncob, ground		5.7	5.6	4.7	3.8
Peanut hulls, ground		5.8	4.3	2.2	2.2
Redwood fibre, ground		14.7	11.8	6.5	6.4
Sawdust		3.0	3.7	3.6	2.8
Wheat straw		5.8	6.4	2.4	1.8
Wood cellulose fibre		18.6	17.3	11.4	9.0
Mineral sorbents					
Perlite		4.6	4.0	3.3	3.0
Vermiculite		4.3	3.8	3.3	3.6
Volcanic ash		21.2	18.1	7.2	5.0
Synthetic sorbents					
Polyurethane foams					
A. Polyether type, shredded		72.7	74.8	70.0	48.7
B. Polyester type, reticul.		30.3	24.5	30.6	27.5
C. Polyether type, 1/2" cubes		72.7	71.7	66.1	64.9
Urea formaldehyde foam		72.7	52.4	50.3	47.6
Polyethylene fibres					
A. Wool type		37.0	27.8	19.7	16.1
B. Sheet, matted		18.6	17.6	11.9	10.6
C. Continuous element, nonwoven		46.0	36.7	45.4	36.2
Polypropylene fibre, nonwoven		21.7	18.1	6.9	4.8
Polystyrene powder		23.4	21.7	20.4	5.8
Polyester shavings		8.8	7.4	6.6	4.7

Sorbents are available in a number of physical forms, such as

belts, chips, mops, pillows, powder, granular fibres and pallet material, ropes and booms, and sheets. Belts must be oilsorbent and sufficiently resilient to withstand repeated squeezing. No sorbent material can be endlessly reused. The number of cycles (its life) will depend upon the oil type, the care with which the material is handled, squeeze pressure, etc. As oils become more viscous, the sorption time increases until finally the oil coats only the surface of the foam. In such cases an oleophilic belt is more suitable, but it should not be more than about 2 m wide. The rate of propulsion must be low and as the speed is small, the rate of collection is relatively low.
Table 2.2 gives information on the oil sorption capacity of some sorbents.

2.2.4 Application of Sorbents

Application of sorbents includes the following operations:
- sorbent distribution;
- sorbent collection or harvesting;
- sorbent separation from contaminants;
- contaminant storage and/or disposal;
- oil/sorbent reuse and/or disposal.

Sorbent distribution depends on the physical form of the sorbent to be employed, its quality, the circumstances and the size of the discharge of contaminant. For small spills and final clean-up near the shoreline sorbents are applied manually, but for large spills mechanical methods such as blowers should be used for distribution. Sorbent mops could be used from ships or shore in inaccessible areas and for the final clean-up of residues, that have escaped heavier clean-up equipment. Sorbent foams and threads, generated in situ, can also be sprayed on water.
Sorbent sheets can be deployed from small boats. The sorbent material may be contained in an inert material to form a pillow, which must have a large size mesh to provide free access to the sorbent. Very elongated pillows can be connected to one another and used as sorbent boom. Another type of sorbent boom is a rope mop as shown in fig.2.5.

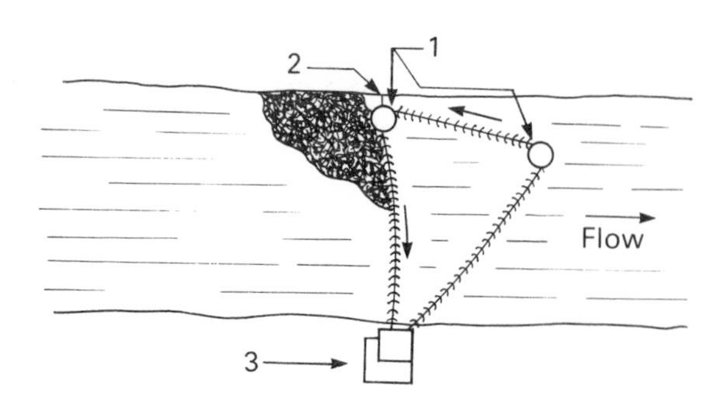

FIG.2.5 Rope type oil mop; (1) self-adjusting tail pulleys for level fluctuations; (2) baffle; (3) mop engine.

Sorbent collection is essential. The sorbent should float when saturated with recovered oil and should not become waterlogged before recovery. Floating sorbents can be contained on the moving water surface and their movement can be controlled by the use of nets, booms etc., but when the current is fast, then the oil-saturated sorbents may be carried under the boom. Floating sorbents may present problems during recovery by mechanical means and they may penetrate the seawater intakes of recovery vessels. Ice may hinder effective sorbent deployment and make pick-up difficult or even impossible. Manual tools should be used such as nets, rakes, forks, shovels, etc. for this purpose, but these operations are very labour intensive and thus very expensive. In case the sorbent can not be recovered, it should at least be inert and preferably biodegradable.

Sorbent separation from contaminant can be quite easily achieved by squeezing. For oil mops a single roller wringer can be used. Sorbents tend to release some oil by drainage (about 20% oil in 24 hours) and centrifugation can also be used as a separation technique. A commercial laundry extractor can separate about 80% of the oil from the sorbent.

Disposal and/or reuse of recovered oil and/or sorbent is a major problem. If the oil and sorbent are separated, then facilities must be provided for temporary and possibly long-range oil storage and final disposal. When this is impossible, then the problem of storage and final disposal is considerably more difficult. Barrel storage, burning or chemical treatment might be the final remedy.

2.3 Dispersion

2.3.1 Chemical Dispersants

The key components of dispersants are surface-active agents (surfactants), which are partly oil soluble and partly water soluble. They reduce the interfacial tension between oil and water, which, when added to wave and wind energy, breaks up a surface oil slick into small droplets, dispersing into the upper few metres of water column. Thus the slick no longer moves with winds and surface currents, but enters the water column, where it is rapidly diluted, quickly losing its lower molecular weight fractions and moves with the subsurface currents. But dispersing slicks introduces a large amount of oil into the upper water column which is generally recognised as a disadvantage of this method, because some impact to water column organisms at the site of dispersion would be expected.

Using dispersants to mitigate oil spills has remained controversial, despite considerable testing in both the laboratory and the field. One of the major concerns, which has not been satisfactorily resolved, is how effective are various dispersants on different oils over a range of environmental conditions. Laboratory experiments can not accurately simulate the real world, while field experiments are difficult to monitor and control.

Concentration test results show that the amount of oil entering the water column from a treated surface slick is dramatically higher than from an untreated slick under the same wave conditions with concentrations up to 70 p.p.m. compared to insignificant amounts in natural dispersion. Even when no agitation was applied to a treated slick, higher concentrations were observed in the water under the slick than under an untreated slick in 28 cm waves. In the case of an untreated slick, the formation of water-in-oil emulsions appeared to inhibit dispersion. When dispersant was applied, 10 cm nonbreaking waves deposited a significant amount of oil in the water column, though the exact amount could not be computed due to undetermined losses to the rest of the basin. In the first hour about 16% of oil entered the water column. Increasing the wave height increased the rate of entry of oil into the water column.

The advantages of dispersion in removing oil from the water surface are lost if the oil droplets do not stay in the water column. Once the oil droplets have entered the water, the stability of the oil-in-water emulsion formed depends on both the turbulence maintaining the drops in the emulsion and the size of the drops. If the level of turbulence is low, the size of droplets is particularly important. Stokes law for the terminal rise velocity of drops is given by

$$V_t = \frac{2r^2 g}{9\mu(\varrho_O - \varrho_C)} \tag{2.1}$$

where: r – radius of the drop;
g – acceleration due to gravity;
ϱ_O and ϱ_C – density of oil and water respectively;
μ – viscosity of water.

For example in 160 minutes a 50 μm drop would rise more than 4 m, whereas a 5 μm drop would rise only 4 cm. Hence an emulsion of 50 μm drops would be very much less stable than one of 5 μm drops. Fig.2.6 shows a typical population distribution and a typical volume distribution for a dispersant test, measured 50 cm beneath the surface. Most particles range between 1 and 8 μm in diameter with a mean particle size of 4.3 μm. The range for all dispersant tests is from 3.3 to 4.6 μm. In the volume distribution most of the oil volume fell into the volume range 3 to 30 μm with a well-defined peak at 6 to 7 μm. When dispersant is used, drop sizes tend to be smaller. In an untreated slick there is an increase in the percentage of particles in the larger 8 to 25 μm range over the background distribution and a decrease in the smaller particle range. The large droplets tend to rise

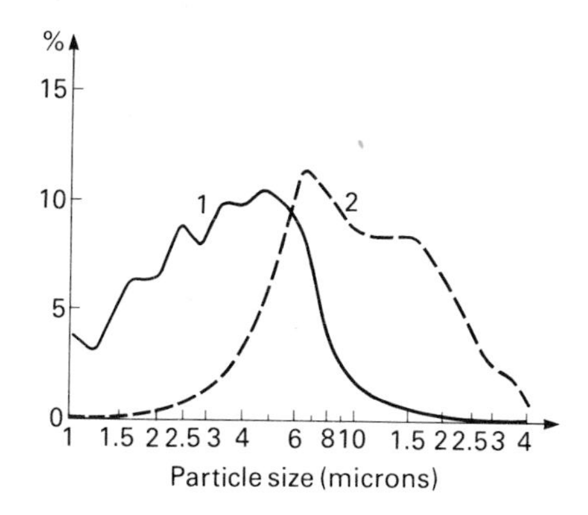

FIG.2.6 Distribution of particle size in 30 cm waves in 15 minutes after application of dispersant, measured at a depth of 50 cm; (1) population distribution; (2) volume distribution.

and coalesce rapidly with the surface slick after being dispersed, especially when no agitating force keeps them in suspension.

2.3.2 Effectiveness, Behaviour and Toxicity of Dispersants

The use of chemical dispersants for lessening the adverse effects of oil spills will depend on the following factors:

- effectiveness of a given dosage of dispersant on a given oil slick;
- diffusion of dispersed oil and dispersant into the water column and subsequent processes including dissolution, volatilisation, degradation and interaction with suspended and bottom sediments;
- effects of dissolved and particulate oil and dispersant on the water column and benthic biota.

The toxicological issue is of importance as it may have a detrimental effect upon marine.

When a dispersant is applied to oil slick the following processes will take place:

- dispersion processes in which the surfactants penetrate the oil surface and reduce the oil-water interfacial tension, thus promoting the formation of small oil droplets, which diffuse down into the water column (fig.2.7.). Larger droplets may rise to the water surface reforming the slick;

- hydrocarbons and surfactants in the oil drops will partition into the water, approaching equilibrium dependent on their partition coefficients and the oil-water volume ratios;
- dissolved organics may evaporate from the solution into the atmosphere;
- dissolved and particulate oil will diffuse vertically and horizontally, yielding continuously changing concentrations with depth;
- oil particles may associate with mineral and organic particulate matter in the water column and sediment to the bottom. Zooplankton may contribute substantially to this process. Sorption of dissolved hydrocarbons may also occur;
- each organism in the water column of a given depth will be exposed to a pulse of dissolved and particulate hydrocarbon and dispersant and will potentially suffer a toxic effect, which is dependent on the combined toxicity of dispersant, the particulate oil and the dissolved hydrocarbons.

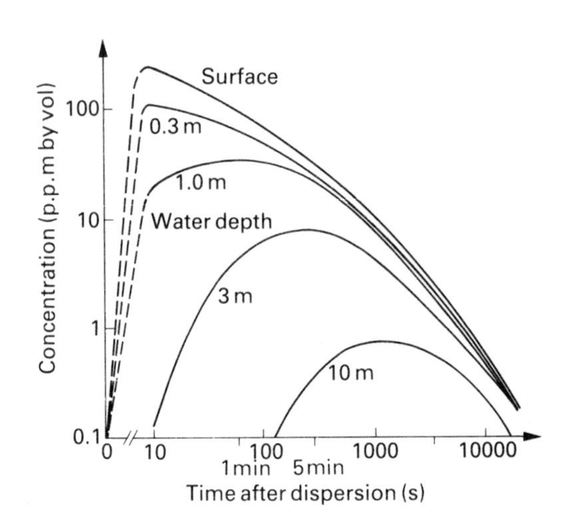

FIG.2.7 Total concentration of oil under the slick at given depth as a function of time.

The parameters to be considered when selecting a dispersant are as follows:

- toxicity to marine life. Consideration should be given to oil toxicity and an increased tendency for the dispersant and oil mixture made available to filter organisms;
- persistency in the marine environment. The persistency of the dispersant should be of the same order of magnitude as that of the oil to be dispersed;
- efficiency under given conditions of use. Materials efficient under given conditions may change their efficiency under changed conditions, i.e. temperature, salinity of water, etc.;
- cost - a low-efficiency material may be more costly in the long run than a more expensive one of high efficiency;
- convenience in use - this includes human toxicity, flammability, viscosity under cold conditions, etc.

2.3.3 Parameters Limiting the Application of Dispersants

Increasing oil viscosity (μ) decreases the effectiveness of the chemical dispersant. The viscosity increase is associated with the emulsification of seawater into the crude oil. There is an important distinction to be made between a viscosity increase of an oil due to weathering and that due to water-in-oil emulsification. This latter is very much more difficult to disperse - if it can be done at all - due to tying up of the dispersant at the many water droplet/oil interfaces. The dispersant is thus kept from the seawater/oil spill interface, which is vital for the chemical dispersion process. When oil becomes semi-solid (10,000 cP), it is difficult to have this mass combine with the relatively low viscosity fluid dispersant. The major portion of the dispersant rolls off the highly viscous oil and is lost into the water. This phenomenon is physical rather than chemical and therefore the use of hydrocarbon solvents in dispersant formulations may not improve the situation.

If the oil viscosity is increased in the lower range (less than 1,000 cP) other factors tend to offset the increase in viscosity and the effectiveness is not always reduced. By viscosities around 100 cP the increase in oil viscosity improves the retention of the dispersant by oil and causes an increase in effectiveness. However, there is a level at which the oil becomes quite shear resistant and effectiveness decreases drastically.

If the water temperature is below the crude oil's pour point, then the oil has solidified and chemical dispersants are not effective. There is no evidence that high pour point oils are not dispersable as long as prevailing sea temperature is above that pour point. There is a popular belief that water temperature influences dispersant performance by changing the oil viscosity, but the solubility of some compounds such as ethoxylated surfactants increases at lower temperature and its effectiveness improves.

The effect of water salinity on dispersant performance is very pronounced for some dispersant formulations. The experiments have shown that at a fixed dispersant treat rate, wave energy, etc. the amount of oil dispersed is approximately 58% at normal sea salinity (33 parts per thousand) compared to 1% at 0% salinity. One reason for such sensivity is due to the hydrophilic-lipophilic balance (HLB) of the dispersant formulation. A formulation balanced for seawater would be too hydrophilic for freshwater use and would not have a similar impact at the oil/freshwater interface.

The chemical composition of various crude oils influences dispersant behaviour in a very subtle manner. If the oil were pure hydrocarbon, then the relative effectiveness of various dispersants could be established more predictably. However, crude oils vary, because they contain varying amounts of dissimilar, indigenous surface active agents. Because of their molecular structure, these surface active agents or surfactants orient themselves at the crude/water interface and stabilize water-in-oil emulsions. In a mixture as complex as crude oil numerous surface active components are present. Studies of Kuwait oil have indicated that porphyrin and metalloporphyrin complexes are among the major components of naturally occurring emulsifiers. A molecular study of these compounds indicated that they were of flat plate-like structure. Hence, they orient at the oil/water interface in such a way that their packing produces a dense cohesive surfactant film at the interface. This actually resembles a plastic

film that presents a coalescence barrier for the emulsified water droplets.
To destabilize crude oil emulsions, a demulsifier (based on surfactants) is added to displace the indigenous stabilizing surfactant from the interface. In order for the demulsifier to displace the emulsifying agent from the interface it must have a higher free energy of desorption. The effectiveness of a demulsifier on a particular crude oil depends on the composition of the demulsifier.
In oil spill situations, dispersants should be tested on the particular crude oil, even if the crude oil's physical properties, such as high viscosity, might suggest that dispersant treatment would not be effective.

2.3.4 Dose Rates

Laboratory tests show that dose rates of about 20:1 oil to dispersant are sufficient to ensure full dispersion using concentrated dispersants for an estimated average slick thickness. However, measurements carried out at sea have shown that the thickness within a single slick span varies more than five orders of magnitude and that about 90% of the oil volume lies within 10% of the slick area. Consequently uniform spraying over the whole slick area led to overdosing of the thinner regions and insufficient dispersant being applied to the thicker areas. In addition airborne losses of up to 50% have been estimated for application from aircraft.
When attacking a large spill by spraying from a boat, it should be attacked continuously around its outer edge, working towards the centre. If the oil is near land, it should be attacked along its landward side, parallel to the land.
The best mechanical recovery can cover a maximum of 1/2 square mile/day. The improved boat spray system can cover 1.5 square mile/day and the largest aircraft system can cover approximately 12.5 square mile/day. The improved boat spray system will apply dispersants at three times the rate of the best mechanical recovery systems and the largest aircraft system can apply dispersant at 25 times the rate of mechanical recovery.

In operations carried out at sea the total quantity of a dispersant to the amount of oil spilled gives an overall oil-to-dispersant ratio of about 5:1, if it is assumed that 30% of oil evaporated in the first 12 hours. In other spills this ratio was 3:1 and even 1:1. This may be due to lack of control of the spraying operation when spraying ships are operating outside the main concentrations of oil and with incorrectly rigged spraying equipment. Some of the vessels of convenience are often unsuitable for spraying operations due to bow wave effects pushing the oil away from the dispersant spray and lack of manoeuvrability.

The major practical difficulty would appear to be in ensuring that the dispersant comes into contact with the oil at the desired dose rate. Experimental work on dispersant effectiveness should be aimed ultimately at producing a matrix that would define the operational conditions for dispersant use, correlating the type of oil, its amenability to treatment, weathering, ambient temperatures, sea state, type of dispersant, means of application and application rates.

2.3.5 Spraying Dispersants From Boats

In order to get the maximum effect the correct amount must be applied to the floating oil and the floating oil must then be stirred or agitated to form a dispersion in the upper layer of the seawater. The natural action of wind and waves provides sufficient agitation when there are even relatively small waves, but sufficient wind must be available to break the tops of the waves. A swell alone, even with large waves, results in little surface mixing and this type of wave motion does little to promote adequate dispersion. Therefore it is usually necessary to mix the oil with the water by mechanical means after spraying the chemical dispersant on the floating oil. For very small spills in harbours or close to the shore in confined spaces it may be necessary to use fire hoses for agitation.

The dispersing chemical is discharged onto the water surface from booms mounted on each side of a boat. The boom is fitted with spray nozzles, which must be specially selected to produce a uniform spray pattern of droplets, which are of such a size that they are little affected by the wind. The same spray booms tow the mixers or surface breaking equipment, which are slotted wooden devices, effectively mixing the treated oil into the upper 25 cm of the sea. The towing speed could be 5 to 10 knots.

The dispersing effect could also be achieved by jet application. This has proved to be effective when using water-based dispersants for dealing with oil spills in brackish waters. This alternative calls for a mixing device that permits effective mixing of the undiluted dispersant with seawater before the mixture is discharged and that permits suitable dosing of dispersant according to the actual circumstances (2 to 5%). The jet application consists of booms with specially fitted nozzles and the dispersant mixture is discharged under high pressure, thus giving the surface mixing energy required.

Dispersants are usually supplied in steel drums of 42 and 55 gallon capacity. These drums can be difficult and even dangerous while loading, unloading and using these containers at sea. Therefore for storage below deck, pillow tanks of 2250 to 4500 litres capacity are now recognised as the most acceptable means of storing. The dispersant can be pump-transferred from the shipment containers into the dispersal ship's tanks. The dispersant can then be pumped from storage tanks below deck to the gunwale-mounted spray nozzles. A more desirable variant of this procedure involves the use of flexible pillow tanks, which could be manually handled and folded for transport and storage. These tanks eliminate contamination of the ship's tanks. They must be harnessed in position at deck level using webbing straps with attached rings or snap hooks.

Most systems designed for boats do not allow a very large swathe width and the speed of the vessel is typically 5 to 10 knots. Offshore supply vessels are capable of carrying large payloads of dispersant, enough to disperse a spill of 10,000 barrels (bbl). Thus the coverage rate of the boat system is its limiting factor. The slow speed of the boat system can make response to remote spills untimely. With aerial application the limiting factor is the relatively small payload.

Thus hydrofoils may be considered to be particularly suitable for oil clean-up at sea for the following reasons:

- upon reaching take-off speed it raises itself above the water and all its weight is carried by submerged parts of the foils and there is no problem with the bow wave. It is possible to fully exploit its very high operational speed, which can be sustained under most conditions for an extended period of time;
- the stability of hydrofoils is very good and long spraying booms could be used;
- high transit speeds allow the hydrofoils to be operating in less than half the time of conventional vessels;
- it does not need to tow breaker boards in order to mix dispersant and oil, because the high-speed sailing structure generates a cloud of water, over which the keel flies. This also avoids evaporation of the dispersant, because spraying occurs in near-saturated moist air;
- the counter-rotating propellers (the starboard propeller rotating clockwise and the port propeller rotating counterclockwise) bring the inner water to the surface at the wake centre line and thus open the wake sideways.

To take full advantage of the large payload carrying capacity of a supply vessel, the swathe width can be increased by using a dispersant fan sprayer, which could increase the coverage rates up to 4 square miles/day, i.e. three to four times. A boat spray system available today consists of spray arms extending out from the bow approximately 9 m. Longer arms are very seldom used as they are difficult to handle and may cause navigational problems in close quarters and when docking. The potential solution to the spraying problem is a high speed fan with a spray manifold installed in the discharge side of the fan shroud. This arrangement incorporates a mist of dispersant into the propeller air column.

An experimental system has been developed using a 36" high velocity, 9 bladed variable pitch fan, capable of delivering 1.000

m^3/min at 1,950 rpm. The fan discharge has an inverted pear-shaped shroud, which helps focus the air column. Within this shroud is the spraying manifold containing 17 discharge orifices. The fan is hydraulically driven and is designed to be mounted on the bow of a supply vessel. Approximately 35 HP is required to drive this fan. Brackets and clamps are fitted to the fan, so that no permanent fixtures are required for attachment to the vessel. The air velocity exiting the fan is approximately 45 m/s and decreases to approximately 6 m/s at 25 m distance (fig.2.8.)

The design of a pumping manifold for the dispersant or diluted dispersant requires special attention. If a large volume of diluted dispersant is to be sprayed, the momentum of the liquid phase will carry the dispersant a significant distance, even without the air stream. However, the air stream does assist in carrying the liquid phase and it does break the water stream exiting from the nozzles into small droplets to form a reasonably uniform pattern on the water surface. A large diluted spray volume of 150 gallons/min or less would appear to be more desirable for windy conditions offshore, where blowing should be minimized. Orifices used for these tests (large volume) ranged from 0.11" to 0.18" dia. If low volumes of dispersant are to be sprayed, then the momentum of the liquid phase alone can not be relied upon to propel the dispersant. Also the air column can not carry large drops. Thus a fine mist concentrate spray carried by a high speed air column can extend the spray pattern further than is possible with the high volume spray. Mist carried by the air column in still ambient conditions, consisting of droplets of 200 to 700 micrometres, can be airborne at 30 m from the sprayer. Focused air stream and dispersant strike the water surface in an approximately straight line. The water surface is gently agitated by the air stream and liquid impact. The oil is gently agitated, and thus the amount of dispersant which reaches the oil/water surface, is enhanced.

Two fans should be mounted on the ship's bow, one directed to each side, allowing spraying from both sides simultaneously in calm wind conditions. In windy conditions the boat should

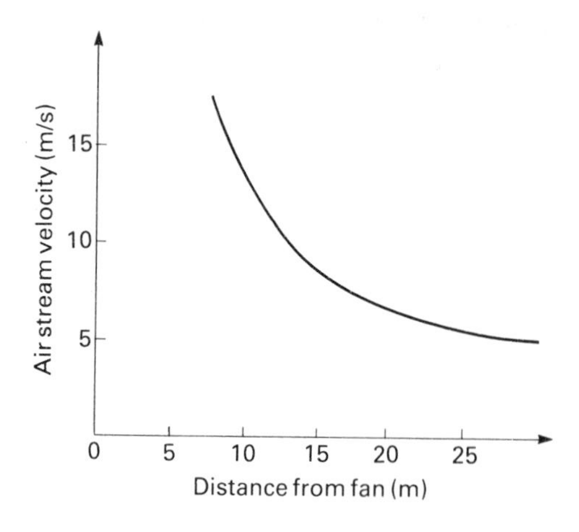

FIG.2.8 Fan air velocity versus distance.

travel with or into the wind, otherwise one fan could be used while travelling normal to the direction of the wind. The fan could also be used for herding the oil by an air stream alone from beneath the docks and through areas of thick undergrowth or other inaccessible places.

Seaborne craft seem to be a most suitable means of applying dispersant with spray booms due to their large carrying capacity and ability to navigate and operate even under bad weather conditions and at night. But ships present disadvantages such as the bow wave causing speed limitations (below 10 knots), breaking and opening up the oil film and not allowing effective contact between dispersant droplets and the oil. Booms are usually placed at the extreme bow in order to avoid or to reduce this disadvantage but with poor results. Spray booms fitted on conventional craft can not be as long as required, because of their distance from the centre of buoyancy making damage likely to result from rolling and pitching.

2.3.6 Dispersant Spraying from Aircraft

Aerial application of dispersants differs from agricultural spraying in the physical properties of fluid spread, the dosage generally required and operational factors such as altitude, speed, swathe and total area to be covered. Consideration of these factors may result in the need for special mechanical designs or alterations in the spray system.

Effective treatment of large spills can be carried out using high speed and high payload aircraft with wing-mounted booms, capable of providing high deposition efficiency, which is warranted by:

- droplet size distribution, which is neither too small, causing off-target drift of chemicals in the wind, nor too large, causing inefficient use of chemical as the drop breaks through the oil slick and is diluted in the water;
- a controllable and uniform application rate of chemical across a desired swathe.

The production of appropriate droplet size minimises misting and results in a more uniform swathe. Also, the energy impact on the slick is related to chemical density and diameter of the drop.

The following three physical characteristics of fluid sprayed affect the atomisation of fluid streams: dynamic viscosity, density and surface tension. Other parameters pertain to the application system and include the velocity of the liquid relative to the air stream and the nozzle diameter. Volatility of the chemicals also affects the droplet size. Rapid evaporation of the sorbent or surfactant components would result in the oil slick being treated by dispersant droplets with reduced diameter and performance.

The literature contains a number of equations for sprayed droplet diameter, which can be reduced to the following form:

$$VMD = k\ (\mu^a\ \delta^b\ \gamma^c\ V^d\ D^e) \qquad (2.2)$$

where: VMD – volume mean diameter, i.e. the diameter of a droplet, whose volume equals the total discharged volume dissolved by the total number of droplets;

k, a, b, c, d, e – empirical parameters based on experimental studies (Table 2.3);

μ – dynamic viscosity;

δ – surface tension;
γ – density;
V – exit velocity, relative to the surrounding air stream;
D – nozzle diameter.

TABLE 2.3 Comparison of parameter estimates for several models of volume mean diameter

Investigator	a	b	c	d	e
1. Ingebo/Foster	0.25	0.25	-0.50	-0.75	-0.50
2. Mayer	0.66	0.33	-0.33	-1.33	1.00
3. Mugele	0.33	0.16	-0.50	-0.66	0.50
4. Weiss/Worsham I	0.16	0.00	0.00	-1.33	0.16
5. Weiss/Worsham II	0.33	0.33	0.16	-1.33	0.16
6. Wolfe/Anderson	0.33	0.50	-0.16	-1.33	0.16

With apropriate grouping of the five variables and exponents, the expression becomes a function of Reynold's number, a dimensionless value relating viscous forces to inertial forces, and the Weber number, a dimensionless value relating interfacial forces to inertial forces. These are two major forces, which control liquid stream break-up and subsequent atomisation during aerial spraying (fig.2.9.). Viscosity, relative velocity and nozzle diameter have the most effect. Surface tension and density vary little for most of the effective chemical dispersants and therefore offer no real control over drop size. (Vertical lines on fig.2.9. show the approximate limits for current dispersants.)

Nozzle orifice diameter and relative velocity have the greatest effect on the droplet size spectrum by generating two types of shear regimes. The mechanical shear at the nozzle outlet is directly related to the rate at which fluid passes the orifice and inversely proportional to the cube of the orifice diameter. A higher shear rate results in more atomisation and a smaller median droplet diameter. The shear rate is given by the formula:

$$\text{Shear rate } (s^{-1}) = \frac{32q}{\pi D^3} \tag{2.3}$$

where: q – flow rate (ml/s)
(if flow given g/min - q = 63.08 $*$ g/min);
D– orifice diameter (cm).

An even stronger effect occurs when the drops enter the air shear regime, which is defined as the difference between the exit velocity of the fluid stream and the speed of the aircraft. This factor increases with aircraft speed, but little effect is seen above about 100 mph. Thus the only real control of maximising droplet size in aerial spraying is the proper selection and use of nozzles.

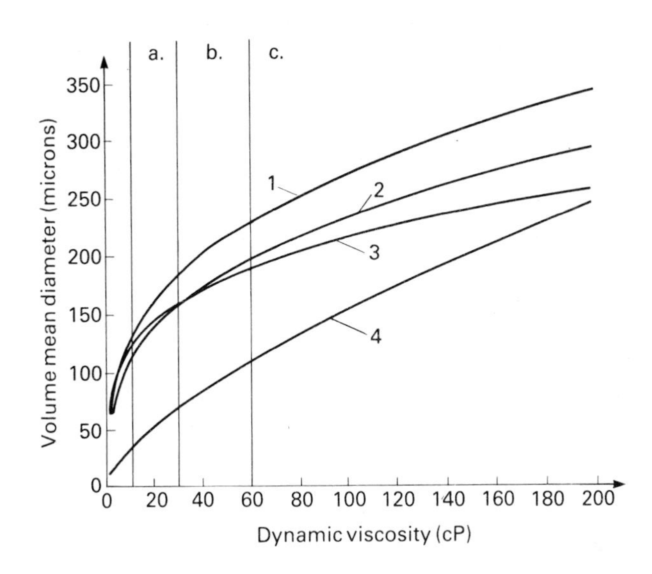

FIG.2.9 Comparison of various droplet size models - volume mean diameter versus dynamic viscosity at density 1,0 g/cm^3, relative velocity 120 knots; nozzle diameter 2 mm and surface tension 30 dyne/cm; a-conventional hydrocarbon; b-concentrated hydrocarbon; c- concentrate; 1-Weiss and Worsham; 2-Mugele; 3-Ingeba; 4-Mayer.

The nozzle should carry 15 to 20% less than its rated capacity at the operating pressures, as extremely high pressures increase the flow rate and thus the shear rate, which should be avoided. Aerial spray pressures rarely exceed 3 atm.
Field tests were conducted in order to compare depositional performance of several products under similar conditions (fig.2.10.). Chemical "A" is a light, low viscosity hydrocarbon-based dispersant, which was seen to "hang" in a cloud in the air, settling only slowly (in up to 20 to 30 seconds) and often blowing off target. The other dispersants settled more rapidly, usually in 5 to 7 seconds. Dispersants whose viscosity is at least 60 cS at

15.5 °C are best for aerial application, although those with viscosity as low as 40 cS are adequate, if the aircraft's ground speed is less than 100 mph. For optimum results, nozzle orifices can not be too small in relation to the fluid flow and throughputs should not exceed 80 to 85% of rated capacity at the operating pressure.

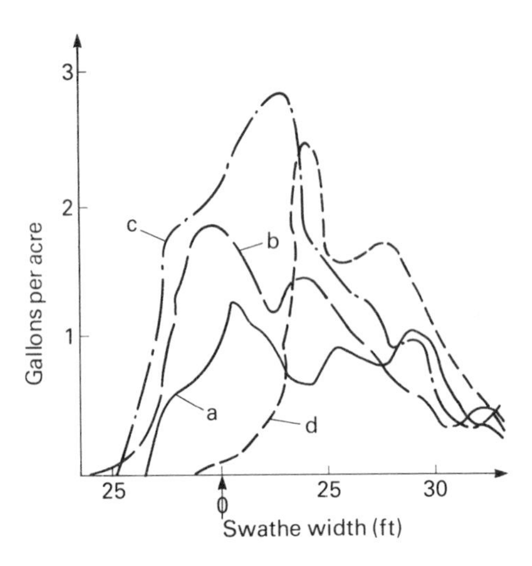

FIG.2.10 Swathe pattern from chemical track for different chemicals: a - Chemical "A" - conventional HC-based dispersant, viscosity (15.5 °C) 9.5 cS; b - Chemical "B" - HC-based dispersant 40 cS; c - Chemical "C" - HC-based high viscosity dispersant 118 cS; d - Chemical "D" standard concentrate (non-HC) dispersant 60 cS; arrow indicates flight centre line.

2.3.7 Application Decisions

Decisions regarding dispersant applications involve the following questions:

- is dispersant use allowed by current government regulations and, if so, conditions must be met;
- do the conditions of the particular spill indicate that dispersants should be used as against natural dispersion or mechanical pick-up;
- what rate of dispersant application should be used;
- are there any locations where the rate of dispersant application should be limited.

All the above questions could be answered well before any spill, but a number of possible solutions should be worked out for the second question. If the conditions for dispersant use and the limitations are planned well in advance of a spill, then the decisions to use or not to use disperants at the time of spill will be expedited.

Despite the operational advantages of aircraft for applying dispersants, ships are considered to be complementary in many cases. It is useful to make comparisons between aircraft and seaborne craft that may be used in similar operations. These data are theoretical and they do not take into account meteorological conditions, the shape of oil slicks, percentage overlap and the necessity of repetition (Table 2.4.).

TABLE 2.4 Aircraft and seaborne craft comparison data.

	Carrier			
	Aircraft CL 215	Aircraft G 222	Hydrofoil RHS 160	Seaborne craft
Swathe width [m]	21	25	30	13
Operating speed [km/h]	110	140	32	10
Length of one ha [m]	476	400	333	769
Time to cover 1 ha [s]	8.4	5.5	20.3	149.5
Surface covered [ha/min]	7.1	10.8	2.9	0.4
Dispersion flow rate [l/min]	1928	2916	593	80
Carrying capacity [l]	5310	7500	26000	30000
Surface covered with one load [ha]	19.60	27.28	130.00	150.00
Time for pumping [min]	2.8	2.6	43.9	3750

For planning purposes, the maximum allowable concentration of oil + dispersant in water should be conservative, i.e. at which little or no toxicity to sensitive marine organisms is observed using either the dispersant or dispersant + oil mixtures. Fortunately most modern dispersants and dispersed oils show relatively low toxicity in static bioassay tests. For most modern dispersants, typical LC50 (96 hours) values are of the order of at least a few hundred p.p.m. either for dispersant alone or for dispersant + oil.

The concentration of oil which would result if it was mixed uniformly into water with the aid of dispersants is given in Table 2.5. A rough guide to visual estimation of the corresponding slick thickness on water prior to dispersion is also included.

TABLE 2.5 Concentrations of dispersed oil in water as functions of oil thickness and water depth

Appearance of oil on water	Approximate oil thickness [mm]	Concentration of dispersed oil in water [p.p.m.] at depth				
		1 m	2 m	5 m	10 m	20 m
Barely visible	4 x 10^{-5}	0.05				
Silvery sheen	8 x 10^{-5}	0.1	0.05			
First trace of colour	1.5 x 10^{-4}	0.2	0.1	0.04		
Bright bands of colour, irridescent	3 x 10^{-4}	0.4	0.2	0.07	0.04	
Colours tend to be dull	1 x 10^{-3}	1.2	0.6	0.2	0.1	0.06
Colours fairly dark, rainbow tints	2 x 10^{-3}	2.4	1.2	0.5	0.2	0.1
Brown or black	0.01	12.0	6.0	2.4	1.2	0.6
Brown - dark brown	0.1	120.0	60.0	24.0	12.0	6.0

The final decision on whether or not to use dispersants must be made on a case-by-case study. Advance planning might indicate that dispersant use is acceptable but if at the time of spill the marine fauna are spawning, dispersed oil would adversely affect the newly spawned organisms. Thus the on-scene decision should be not to use dispersants.

TABLE 2.6 Allowable quantities of dispersant

Water depth [m]	Maximum allowable dispersant application rate [gallons/acre]
<1	<1
1-2	1
2-5	2
5-10	5
10-20	10
>20	Dispersant rate governed by amount of oil on the water surface

For estuaries, bays and confined waters the criterion for accepting the use of dispersants should be that the exchange rate of water will result in reasonably complete flushing, i.e. replacement of a substantial fraction of water volume (90%) in less than one day. If the water surface area covered by the the oil is less than about 1% of the total surface area of the water body this criterion could be waived.

For the waters which are less than 20 m deep, assuming that dispersant usage is ecologically acceptable, the maximum rate of dispersant should be governed by water depth. The allowable rate will usually be such that the total concentration of oil + dispersant introduced into the water column will be no more than 10 p.p.m. if mixed uniformly from top to bottom. If the effective ratio of dispersant:oil is 1:10, the maximum allowable dispersant application rate in gallons per acre will be roughly equal to the water depth in metres as shown in Table 2.6.

2.4 Burning of Oil at Sea

2.4.1 Flames and Oil Spreading on Water Surface

Burning has been used for combating oil slicks on open waters and stranded oil on shorelines. It may require addition of burning agents, such as gasoline or kerosine, or wicking agents, such as straw, moss or sawdust. Burning is considered a primary method for Arctic spills on ice. For shoreline habitats it is not recommended, because it destroys plants and animals, leaves behind toxic residues and may result in increased penetration of oil into sediments.

Burning oil in situ as a countermeasure for oil spreading on open water has theoretically the potential to remove a considerable percentage of the oil from such spills provided ignition could be effected within a few hours of the incident. Delaying ignition decreases the combustion efficiency of burning, which is the difference between the volume of the oil spilled and the volume of the remaining layer of unburned residue. For any given spill volume, the combustion efficiency is maximum when the slick is ignited immediately. If ignition is delayed until the slick thickness is less than 0.8 mm, none of the oil can burn and the combustion efficiency is zero.

The spilled oil is assumed to spread according to the laws formulated by Fay and explained in Chapter 1., initially a gravity-inertial spread with slick radius proportional to $t^{1/2}$, followed by a gravity- viscous spread with the slick radius proportional to $t^{1/4}$. A subsequent surface tension-viscous spread may occur when the slick is already too thin to burn. The combustion process is assumed to affect the spread of the slick in one way only through air flowing into the flames and inducing a water surface current that opposes the spreading of the slick. This is a self-generated "wind-herding" phenomenon. Induced current increases the burning time and combustion efficiency, but decreases the size of the slick when burning ceases.

Flame velocity is constant for a given wind speed and oil type (fig.2.11). The flame flashing velocity, i.e. the velocity at which flame propagates through the combustible mixture of vapours, was measured at 1.3 m/s. Burning oil does not spread appreciably faster nor farther than cold oil. Only in the case of diesel oil do the flames not keep up with the oil and the oil spreads faster than the flames. Slicks with a thickness above 5 mm burn at a rate independent of the slick thickness, but small slicks (below 2 m diameter) burn slightly more slowly than slicks above 2 m, where the burning rate does not seem to be a strong function of slick size. There is a definite impact of ambient temperature on the rate of burning. The slicks burn faster at high ambient temperatures (about 15% faster for an average 15 °C rise in temperature).

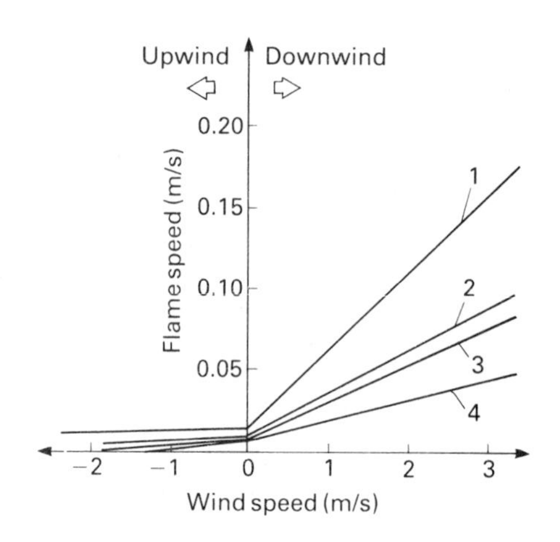

FIG.2.11 Flame spreading velocity of aged crude oils: (1) 1 hour; (2) 4 hours; (3) 8 hours; (4) diesel oil.

The computed combustion efficiency (η_{comb}) as a function of the ignition delay (τ_d) is shown in fig.2.12. for four pairs of curves, one for each of the following values of $V_S = 10^{-2}$, 1, 10^2 and 10^4 m^3. One curve in each pair was calculated for zero induced current (Uc = 0) and the second one for Uc= 0.01 m/s. In each case, the combustion efficiency decreases from maximum value at the smallest delay. The values shown for $\tau_d = 10$ are practically the same as those for $\tau_d = 0$.

TABLE 2.7 Maximum ignition delay τ_{dmax} and corresponding τ_q, x_q and η_{comb}

Volume of spill V_S $[m^3]$	Maximum permissible delay between spill and ignition τ_{dmax} [s]	Burn time τ_q [s]	Dimensionless radius at extinction x_q	Combustion efficiency η_{comb} [%]
10^{-2}	125	190	8,322	19.3
1	500	665	14,127	49.8
10^2	1,950	2,312	24,392	68,0
10^4	7,150	7,830	41,555	79.8

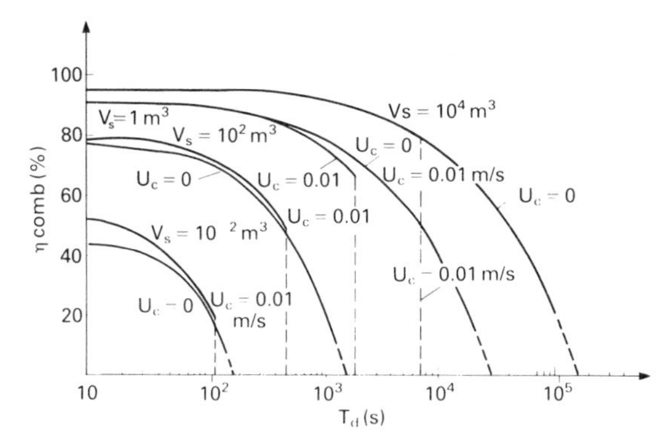

FIG.2.12 Combustion efficiency as a function of ignition delay with induced current Uc = 0 and Uc = 0.01 m/s and for different volumes of spill.

In the case of zero surface current the curves of η_{comb} approach zero continuously, but with Uc = 0.01 m/s the equations beyond certain values of τ_d could not have been solved as the burning rate began to decrease rapidly. The threshold value of τ_d has been reached, which is the maximum possible delay in igniting the slick, beyond which burning may be ineffective. A rough approximation of theoretical combustion efficiency for instantly ignited slick is:

$$\eta_{comb} = (1 - \frac{1}{3} V_S^{-\frac{1}{5}}) * 100\ [\%] \qquad \text{where } V_S \text{ is in m}^3. \quad (2.4)$$

The critical maximum permitted ignition delay could be estimated from the formula:

$$\tau_{dmax} = 0.1 * V_S^{\frac{1}{2}} \quad \text{where } \tau_{dmax} \text{ is in hours, } V_S \text{ is in m}^3. \quad (2.5)$$

Table 2.7. shows the maximum ignition delay depending on the volume of the spill.
Ignition of the periphery of the slick results in combustion efficiencies almost as high as those for ignition of the entire surface area. The required number of conventional igniters spaced 3 m apart at the maximum ignition delay (Table 2.8) can be estimated by:

$$N = 31 * V_S^{0.45} \qquad \text{where } V_S \text{ is in m}^3. \qquad (2.6)$$

TABLE 2.8 Number of igniters needed at the maximum ignition delay

Volume of spill [m^3]	Number of igniters needed
10^{-2}	4
1	30
10^2	238
10^4	1,875

Air entrained by combustion of the oil slick at a velocity of about 0.14 m/s induces an inward surface current that inhibits and finally stops the spread of oil. The slick thickness at which this occurs is related to the size of the fire and could be estimated by:

$$h = 7 * 10^{-4} * r^{\frac{1}{2}} \qquad \text{where } h \text{ and } r \text{ are in [m].} \qquad (2.7)$$

2.4.2 Ignition of Oil Spills

A liquid will burn if the following criteria are satisfied:

- the surface of the liquid must be hot enough to provide a vapour that is combustible when mixed with ambient air;
- the vapour/air mixture must be raised to a temperature sufficiently high to trigger combustion;
- the liquid fuel must establish temperature and flow fields that transport less power into the liquid's volume than the nascent flame is capable of supplying to the liquid surface.

Cold, aged crude oils fail to satisfy any of the above criteria. To generate a combustible mixture, the surface of the liquid must be heated to its flash point (about 100 °C). Igniting the vapour/air mixture requires exposure to a heat source of 700 °C or more. The third criterion requires transfer of a large amount of heat into the liquid before ignition, but this energy must not be removed by the essentially infinite sink of underlying cool oil, ice and water.

The various products which assist burning of oil on water can be divided into two groups:
- igniters;
- wicking agents.

Experiments with igniters have shown that if the oil will burn, the use of an igniter is unnecessary. If the layer is too thin or if the oil is unsuitable, the igniter has very little useful effect. Wicking devices are based on the theory that if the oil can be raised above the water surface by capillary action, then the cooling effect is reduced. Thus various non-combustible, porous floating materials have been developed to employ this idea, but there are two general limitations to their effectiveness:
- the oil surface must be almost completely covered with wicking material;
- the oil must have a viscosity low enough to permit the oil to rise into the wicking material.

Under favourable conditions wicking agents will increase the amount of oil that can be burnt, but their use is limited to special cases only. A fumed silica wicking agent added to oil in a mass ratio of 1:10 causes an increase in combustion efficiency in warm water tests (above 15 °C), but in cold water (below 6.5 °C) it causes formation of a residue comprising balls of oil residue and wicking agent. Considering these effects and the difficulties in using wicking agents, especially in open seas and ice infested waters, where transportation problems are not easy to solve, the use of wicking agents is very limited.

Other igniting techniques include hand-held drip torches, aerial and electrical incendiary devices and gun operated incendiaries. All these techniques present access or safety problems or both. In recent years they have led to a significant number of disasters or near-disasters. Furthermore, many are costly, both in capital and in man-power.

Modern technology presents the possibility of igniting the oil spill using a laser beam focused to give a small, intensive spot of heat at distances ranging from approximately 50 to 1500 m. Target acquisition and precise focus control are achieved by the use of a range finder and an acquisition telescope. The device is fully steerable, can be ground based, airborne or mounted on a seacraft. There are two laser igniting systems, based on a single or dual laser.

The single laser device consists of a one kilowatt carbon dioxide (CO_2) laser operating at a wavelength of 10.6 μm in a continuous-wave mode. Power densities greater than 30 W/cm^2 are required for rapid ignition and thus for longer distances, due to beam divergence, focusing optics are necessary, which allow an increase in energy output and minimise the losses. To vary the output of convergence angle the focusing optics could be solved in two alternative ways:

- a two mirror telescope as the projection system with variable mirror separation;
- auxiliary optics with fixed mirrors.

Due to the mechanical difficulties in moving mirrors and associated optical problems, the second option with two optical subsystems that move relative to each other is more practicable.

Laser ignition devices require an exact measurement of the distance from the target with an accuracy of ± 2 m over a distance of 1000 m for successful and safe operation. A laser range finder is therefore employed with automatic built-in calibration and checks. The laser beam is projected onto the target by a single or two mirror telescope, which has motions of ± 45 ° in altitude and ± 180 ° in azimuth. The device pans at a rate of approximately 0.5 m/s at the target. Faster or lower speeds are available, controlled by use of a joystick.

Acquisition of ignition sites is effected by means of an auxiliary optical telescope and/or television monitor located adjacent to the laser range finder. Once a site has been selected, ignition is activated by a push-button control on the joystick hand paddle. Power densities are such that adequate heating for ignition is possible over a wide range of distances and a wide range of physical conditions such as temperature, wind and water content. If the laser ignition device has inadequate power density for oil ignition within a few seconds, when operated in a CW-

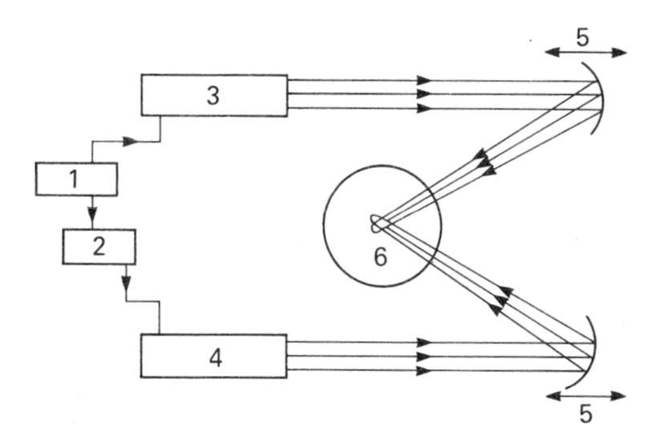

FIG.2.13 Set up of dual lasers; (1) trigger; (2) delay; (3) laser A: (4) laser B; (5) moving mirrors; (6) oil pool.

mode, it can be rapidly switched to pulse mode. In this mode intense energy pulses of microsecond-class width are produced, causing instantaneous ignition of the oil vapour. Switch-over times between CW and pulse modes of operation are less than one second, enabling the oil vapour phase to be maintained.

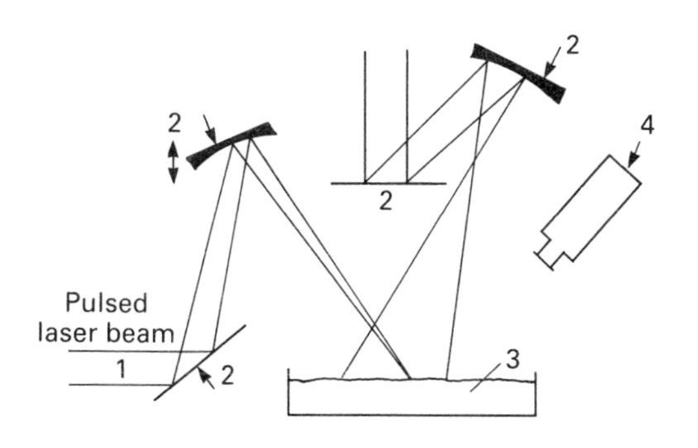

FIG.2.14 Pulsed laser set up; (1) pulsed laser beam; (2) mirrors; (3) oil pool; (4) video cameras.

This means of safe ignition of oil spills from a distance will be extremely cost-effective compared to current conventional methods.

In the dual laser device, which is intended for helicopter installation two lasers are employed, each focused by a single 760 mm focal-length concave mirror onto the surface of cold crude oil (fig.2.13.) The area irradiated by each laser and thus the fluence, is altered by moving the focusing mirrors toward or away from the oil. A CW (CO_2) laser producing up to 1.000 W of power is optically configured to irradiate an oil pool with the required preheating intensity (fig.2.14.). The second laser is a pulse laser, which is focused to provide sufficient fluence to ignite fires remotely on preheated spots in oil pools. The device has the means to align and hold the laser beams on target despite helicopter movements and vibrations.

2.4.3 Combustibility and Incineration of Beaufort Crude/Seawater Emulsions

The highest possible water content in an emulsion is 70% of the total emulsion by volume. At 70% water, no further emulsification is found. Microscopic observations showed all emulsions to be of water-in-oil type. With decreasing water content, the emulsions become more homogeneous and the mean diameters of the water droplets become smaller. For a 50% emulsion, drops ranged up to 50 μm in diameter, with 30% water very few drops above 30 μm were found and with 10% water only a few droplets above 10 μm. For all emulsions the minimum drop size is below 10 μm.

The stability of emulsions is quite variable - emulsions eventually separate. Only a thin layer of emulsion with a very low water content remained at the interface under quiescent conditions, but it was rapidly destroyed by slight agitation. All emulsions with more than 50% water start to separate immediately after stopping the shaker. At room temperature all emulsions with 50% or less water were relatively stable. The smaller the water content, the longer it took for water to come out of the emulsion. Many experiments were conducted on the burning of water-in-oil emulsions. High speed photography of single emulsion droplets introduced into a hot chamber showed that the droplets went through a faster and more disruptive combustion process than pure fuel droplets, because the rapid evaporation of the water in the droplets ruptured the droplets in microexplosions. Water in emulsified fuels did not impair, but improved, combustion due to breaking of droplets, which increased the evaporation surface and improved the mixing of fuel with air. The ensuing reduction of combustion time had a favourable influence on the burn-out of sooty residue and NO_x emissions. Barret reported a substantial reduction in smoke levels and a reduction in NO_x from about 400 p.p.m. to about 100 p.p.m., when using 20% water in No.6 fuel oil. It is accompanied by a substantial reduction in the transformation of SO_2 into SO_3.

The limit on the amount of water in the emulsion depends on the type of fuel and burning conditions. With a light fuel and lean mixtures, an optimum seems to be reached at 5% water - this limit increases with increasing residual content of fuel and increasing richness of the mixture. A limit is reached in all cases at about 50% (mass) of water in the emulsion. Increasing the water content beyond this level leads to excessive quenching effects and a drastic increase in CO and unburned hydrocarbons and eventually to flame extinction.

Combustion tests of oil emulsions were preceeded by atomisation tests, which have shown that adequate atomisation could be achieved for all emulsions (close to 0 °C) with gauge pressures of about 1.3 to 1.5 mPa. Ratings of nozzles tested ranged from 0.75 to 1.5 US gallons per hour at 100 psig. The nozzle used for the subsequent combustion tests had a rating of 0.75. The pump used was a standard gear pump for hydraulic applications. The tests were carried out at temperatures between 5 and -15 °C. The combustion chamber was cylindrical, with internal diameter 150 mm, and 250 mm long. The end was lined with 25 mm of castable ceramic cement (outer diameter 200 mm). The end was closed by a ring baffle with an opening of 90 mm. Combustion tests have shown that emulsions containing up to 60% water could be burned successfully after preheating the ceramic lining either using diesel fuel oil or emulsions with a low water content. Emulsions with a higher water content burned quite well initially. The low heat release was insufficient to keep the ceramic lining hot and the flame rapidly deteriorated and eventually was extinguished.

Although the air-fuel mixtures were fuel rich (with equivalence ratios of about 1.1 to 1.8) all emulsion flames were without visible smoke in contrast to the diesel flame. All the flames were fairly luminous. Spectroscopic observations have shown that this luminosity was not due to carbon, but to sodium from the seawater used.

Emulsions containing 30% or less water could be ignited directly using the standard incorporated igniter when fuel pressures were above 3 mPa. The pressure needed increased with increasing water content. Very satisfactory ignition was achieved for all emulsions by preheating the combustor (usually with diesel fuel) and then switching over to the emulsion. Air was supplied from an external blower and metered through an orifice plate. Unburned hydrocarbons were more or less constant and very low (about 30 p.p.m. as propane) which is considered to be insignificant. Evaluation of exhaust gases by smell showed the obvious presence of hydrocarbons. The intensity of smell was in keeping with the low levels measured. Carbon monoxide levels were constantly below the detection limit (on a 0.0 to 0.5% scale).

In conclusion it can be accepted that emulsions of crude oils can be burned successfully with a minimum of pollution. Emulsions with a water content greater than 50 % can be expected to separate so rapidly that they do not constitute any problem as long as the emulsions are fed to the burner from the top of the retaining reservoir.

2.4.4 In-place Burning of Crude Oil in Broken Ice

If oil is spilled in the Arctic when ice is present or forming, it very rapidly becomes encapsuled in the ice, where it remains until the onset of break-up. Oil then surfaces through a combination of ablation of ice and upward migration of the oil via brine channels. The oil, which is relatively fresh, floats on the melt pools that cover the surface of the disintegrating ice and is concentrated by wind herding between the ice floes.

In broken ice conditions, where oil spill clean-up operations are hindered and mechanical means of clean-up may not be possible, in-place burning of the slick may be the only alternative. The behaviour of oil spilled in pack ice is quite different compared with its behaviour in open waters. The following conclusions can made from experiments and observations made so far:

- spreading of oil in pack ice is dramatically reduced over that in open waters by the presence of ice forms;
- evaporation and subsequent oil property changes can be adequately predicted using the evaporative exposure approach;
- emulsification and natural dispersion do not seem to play as important a role in determining the fate of an oil spill in pack ice as they do for spills in open waters;
- in-situ burning is an effective countermeasure for oil spills in broken ice concentrations. Combustion efficiencies up to 80% could be achieved.

In-place burning of Prudhoe Bay crude oil performed during a Shell Company demonstration yielded combustion efficiencies of 55 to 85%. In a subsequent study, burning efficiencies for pooled oil in ice were predicted to range between 80 and 90%. In evaluating in-place burning of pooled oil, combustion efficiencies obtained were 90 to 95% and even >95%. The regular waves generated did not significantly inhibit flames from spreading to the entire test area. The minimum slick thickness on cold (<6.5 °C) water was 2.5 mm for Prudhoe Bay crude oil.

2.4.5 Air Curtain Incinerator

The incinerator should be able to burn pure oil, emulsions of any oil/water ratio and debris of all types. The incineration should be smokeless or nearly smokeless. The incinerator should be portable and simple to set up and operate in the field. An open-pit air curtain incinerator provides this capability (fig.2.15.). A large volume of air is forced down a plenum and out of the long nozzle. The diagram shows the nozzle placed along the top edge of an excavated earthen pit that contains the waste material. In permafrost areas, an above ground open top chamber would be used. A strong air current directed at an angle into the pit

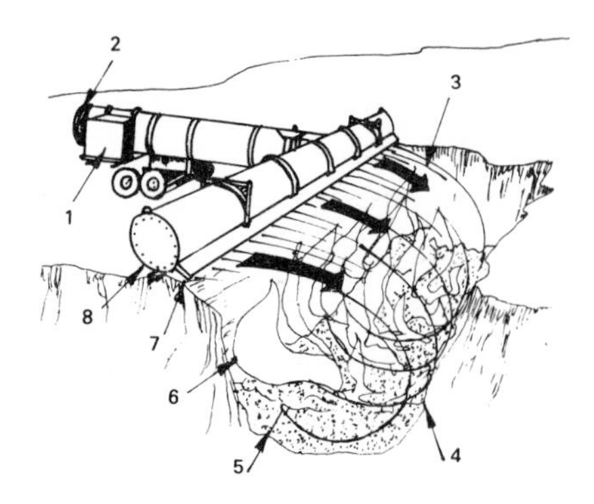

FIG.2.15 Air curtain incinerator; (1) prime mover; (2) blower; (3) circulating air flow; (4) excavated trench; (5) debris; (6) fire; (7) air nozzle; (8) air plenum.

supplies oxygen to the incineration process and recirculates unburned particulates back into the combustion zone. With this type of incinerator the waste material itself is the fuel for the combustion process. The system consists of:

- an air curtain blower and plenum situated on a trailer in such a manner that it could be elevated to the top of an above-ground combustion chamber. The air curtain is generated by a 34 inch diameter two stage axial vane fan. The air velocity at the air nozzle can reach 125 mph. The fan is powered by a 75 BHP diesel engine at 1,900 rpm with a 8,200 ft^3/min air delivery. The air curtain impinges the front wall of the 10 feet wide combustion chamber at about 50 mph. The nominal angle of the air stream is 15 ° from the horizontal, so that air impinges the front wall of the combustion chamber about 3 ft from the top. An adjustable air deflector is installed over the nozzle so that the direction of the air can be changed to impinge the front wall lower into the combustion chamber;
- the combustion chamber consists of a 10 ft high portion above the ground and a 4 ft below-ground portion 10 ft wide and 10 ft long. The above-ground structure is made of four high temperature refractory walls bolted onto a steel structural framework. The chamber lining consists of a 6" thick blanket fibre refractory capable of enduring thermal shock. The front wall opposite air nozzle is hinged at the bottom, which allows

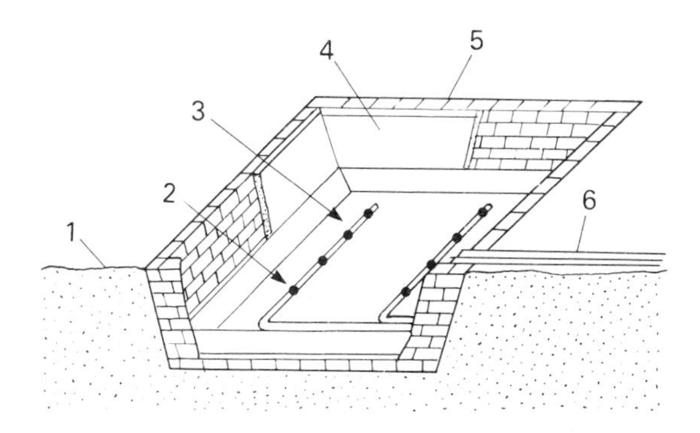

FIG.2.16 Below-ground portion of combustion chamber; (1) soil; (2) manifold with 4 spray nozzles; (3) holding pan; (4) refractory lining; (5) concrete structure; (6) fuel lines.

the wall to be slanted inward about 25 ° from the vertical to change the direction of air impingement on the wall. A 3 ft opening into the chamber side wall on each side is provided for loading the debris into the incinerator. The total weight of the structure above the ground is about 4 tons;

- the below-ground portion of the chamber was made of concrete blocks and covered with a layer of refractory material (fig.2.16.). Concrete blocks provided a footing for the above-ground portion to rest on. A refractory seal is installed between the top and bottom portion, when joined. Inside the bottom portion is a steel holding pan of 1,000 gallon capacity to collect and to contain any unburned oil;
- two manifolds with 4 fan spray nozzles each are located along the bottom of the front and rear wall to distribute oil and emulsion into the incinerator. The oil spray is angled upward toward the centre of the combustion chamber. Atomisation of oil in the combustion chamber is important to obtain fuel/air mixing for maximum combustion efficiency.
- a waste fuel system delivers oil and water from separate holding tanks by means of a set of gear pumps with a total pumping capacity of 30 gallons/min at 100 psi. The amount of water and oil is adjustable, so that any oil/water ratio could

be formulated. Solid debris is added to the incinerator manually either over the top or through a side chute mounted on the incinerator chamber;

- instrumentation has to facilitate easy data acquisition for the whole incineration process. Temperatures are monitored at four locations inside the incineration chamber and at one location on the exterior surface of the refractory with type K thermocouples. Oil and water flow rates are measured before entering the pumps using orifice plates and pressure transducers. The air flow rate is set by adjusting the engine speed and reading the mechanical tachometer. Emissions are monitored visually using a standard American Society for Testing and Materials (ASTM) Rinleman chart. An observer stands about 50 ft from the incinerator, sight the chart towards the plume and compare the smoke colour or density of the plume to the Rinleman chart scale. A reading of 5 is 100% smoke density, 0 is no black smoke.

The experiments have verified that there is an optimum air flow rate (7.000 ft^3/min) and an angle of 30 ° from the horizontal. Water overspray helped to reduce emissions. Optional auxiliary components can include a conveyor system to load debris at a rate of 30 to 50 pounds per minute, water spray for emergency shut-down and lights to illuminate the area around the system for night operation.

In Sweden a burning facility has been developed for burning oil-contaminated waste without having it transported from the scene of damage. The endevour to recover the oil by separating it from solid impurities is not an economic proposition. The material to be disposed of may be oil-drenched sand, stone, weeds, reeds, branches, birds, boards etc. An air-blown incineration furnace is the most economical and also the technically most viable proposition for a mobile system of the type needed for Swedish conditions. To make it easily transportable, the outside dimensions of the entire incineration facility do not exceed 3 m x 1.8 m (fig.2.17.). It is capable of being transported by standard trucks on public roads, by air and at sea and of being

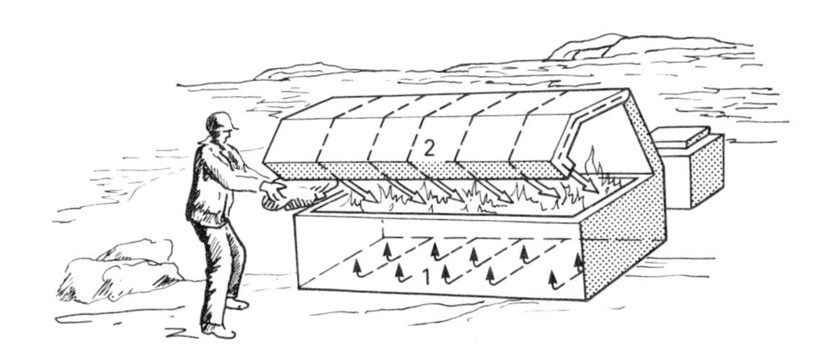

FIG.2.17 Swedish incineration facility; (1) for primary combustion, oil is blown into the bottom of the combustion chamber; (2) secondary combustion is supported by air via the edge of the cover.

lifted by truck, crane or winch. The weight of the incineration container and of the motor with the fan does not exceed 500 kg. During operation, the incineration container is able to stand on the beach, lie in the water and, in favourable circumstances, can be left permanently on a working boat or barge. All the equipment included in the facility functions in all weather situations encountered in Sweden when efficient oil clearance work is likely to be carried out. It is powered only by one type of energy and it is arranged so that the risk of fire is minimised. The incineration capacity of the facility is at least 3 m^3/hour of oil-polluted material collected from beaches. The combustion gases due to excess air supply are acceptable environmentally for temporary use of the incineration facility in the shore zone. After incineration, the ash and sediment is environmentally acceptable when left on the beach. The incineration container is capable of continuous operation for 12 hours.

2.4.6 Testing of a Prototype Waste Oil Flaring System

A new portable system has been developed capable of flaring 180 gallons/minute of light oil or 90 gallons/min of oil with viscosity up to 1,600 cS. The system can be transported in a C-130 aircraft or broken down into modules light enough to be carried by helicopter.

The system consists of the following functional areas (fig.2.18.):

- the burner consists of 12 independent burner modules each having an atomising oil nozzle, a pilot light and two suppression water spray nozzles. Different tips can be installed to accommodate a range of oil types - a large orifice (A-tip) with a nominal capacity of 15 gpm of light oil and small orifice (C-tip) with a nominal capacity of 7.5 gpm of heavy oil. Air consumption is the same for each. Individual tips can be blanked off easily, if a low overall burning rate at high atomising efficiency is necessary to keep thermal radiation to a minimum in close quarters such as burning from a barge deck;
- the burner support consists of a boom made of four tubular longitudinal members adequately braced. Each of the four tubes carries one of the fluids for the system: oil, air, smoke suppression water or radiation suppression water. Pilot light connections are made of separate 1" aluminium tubes attached to the boom. Just at the back of the intermediate boom flanges an array of water spray nozzles is installed to provide a fan-shaped radiation suppression screen between the flame and the operation area. The boom is 78 ft long made in two parts. The burner is attached to the forward end. The boom assembly further consists of a lifting bridle, supporting A-frame legs, an alignment cable and support, and a base assembly weighted down with two compressors (6,000 pounds);
- the oil pump module consists of a triple screw pump, a diesel air-cooled engine with four speed transmissions, a butterfly valve, a flow meter and a skid. A propane vaporiser system is also included in the module, as the engine exhaust is used as a heat source in the vaporiser;
- three identical air compressors are used to provide atomising air. The rated capacity of each unit is 250 ft^3/min at 105 psig. The compressors are rotary screw types driven by diesel engines of the same size as oil engine;
- the water pump module consists of a centrifugal pump, a diesel engine, a 3" basket strainer, a butterfly valve, a pressure

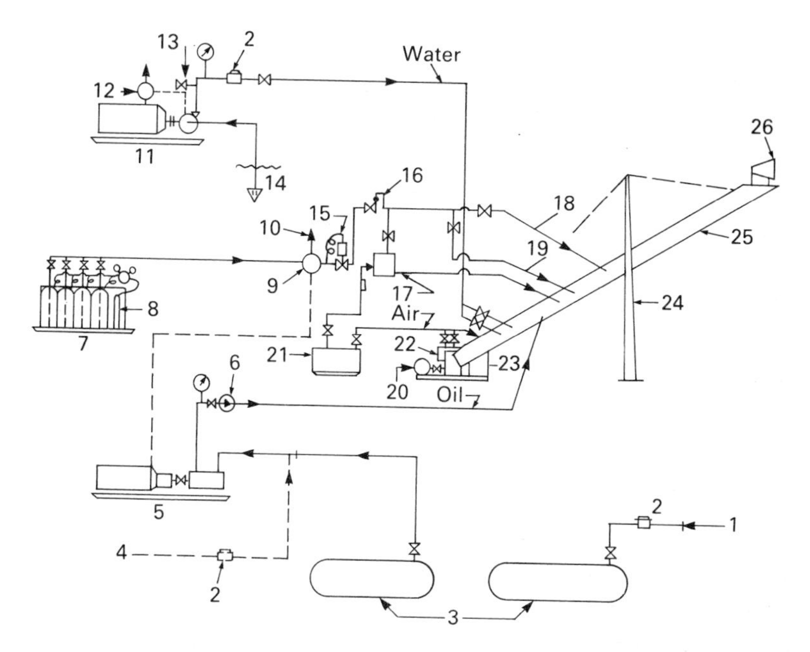

FIG.2.18 Schematic diagram of waste oil flaring burner system: (1) oil from trash pump; (2) strainer; (3) pillow tanks; (4) optional input from adapts or from Dracome; (5) oil pump module; (6) flow meter; (7) propane cylinder module; (8) N_2 cylinder; (9) vaporiser; (10) engine exhaust; (11) water pump module; (12) exhaust primer; (13) fire water; (14) foot valve; (15) temperature regulator; (16) pressure regulator; (17) ignition components; (18) main pilot gas supply; (19) ignition runner line; (20) winch; (21) AC-3; (22) AC-2; (23) AC-1; (24) A-frame; (25) boom; (26) burner.

gauge and a skid;

- the pilot gas system consists of a supply of liquid propane, a vaporiser for use in cold weather and an igniter unit for the pilots. In liquid feed conditions, a coil-type heat exchanger using diesel exhaust (oil pump engine) as the heating medium is used to vaporize the liquid propane. To ignite all 12 pilots simultaneously, an "ignition runner" is used. This device is essentially a small propane burner, which provides a continuous ring of flame from a perforated pipe, which intercepts all of the pilot flames. To ignite the ignition runner, a flame front generator (FFG) is used. This apparatus provides a flammable mixture (propane/air) into a 1" pipe leading to the ignition runner. A manually-activated electrical sparker is then used to ignite the remote end of the pipe mixture. The flame front propagates to the other open end of the pipe to light the ignition runner;

- oil can be pumped from several sources like Dracome barges, trash pumps, steel barges or portable tanks.

The tests have confirmed the design parameters. Low viscosity oils (light fuel oil, crude oil) can be burned with no smoke production by spraying water into the flame. Water spray does not eliminate smoke at high viscosities. The smoke intensity is usually in the light to moderate range until viscosities of approximately 800 cS and higher are burned. The maximum noise level was measured at approximately 96 dB(A). The average measurement for all runs was 87 dB(A).

CHAPTER 3

Mechanical Response Technology to an Oil Spill

3.1 Containment Systems

3.1.1 Physical Fundamentals of Retention

Mechanical methods of oil removal from the sea surface serve to contain and to recover spilled oil. It is ecologically essential and technically and economically feasible only as long as the oil is floating on the water surface and has not reached a state of total dispersion. Thus mechanical response should start as soon as possible after a spill in order to contain and collect most of the oil.

Booms provide a mechanical barrier to floating oil as they extend above and below the water line. The purpose of their application is:

- to contain the oil slick, to reduce spreading and to facilitate recovery;
- to divert oil to areas where recovery is possible;
- to protect specific areas, such as biologically sensitive areas.

To accomplish these tasks one of the following measures needs to be taken:

- to act only on the oil slick without affecting the water body e.g. by using mechanical barriers of small draft, by using wind or air jets (fig.3.1a.);
- to induce a retention current against the spreading oil with special effects concerning the water body, i.e. by hydraulic or pneumatic jets (fig.3.1b and c);

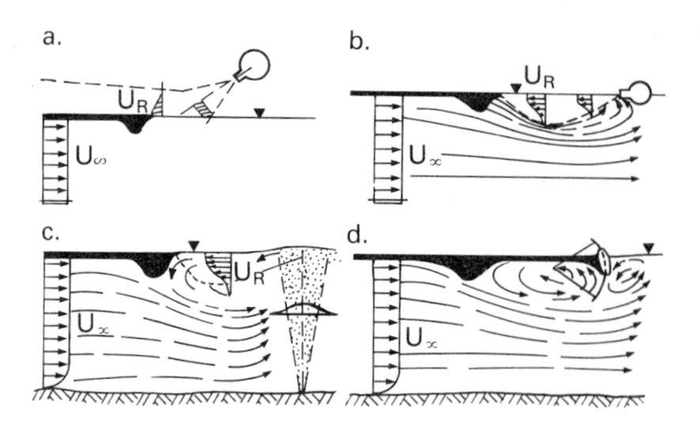

FIG.3.1 Types of oil barriers; (a) surface acting air jet; (b) hydraulic jet; (c) underwater pneumatic jet; (d) mechanical barrier.

- to retain the upper oil transporting layer of the water body by mechanical means (fig.3.1d.).

Oil barriers in calm and stationary waters have only to overcome the oil spreading forces. Their physical background is a surface current with a maximum velocity deductible from the equilibrium condition in the stagnation zone:

$$U_{Rmin} = \left(g * h_0 \left[1 - \frac{\varrho_0}{\varrho_F} \right] \right) - \left(2 \frac{\delta_A + \delta_F}{\varrho_0 * g * h_0^2} \right)^{\frac{1}{2}} \quad (3.1)$$

where: U_{Rmin} – minimum retention velocity;
g – gravity;
h_0 – oil slick thickness in stagnation zone;
ϱ_0 – oil density;
ϱ_F – water density;
δ_A – surface tension in the oil/air interface;
δ_F – surface tension in the oil/water interface.

Equation (3.1) is valid for all waters with a surface current $U_\infty \leq 0.2$ m/s and without waves. If the surface current exceeds 0.3 m/s, surface active booms as well as hydrostatic booms become ineffective due to oil slick being accumulated in the stagnation zone and from there transported by recirculation flow into the jet area. Here the horizontal momentum flux of the highly turbulent jets provokes the oil to penetrate underneath the barrier (fig.3.2.). All booms in the current establish a front area (fig.3.2a.) in the neutral area between the oncoming and recirculation flow.

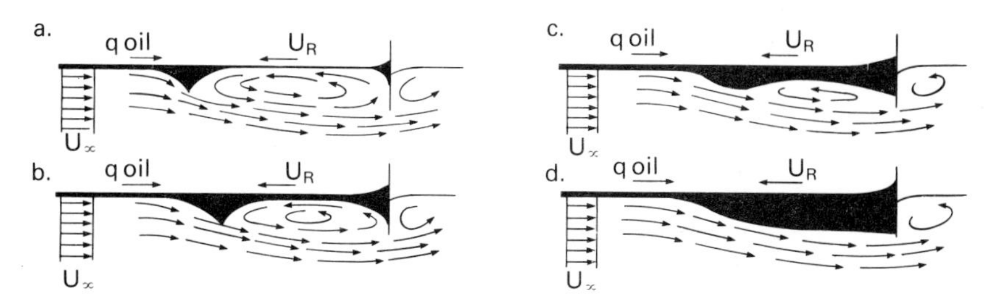

FIG.3.2 Stages of boom failure.

Oil losses can arise in the form of oil droplets, which are formed due to Kelvin-Helmholz instabilities of the leeside of the front zone (entrainment failure) and which are transported to and under the barrier if the downward momentum of these droplets is sufficiently high. If their buoyancy dominates, they will enter the recirculation current in front of the barrier (Fig.3.2 b-d.) until a quarter elliptical body with $h_{omax} \approx h_T$ will be established. If supercritical currents now occur, oil will start to drain behind the barrier.

The existence of a front zone is dependent on inertial and gravitational forces. The equilibrium of the critical containment volume is governed by viscous and surface tension forces.

In the presence of waves additional forces occur due to orbital motion of the water particles, which affect the oil slick, i.e. compress it or draw it apart, promote emulsification by breaking, disrupt the induced current of pneumatic and hydrodynamic barriers and excite natural oscillations of mechanical barriers. The containment process in waves is not fully understood as yet. It is stated that with increasing wave steepness i.e. with decreasing wave length, the oil slick thickness as well as the critical drainage velocity will increase. Because of the complexity of the motion of boom segments and the existence of non-linear effects, an exact prediction of the sea keeping response of boom configurations and of their influence upon the retention process is impossible. However, boom elements should have maximum elasticity to follow the waves and also sufficient elasticity to prevent peak tensions. Of great importance is the tension in mooring lines, which can deform the shape of the boom curtain below the water surface considerably (fig.3.3.). When the angle varies, it will certainly have an impact upon the containment capability of the boom.

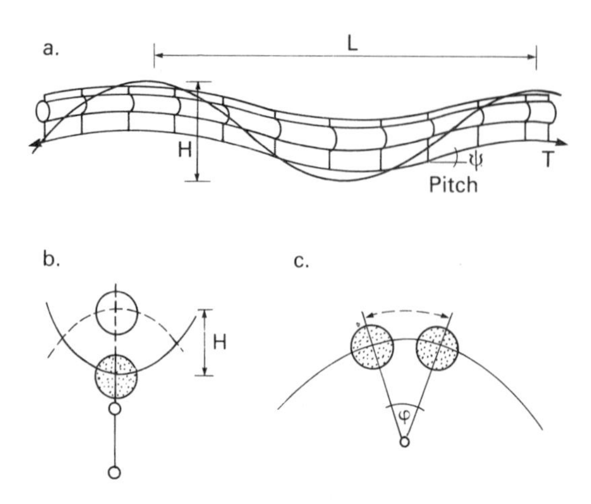

FIG.3.3 Behaviour of oil booms in waves; (a) boom traverse the waves; (b) heaving and pitching in longitudinal waves; (c) rolling of booms in longitudinal waves.

A large flat plate barrier 3.5 m long, 1.0 m high and of 0.6 m draft was placed in a circulating water channel and its oil containment performance was observed in calm water. The relation of plate inclination angle to the direction of oil leakage gives a different flow in each case. The eddy current just in front of the plate starts oil leakage under the barrier depending on the speed of flow (fig.3.4.).

3.1.2 Boom Structure

Floating booms consist of four basic components (fig.3.5.):

- a means of flotation, supporting the boom on the water. Flotation material is attached in a portable manner to the boom or may be integrated with it. It consists of expanded plastic foams such as polyethylene or polyurethane, of natural cork or wood or of gases such as air or carbon dioxide. When gases are used for flotation, then the booms can

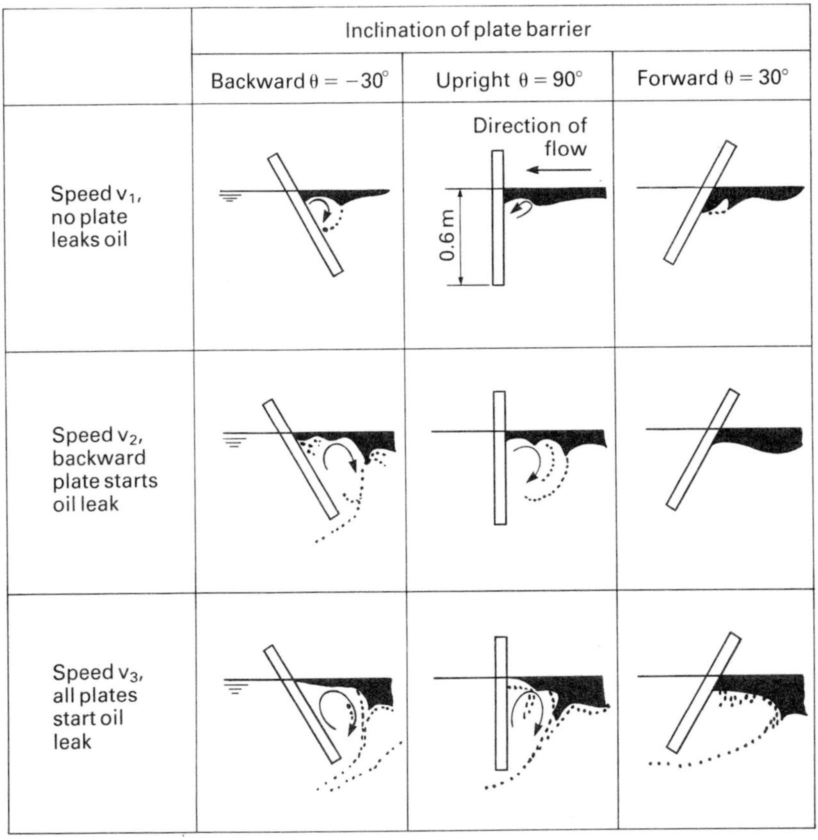

	Inclination of plate barrier		
	Backward $\theta = -30^\circ$	Upright $\theta = 90^\circ$	Forward $\theta = 30^\circ$
Speed v_1, no plate leaks oil		Direction of flow 0.6 m	
Speed v_2, backward plate starts oil leak			
Speed v_3, all plates start oil leak			

FIG.3.4 Schematic illustration of oil leakage around a flat plate barrier in calm water; Θ - angle between the plate barrier and water surface.

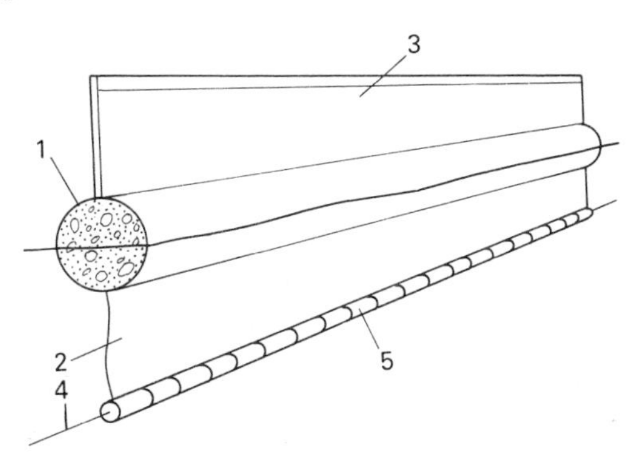

FIG.3.5 Basic components of flotation booms; (1) means of flotation; (2) skirt (3) freeboard; (4) tension member; (5) weighted ballast.

be easily folded for storage and they require little room, but they must have sufficient resistance to punctures. The most modern booms are selfinflating after being dropped into the water and thus they are very easy and safe to use;

- the skirt, which provides the basic barrier to the spread of oil. There is a great variety in the design of boom skirts and in materials being used. Normally a strong fabric coated with nylon or fibreglass is employed for this purpose, sometimes reinforced with metal cloth. The deeper the skirt, the greater will be the load exerted by water and currents on the entire boom;
- the freeboard, preventing the waves from washing oil over the top. It offers considerable resistance to wind and waves, thus it should not be too high, as it might be prone to hydroplane the boom under even moderate wind conditions;
- the tension member, extending the whole length of the boom and carrying the load imposed on the barrier by wind, wave and current forces, which depend on the weather and current and in unfavourable conditions will be quite high. If there is no tension member, then the whole load must be taken by the skirt. Such booms could be used only in very calm waters. The tension cables may be located at, below or below and above the water line. They are generally made of wire cable, although in some cases chain or manilla cable is used. All booms have anchor points located at intervals along their entire length attached directly to the tension member. The distance between anchor points should be no more than 30 m to prevent the formation of deep cusps between them when the boom is exposed at right angles to the flow. At these sections oil may escape when the flow exceeds the limiting value.

Most booms have some form of weight or ballast to keep the skirt vertical in the water. This ballast is generally placed along the lower edge of the skirt. The material used for this purpose is lead chain or lead polyvinyl chloride rods. It should be attached in such a manner as to avoid any chafing.

There are a large number of commercially available booms designed for different current conditions, sea states and locations, e.g. open sea, river, harbour, etc. Generally booms are most effective in calm waters, low currents and sea states. The oldest type of boom is an interconnection of wooden pilings or construction beams without any freeboard structure or supporting skirt. There is a group of booms, in which flotation is secured by means of plastic foams (fig.3.6.) and a group with flotation secured by gas. Some booms are one sided, i.e. not symmetrical, and consequently not effective when improperly deployed.

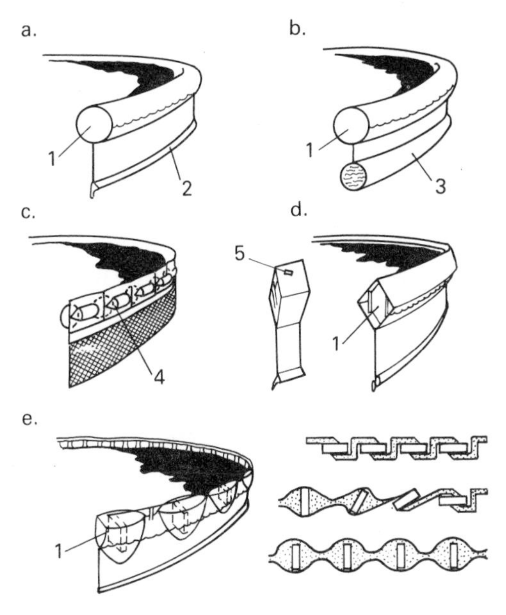

FIG.3.6 Typical commercial booms; (a) boom with air flotation body & lead ballast; (b) boom with air flotation body & water ballast; (c) boom with indiv. floaers; (d) boom with spring loaded air filling; (e) boom with air filling floaters; (1) air; (2) lead ballast; (3) water ballast; (4) air floaters; (5) air valve.

When several booms have to be joined together to form a containment barrier, the ends must be compatible in order to obtain a tight joint which will be strong enough to carry the tension forces, will not form a dip in the skirt nor in the freeboard, through which oil would easily penetrate, nor cause deeper draft of the boom in way of joint and thus lower the freeboard. This joint should be easy to make as very often it has to be made from a small swaying craft in adverse weather conditions. Use of light weight clamps is highly desired, but they must be floating in the same position as the barrier. A splice in the skirt should not make any hard spots in the flexible skirt, which is undergoing cyclic loading for many hours and sometimes days. All the corners and edges should be well rounded off, clamping pressures should be evenly distributed and local stress-raisers should be avoided. In some cases it may be necessary to use a specially designed transition piece between different booms to avoid any improper joints that might restrict any of the joined boom sections.

Some booms operate on the principle of float removal for easy stowage. This entails physical attachment of the floats prior to deployment and use, which is a difficult and lengthy operation and in some circumstances may even be dangerous. Buoyant materials of cellular synthetics are easily damaged by impact or compression and have to be renewed, because they lose their buoyancy. If synthetic beads are used, then in the case of the boom fabric being torn or otherwise failing the beads escape and float on the water with the boom losing its capacity to float.

Some materials of which booms are fabricated require very thorough cleaning after use. This cleaning operation creates environmental problems from disposal of contaminated scrub and rinse water. Some materials deteriorate rapidly while in storage.

3.1.3 Methods of Boom Utilisation

There are many methods of boom utilisation depending on the local circumstances and the spill size:

- encircling boom (containment boom - fig.3.7a.) is used for containment of oil leaking continuously from a point source such as a damaged ship, underwater well or leaking pipe. In the event of a large leak multiple barriers may be required. End moorings and back moors are required to prevent the barrier perimeter from collapsing under wind and current. Part of the perimeter might be formed by the shoreline or by a ship. For terminal anchorage on a steel hull magnetic anchors could be used. If the ship is lying on the ground or the boom is attached to the shoreline in tidal waters, then frequent adjustment of the attachment is required to prevent the boom attachment from hanging high and dry, permitting leakage at the water/shore interface.

- a U-shaped boom is suitable for moving slicks driven by currents or wind in one direction in order to entrap the slick. The boom could be anchored at an appropriate distance to keep the tension stresses in the barrier below the material strength limits. Instead of mooring the boom it could be towed by tow boats at each end to draw it slowly through the slick (Fig.3.7c). The skimmer is always placed at the deep end of the catenary of the boom, but the combating ship might be placed within or outside the boom;

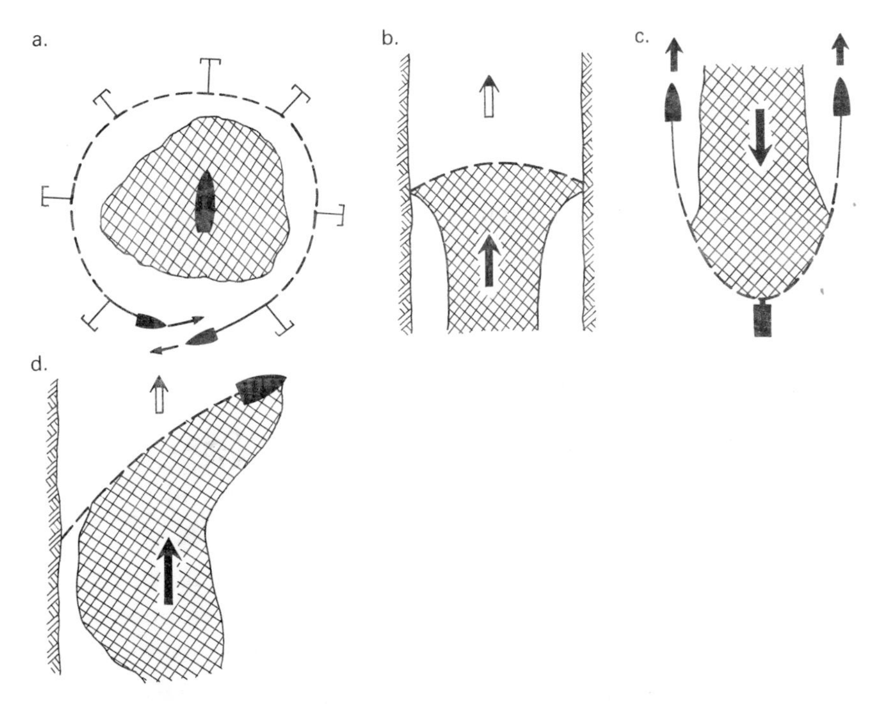

FIG.3.7 Modes of application of oil booms; (a) containment boom; (b) oil barrier; (c) collecting boom; (d) diversion boom.

- diversionary booms are used in cases of currents above the boom retention capability. A boom laid across the river at right angles to the flow will bulge at the centre of the river (fig.3.7b.), where the flow is strongest, and thus the oil will collect where it is most difficult to remove it and where the current is most likely to be high enough to carry it underneath the boom. The recommended alignment is in the shape of a chevron, so that floating matter is deflected to each side of the stream (fig.3.8a.). The boom should be angled at approximately 50 ° to the surface of the water flow in order to reduce the head water pressure and current velocity acting

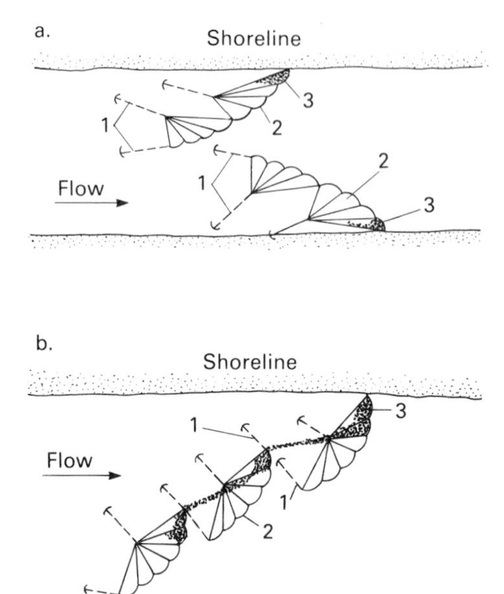

FIG.3.8 Use of floating booms in a diversionary mode; (a) Chevron shape deployment; (b) cascading boom deployment; (1) anchor lines; (2) boom; (3) oil recovery zone.

on the submerged section of the boom. This concentrates the oil and floating debris at the inshore end of the boom, where it could be easily collected. It has been discovered that for such operations a shallow draft boom (0.3 to 0.6 m) will be more stable and efficient than a deeper one. Multiple moorings are required to maintain the barrier in a good line. In instances where strong currents prevent optimum positioning of one boom, several booms could be deployed in a cascading manner (fig.3.8b.) to move oil progressively to one side of the water course. This technique is used particularly effectively in wide rivers, where simple chevron patterns due to strong currents and the width of the river would fail. This mode of application is also useful for protection of shorelines in currents along the coast. Such barriers could be used to force the oil offshore, with one end anchored at the beach and the other one in the stream (fig.3.7d.).

Barriers could be launched from the shore and a small watercraft would be required to carry the moorings into position. Cyclical tidal currents have to be taken into consideration as the directions and velocities of the currents may be changing and the barrier may need resetting at high and low tides, or two settings of moorings could be set out for each direction of current.

3.1.4 Boom Deployment and Mooring

The actual deployment of equipment in rough conditions may expose the workforce to many dangers. Thus at all times the safety of response team should supersede all other considerations. The limitations of the available equipment should be well known. It is useless to deploy equipment under conditions for which it was not designed. Local conditions such as water depth, direction and velocity of current under various tidal states (highest values should be used), ground conditions for mooring the boom or permanent anchoring, the best position for launching the boom (pier, beach, barge or ship), the type and length of the boom - all of these problems should be considered in predeveloped schemes in contingency plans based on probable needs in specific areas. All splicing and adjustment should be performed prior to launching the boom securing the continuity of the skirt and strength of members. Mooring connections must be made available. Provisions should be made for lighting the booms if they could be an obstruction to navigation. The barrier must not be permitted to twist into a cork screw during deployment. One end of the barrier should be moored first in a predetermined position and only then should the other end be towed to a second predetermined position. Towing vanes or paravanes are recommended for both ends of the booms, particularly in strong currents, as they improve the stability of the boom by providing additional buoyancy and an even distribution of tension at the end of the boom.

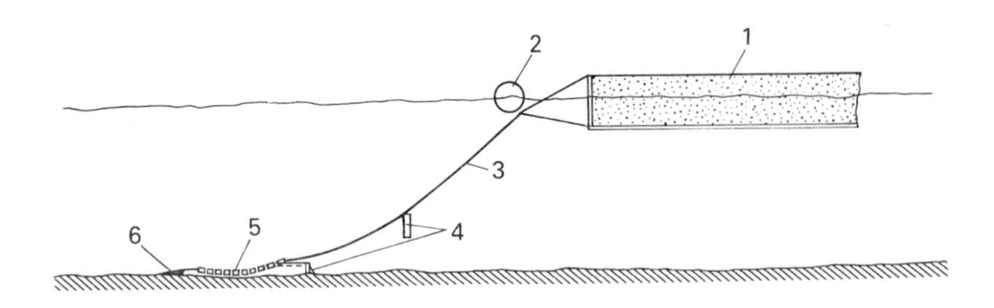

FIG.3.9 Floating boom anchoring system; (1) boom; (2) buoy; (3) anchor line; (4) weights; (5) anchor chain; (6) anchor.

In very shallow water or at low tide, where barrier grounding may occur, the barrier stability will be impaired and the barrier will become ineffective. Thus it must be artificially stabilized, so that leakage past the barrier will not occur. If the water depth is below five times the draft of the barrier, the speed of water below the barrier might increase as a result of the "Venturi effect", by which the confinement capacity of the boom becomes less effective. Thus a small shallow draft boom should be used in such cases.

The mooring forces of the barrier are a function of barrier length, gap or opening, current velocity and sea state. Most barriers have limited reserve buoyancy and they can not support large vertical mooring loads, especially at their ends. Thus it is preferable to place a float or a buoy between the barrier and the mooring with a 100% reserve displacement to absorb any vertical component. Any mooring suitable for the bottom of a river or the sea and for the expected loads, such as patent anchors, concrete blocks, etc. could be used (fig.3.9). A terminal anchor chain might be desirable. If the barrier has to be shifted periodically, it is easier to put moorings at alternate locations and shift the barrier between them than to pick up and reset the moorings each time the barrier is moved. Thus a ship or craft with appropriate equipment to handle the moorings is required. Onshore mooring equipment can be very simple such as the nearest tree, driven stakes or pipes, mushroom or patent-type anchors. They must be appropriate for the expected load and placed in required positions. They may be stayed off in order to increase their holding power.

The barrier must retain flexibility to conform to the wave profile and should be oriented normal to the wave crests, because such a location reduces the stresses imposed and prevents the destruction of the barrier.

When booming the damaged ship, whether the ship is floating or resting on the sea bottom the method of booming should be considered. Approximately three times the length of the ship is required for full encirclement with the appropriate number of moorings (minimum four). A barrier should be emptied of retained oil, because when the current is reversed the U-shaped barrier will lose its contents and the barrier will lose its shape unless some moorings are arranged on the back side. The accumulated oil should be restrained with an additional opposing barrier, which could use the same moorings. A two-sided barrier could move across to a reversed position without switching the ends.

Floating debris in currents and waves present danger to barriers with skirts or components made of light materials and plastics. Rips and tears may be initiated by impact, which can then propagate and destroy the retention capability of the system. Large floating objects such as logs, piles and planks can be towed away, but small debris consisting of seaweed, grasses, straw, kelp, plastic sheathing, etc. must be handled by nets and rakes. To prevent redispersion, it is desirable to remove the debris from the water as soon as it is brought together. Thus small craft with experienced personnel equipped with rakes, nets, tow lines, grapnels, etc. is essential. The floating debris should be collected as it tends to clog the removal systems. Pumps and separators with rotating parts may be severely damaged by these materials. Di-

aphragm pumps are less susceptible.
Protective clothing for the crews should be provided. Special off-loading and transportation facilities for this dripping mass of miscellaneous material should be considered as well as cleaning facilities for boats and equipment. The variations in shape, size and material greatly complicate the required countermeasures. Small items could be macerated, but are difficult to screen, and large items could be screened but are difficult to macerate. The best method of disposal is to burn the debris, but due to its high water content, it is not easy to burn and special ovens should be provided. The alternative is burial at an approved site.

3.1.5 Boom Retrieval

Boom retrieval can take place at the end of a spill combating operation or under the threat of changing environmental conditions. In the first case, it could be done slowly, but in the second case the time element is of the essence. The booms could be towed by a tug to a site where they could be easily retrieved from water, cleaned and folded for storage. Light barriers could be placed manually on a deck or shore, whereas heavier barriers require power assistance (winches, cranes or hoists) and thus are vulnerable to damage. The identification of strong points for lifting and handling should be made by the manufacturers and they should issue instructions for folding and storing the booms. Some manufacturers provide special power driven reels, on which the booms are stored.
The boom equipment will be contaminated with oil. Thus during retrieval it should be thoroughly cleaned by means of high pressure water hoses. This causes some local contamination, which should be dealt with if practicable.

3.1.6 Other Types of Booms and Barriers

Improvised booms can be used for relatively small oil spills in sheltered waters or as temporary measures until more suitable commercial booms arrive at the spill site. These could be constructed from inflated fire hose, telephone poles and logs. They must be properly tied up with no cavities between them, through which oil could easily pass to the other side of the boom. They would have a very low freeboard and oil can easily spill over the top. Improvised booms could be used upstream of conventional booms to divert debris or in very shallow waters, where conventional booms may prove to be ineffective.

Sorbent booms and barriers absorb the moving oil slick in a porous material such as straw or hay or some synthetic product, combining the containment and recovery operations. Thus they could be used only for relatively thin oil slicks, since their recovery efficiency rapidly decreases once the porous material is saturated with oil. They require additional support to avoid breakage under the force of water as well as during handling; care must be taken when they are saturated not to squeeze the oil back into the water. They should not be allowed to turn over the contaminated side, thus recontaminating water downstream. Straw held in place by wire cable or wire mesh in very shallow water may form a very effective sorbent barrier.

Pneumatic barriers use a screen of air bubbles released below the water surface to act as a barrier to the movement of floating liquids. This barrier lying on the bottom of the basin does not impede navigation (fig.3.1c.). It could also be used for diverting debris and oil from a protected area.

When air bubbles are released below the surface, they rise and expand as hydraulic pressure decreases. At the same time they generate an upward water flow, which on reaching the surface is diverted into the horizontal plane. This surface current flows in both directions away from the vertical bubble stream and efficiently retains the oil. Rising bubbles rapidly achieve a limiting velocity of ascent. Release from greater depth does not increase the velocity of approach to the surface nor the velocity of entrained water. For this reason there is no sense in submerging the barrier to depths greater than about 3 to 5 m. Deeper submergence results in increased power requirements. The barrier consists of a pipe with drilled holes, but these are apt to be clogged by sand, mud and marine growth. A pipe filled with air will have positive buoyancy and must be anchored to the bottom. Air bubble screens also tend to suppress small waves and prohibit formation of ice.

Pneumatic barriers are used in the sheltered waters of harbours to contain spills at unloading stations or to protect harbour entrances from oil in the sea. There are very high costs of installation, maintenance and powering of the barrier. Thus they are used to a limited extent as stationary installations.

Surface acting air or water jets may be used to contain or divert an oil slick. A ship's fire equipment may be used for this purpose, but the jet should be as close to the surface of the water as possible and at a low angle of incidence. Portable fire pumps placed on small boats could be used for this purpose. If the water is played directly into the oil slick, oil will be emulsified. A line of sorbents placed along the "up-wind edge" of the slick proved to greatly increase the sail area.

Trawling systems could be used for removing oil from the water column. Once the density of spilled oil increases to that of seawater, the oil may sink, depending on oceanographic conditions. At high latitudes ice melts and creates layers of brackish waters and strong pycnoclines. Oil of intermediate density between the surface water and lower water may sink through the surface layer, but rest on the pycnocline. Experiments are conducted with underwater trawl nets for recovering oil sunk below the water surface. Several types of nets were tried (paravane otter type, kite type) made of different materials, threads, mesh, shapes, etc. and they were towed at different speeds and depths. The results of these tests have shown that it is possible to recover highly viscous underwater oils by means of underwater trawl nets, but further work is required.

3.1.7 High Pressure Water Jet Barrier

Conventional oil booms have limited capabilities. Oil is swept under them when the speed of the current exceeds 1 knot. Sea state 3 with about 1 m waves is about the limit of effective containment of oil at sea. Considerable splash over takes place in choppy conditions when the waves are only 0.4 m high. All conventional booms have to be kept in position by boats, ships or anchors. Reorientation to accommodate changes of wind direction or tide is usually a time-consuming operation. Ships not equipped with thrusters are unable to maintain position at 1 knot limiting speed for holding oil.

Conventional booms operate on the principle of obstructing the flow of the surface layer of water to catch whatever is floating on it. Thus they have to be strong enough to take the associated loads. At the same time they have to be flexible and buoyant enough to conform to the waves, so that gaps do not occur in the troughs and the crests do not wash over the top. Any debris mixed with the oil is also caught and this creates a major problem in areas of heavy ice. The containment area can soon become filled with ice, leaving no room for the oil.

To offset these difficulties the stream of water from a fire hose has also been used for localised control of oil slicks. But when such a stream enters a body of water, it induces a current and surface turbulence to move the oil. If played directly on the oil, the oil is churned up and emulsified and containment is lost. A coherent water jet directed vertically downward onto the surface of the water creates a surface current and wave effect in the water to move floating oil.

The basic idea of using high pressure, low volume water jet technology to contain oil relies on a series of nozzles mounted on floats to emit flat, fan shaped, high velocity jets of water horizontally about 15 to 20 cm above the surface of an oil slick. Each high velocity stream of water would break into a fine spray or mist of water droplets, which would entrain a large amount of air blending the discrete water jets into a thin, horizontal air jet over the surface of the oil. That induced air flow would move the oil over the surface of the water in the same manner as wind does. The underlying water would not move until it becomes exposed by the departing oil. The secondary surface current created in exposed water by the artificial wind would inhibit any tendency for oil to return towards the barrier.

The nozzles have to be spaced so that the sprays would overlap to create a continuous front. To balance the thrust from these jets, a second series of identical jets would be installed back-to-back with the first, pointing in the opposite direction.

An experimental water jet barrier system was designed (fig.3.10.), consisting of four main components:

- power supply unit;
- control manifold;
- tether line connecting the manifold to the barrier;
- 10 interconnected float-mounted nozzle assemblies.

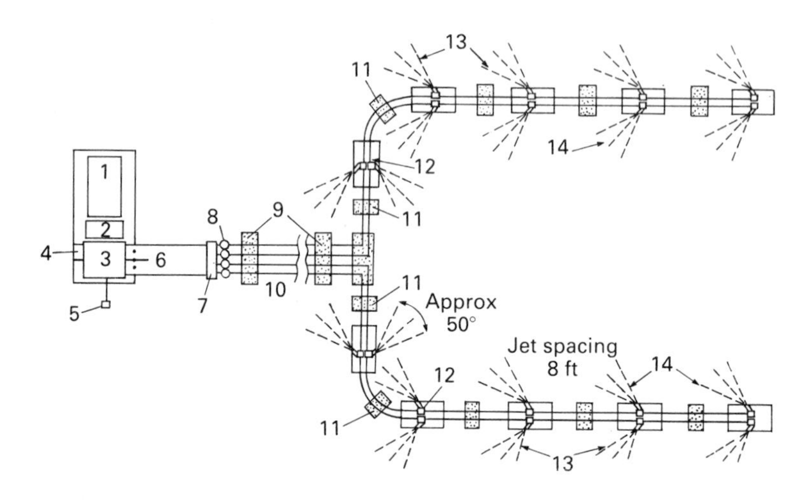

FIG.3.10 High pressure water jet barrier system for oil containment; (1) diesel engine; (2) clutch; (3) triplex pump; (4) priming pump; (5) screened intake; (6) dump for overload; (7) manifold; (8) control valves; (9) tether line floats; (10) tether line 50 ft; (11) hose float; (12) nozzle assembly float; (13) reactive jets floats; (14) oil control jets .

The manifold combines the dual output from the triplex pump to feed two or four separate lines to the barrier. Each line from the manifold is controlled by a valve and supplies flow to nozzles on one side of the barrier arm. Two lines are required for a diverter barrier and four lines are needed for a containment barrier.

The barrier system consists of a 16.5 m tether line, fabricated from 38 mm internal diameter dual-braid steel-reinforced hose with a burst strength of 34,500 kPa (5,000 psi) and nozzle assemblies each containing two jets, spaced 2.4 m (8 ft) apart and supported on styrofoam floats with the connecting hoses supported on separate floats, placed midway between adjacent sets of nozzles. The turning moment (torque) created, when pressure is applied is compensated for by crossing the hoses. Boat fenders are used to support a platform, which keeps the nozzle 10 cm (4") above the water surface.

The barrier was found to be functionally satisfactory and ready for use. It was stable, very manoeuvrable and easy to operate. It could be moved laterally, shaped, advanced and withdrawn merely by adjusting the pressure of opposing jets, and quick to employ. The tests have shown that containment configuration can be held in position in a 1.5 knot current without anchors or tow craft and have enough power in reserve to oppose oil.

3.1.8 Fireproof Boom

A fireproof should:

- withstand flame temperatures of 980 °C for extended periods of time in a salt water environment and be reusable;
- contain oil in a "U" configuration at sea state 4 °B and survive sea state 5 °B;
- be as compact as possible and remain flexible down to -20 °C and storable to -50 °C;
- have good abrasion resistance so as to be able to withstand frequent handling and some contact with ice;
- be easily deployed, handled and recovered using supply vessels and be easily towed at 2 knots;
- have provisions for anchoring and towing;
- have a freeboard of about 300 mm and draft of about 400 mm - larger versions for offshore are 600 mm and 800 mm respectively;
- conform easily to waves for long periods of time;
- have a tensile strength of at least 110,000 Newtons (N);
- be low enough in cost to encourage widespread use.

The boom is constructed of 1.5 m long flotation units (fig.3.11) joined to a 0.75 m long flexible panel to provide wave conforma-

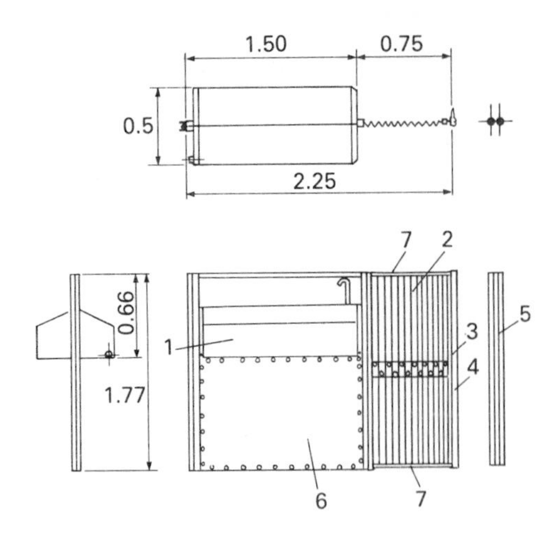

FIG.3.11 Fireproof boom with corrugated steel connectors; (1) flotation unit (2) flexible panel; (3) flexible panel securing; (4) flexible connecting end; (5) joiner; (6) bottom panel; (7) flexible connecting cable.

tion. Each flotation unit (1.77 m high with 0.66 m freeboard) consists of a pentagonal cross section with a "sail" to provide the required freeboard and an underwater skirt to provide draft. In the final design the flexible connecting panel consists of a sheet of pleated light gauge stainless steel, through which a universally jointed box beam is passed to provide tensile strength to resist towing and ice loads. The pleat bends are reinforced by the addition of a tubular backing and are of sufficient radius to reduce stress concentrations to well below the material's endurance limits.

The relative vertical motion of two adjacent flotation sections is restricted by the use of a guide shoe mounted on the universally jointed tongue. All parts are made of stainless steel to ensure high abrasion resistance. The final design of the fireproof boom has the following major characteristics: section length - 2.58 m; section weight - 210 kg (boom and connector); linear weight - 81 kg/m; gross buoyancy - 440 kg; buoyancy/weight ratio 2.1 to 1; draft - 1.2 m; freeboard - 0.57 m.

It was found in towing that the stability of the boom in a straight line and in catenary configurations was satisfactory. The boom could be towed in a straight line at speeds up to 5 knots, but at that speed the propeller wash tended to deflect the first section of the boom. Also a significant bow wave was set up by the first section, which resulted in a high drag force on the boom. The catenary towing trials were held in a short, choppy sea with wave heights of approximately 1 m and winds up to 30 km/h. The boom conformed well to the waves and demonstrated excellent stability.

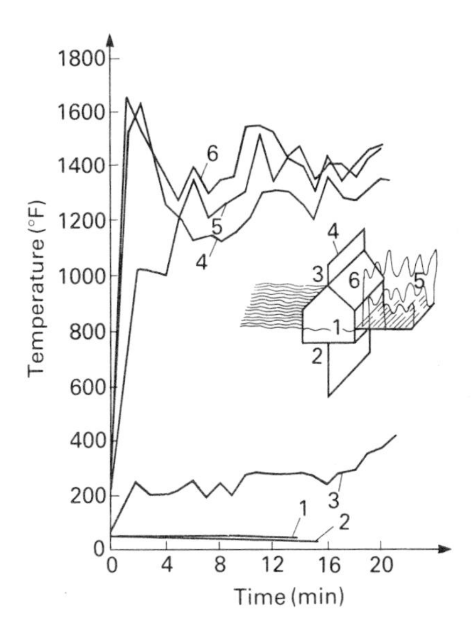

FIG.3.12 Burning trials of fireproof boom with temperatures measured at 6 points.

Thermocouples were mounted at various locations on one section of the boom and were monitored from a barge adjacent to the test site, which served as a logistics and observation platform. The temperatures measured at various points on and around the booms are shown in fig.3.12. The maximum temperature measured was 905 °C, well within the design maximum exposure temperature of 980 °C. The smoke plume generated by the burn rose vertically to a height of approximately 300 m and then dispersed horizontally with visible smoke disappearing within 2 to 3 km downwind. All 1,545 l of oil was burned leaving only 2 l of oil residue, giving a combustion efficiency of 99.87%. The temperature of the skin was consistently higher than the tem-

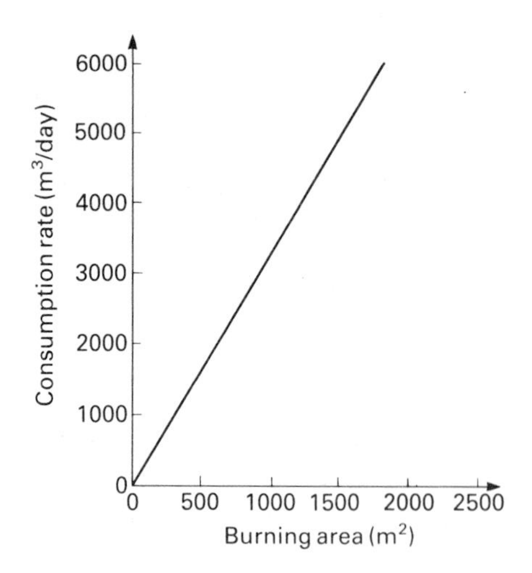

FIG.3.13 In-situ burning oil consumption rate.

perature just above the slick. The combustion process due to oxygen starvation took place at the edges of the boom and in the air space above it. In-situ burning is a one-step removal process that can dispose of large volumes of oil on water. Experiments with the fireproof boom (fig.3.13) show the dependence of oil combustion rate on a burning area.

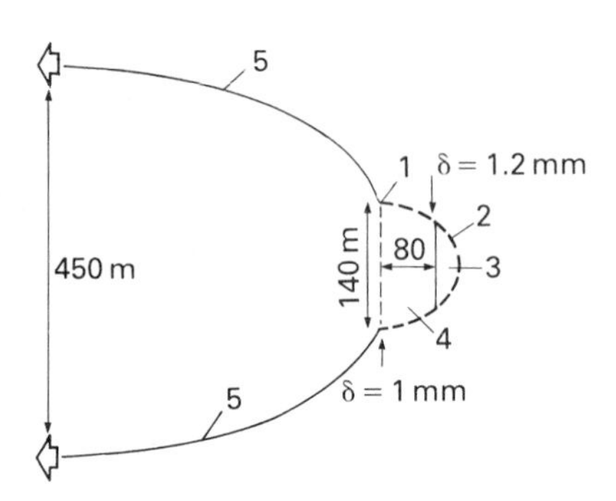

FIG.3.14 Boom configuration; (1) wire bridle; (2) fireproof boom; (3) continuous burning area; (4) safety zone with intermittent burning; (5) conventional boom.

Fig.3.14. illustrates one possible configuration for the use of a boom in response to a blow-out offshore. A conventional offshore boom is held by supply boats or anchored so as to direct the oil

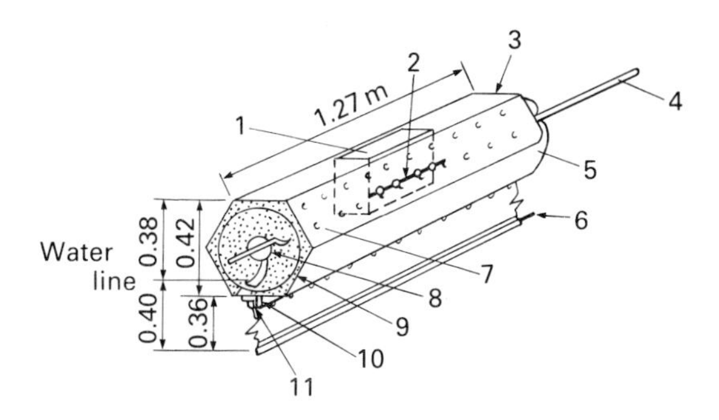

FIG 3.15 Construction details; (1) sliding cover; (2) turnbucle; (3) 4 mm GRP skin over foam glass core; (4) 10 mm steel cable; (5) convex rounded end; (6) towing cable; (7) vent holes (8) 100 mm dia steel tube; (9) seal; (10) twist fastners on 150 mm centres; (11) aluminium angle 40 x 40 x 3 mm.

toward a pocket of the fireproof boom. As the oil moves toward the pocket, it thickens, however the overall slick thickness will not exceed 1 mm until the oil enters the pocket. As the oil moves toward the back of the pocket it continues to thicken until it reaches the area of burning slick and then ignites.

An effective low cost boom combines a foamglass core with a fibre reinforced refractory cement cover and omits the intermediate wicking layer. To obtain oil-tight, yet flexible joints between boom segments a simple ball-and-socket combination has been used (fig.3.15.). Tests have shown good oil retention performance. A small amount of burning oil escaped through the ball-and-socket joints and ignited some of the entrained oil as it surfaced. But all burning in the wake was confined to a small area which did not extend beyond about 1 m from the rear face of the boom.

3.1.9 Portable Oil Booms in Broken Ice

Performance tests on portable oil booms were carried out in Japan. Booms were towed in a catenary configuration with oil and/or simulated ice pieces. Booms were tested for critical tow speeds at which oil began to escape the boom and for the towline tension. The thickness of the ice mass contained in front of the boom increased with increasing water current. The leading edge of the ice mass was a steep gradient at speeds more than 0.4 m/s and the maximum thickness was found at the rear of the leading edge. The ice mass without oil was crushed at speeds more than 0.45 m/s. At this speed the ice pieces began to escape the boom. But the ice mass with oil was stable at speeds of 0.5 m/s. Fig.3.16. shows the ice mass at 0.45 m/s without and with oil. The length of the ice mass with oil will be greater than that without oil. The profile of the ice mass contained in front of the boom resembles to a wedge. The drag of the wedge is less than that of the fence (fig.3.16).

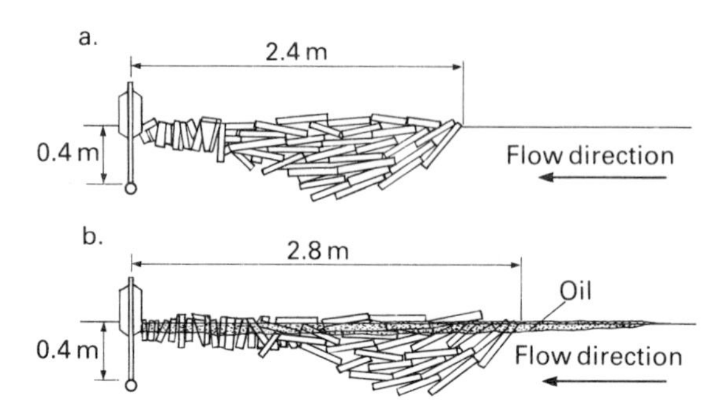

FIG.3.16 Ice mass in calm water with current 0.45 m/s; (a) without oil; (b) with oil.

For a fence type boom without ice it was found that a vortex formed between adjacent flotation units, drawing a small quantity of oil beneath the skirt at 0.26 m/s. This phenomenon was stopped by the presence of ice. The amount of oil droplets in front of the ice mass was less than that without ice at the same speed.

The results of the tow tests could be summarised as follows:

- the critical tow speed of oil with ice was greater than or equal to that without ice;
- the critical tow speed of ice with oil was greater than or equal to that without oil;
- the towline tension of booms with ice was less than or equal to that without ice.

3.1.10 Boom Test Procedure

Booms should be tested in tanks and in open water in order to obtain a clear picture of their adaptability to perform their two basic functions: to contain and/or to divert oil. Boom tests in tanks may have specific and/or generalised goals. Tank tests are limited by tank capabilities and open water tests are limited by the lack of control and high costs of execution. Additional differences are caused by oil distribution methods, volume and type of oil used, boom length and rigging.

A programme for in-tank test procedures must be developed and it should be kept uniform through the tank tests for all booms. The first step is to determine the oil-holding ability of the tested boom, rigged in a uniform manner with the same quantity and type of oil and subjected to the same test conditions. Each boom must be towed to determine the calm-water first-loss tow speed - the speed, at which the first evidence of entrainment could be observed. This must be followed by a test to determine the calm-water first-gross-loss tow speed, i.e. the tow speed at which drainage or other massive oil loss occurs. All these tests are run in:

- calm water;
- 0.2 m high, 2.8 m long regular waves;
- a 0.45 m harbour chop.

An additional test should be made at a speed expected to be outside the operational range of the boom to give the maximum speed, at which the boom could be towed without submarining, planing or mechanical failure, which is called the critical tow speed.

The final test is the endurance test, when the boom is exposed to a 0.5 m harbour chop for 45 minutes to obtain an accelerated cyclic stressing of the boom.

Two tow points are mounted on the main towing bridge with an 11 m separation to attach a 30 m boom. During calm-water tests the tow force is recorded at known incremental tow speeds. These data are used to extrapolate the maximum allowable tow speed in calm water, corresponding to the design strength of the tow point.

The first test may begin at 0.5 knots and may be increased in 0.1 knot increments until loss occurs by entrainment of oil. This test determines first-loss tow speed in calm water and the tow force is recorded. A preload of a certain quantity of heavy oil (say 100 gallons = 0.38 m^3) should be distributed ahead of the boom catenary prior to each oil loss test. Then the tow speed should be further increased by 0.1 knot increments until the first-gross-loss tow speed is recorded.

In the second series of tests a 0.2 m high, 2.8 m long breaking wave is created. Tests start with a speed of 0.25 knots and the speed is progressively increased by 0.1 knots until the first-loss tow speed is reached. The first-loss tow speed and mode of loss are recorded.

The third test determines the first-loss tow speed in a 0.45 m high harbour chop starting at 0.25 knots and increasing by 0.1 knot increments. The mode of loss under or over the boom and the first-loss tow speed are recorded.

In calm-water critical speed tests, the boom is at first towed at 0.5 knots and the tow speed is increased by 0.5 knot increments to 2 knots. Then the speed is increased by 0.25 knot increments until the boom fails to float properly by either planing or submarining.

In durability tests, the boom is towed intermittently at 0.5 knots in a 0.45 m harbour chop. The 45 minutes test did not include periods when waves were building up or subsiding.

The open water test site must be carefully selected. It must have sufficient area, at least 500 x 250 m with a mean low water depth of at least 5.0 m and be exposed to frequent winds. The length of the boom should be at least 150 m with the boom gap at least 50 to 75% of the boom length. Typical mooring facilities for the booms should be provided, enabling placing of the booms in any specific configuration. There should be a high tower provided for visual observation and video-taping, as well as for servicing the booms. The booms should carry light floats to mark their location and to avoid creating hazards for navigation. The floats

should be attached to the tension member. The lantern lamps should be fitted about 1.0 m above the water surface.
Boom motions are measured with both pressure transmitters and coupled accelometers. The accelometers have to be mounted in pairs on the boom as close to the estimated centre of rotation as possible in order to minimize the magnitude of the measured angular accelerations without affecting the measurement of translational accelerations. The translational accelerations can then be double-integrated to obtain boom motions. Pressure transmitters will be attached to the bottom of the boom skirt. This pressure measurement can then be changed to a depth measurement, if the specific gravity of the surrounding water is known. It is important to ensure that the parts of the system's instrumentation and cabling are watertight and to minimise the weight and size of the components. Data should be recorded in high-density digital format using a watertight data logger. The decision to record the information will be made based on sea spectra and boom orientation.
The waves are measured using a wave-rider buoy. Boom tension and orientation will need to be measured in addition to environmental parameters to complete the final computation.

3.2 Clean-up Techniques on Water

3.2.1 General Requirements

Following the containment of an oil spill, the next step in the clean-up operation is the removal of oil from the water surface. Clean-up operations are facilitated by the concentration of oil into a thick layer using a containment operation. These two operations should commence as early as possible after a spill in order to avoid operating in weathered oil and to limit the extent of the spill. Clean-up operations are carried out by means of mechanical devices called skimmers, which are specially designed to remove oil from the water surface without causing major alterations in its physical or chemical properties. Skimmers vary in both recovery efficiency and capacity depending on slick thickness and operating conditions, type of oil spilled, presence of debris in the oil or water, location of the spill, ambient climatic conditions and calmness of the sea. Most skimmers have a high overall rate of recovery and collection efficiency (volume recovered relative to the total volume of liquid collected), when an oil slick is relatively thick. Therefore skimmers should be used in conjunction with containment booms, being placed in the area of greatest oil concentration. All skimmers work best in calm waters with little or no wave action, since waves reduce the mobility of some types of skimmers and decrease recovery efficiency of nearly all types. When the current velocity exceeds 36 cm/s (0.7 knot) oil may be swept under the stationary skimmer. Wind may divert a slick toward a skimmer or away from it. Many types of skimmer will suffer substantial decreases in operating efficiency as soon as seaweed, branches or debris are picked-up. Some skimmers are provided with protective screens around the intake, which must be periodically cleaned to maintain operating efficiency.

Pumping systems in skimmers have tremendous importance for successful operation. Intake of air makes some pumps lose their prime and the skimmer stops operating. Some skimmers have difficulty in handling water-in-oil emulsions and heavy oils in cold waters. Highly viscous, waxy or high pour point oils block the entrance to some skimmers and render them inoperative. Thus weathered oil tends to be more difficult to recover than fresh oil since its viscosity increases. In sorbent surface skimmers, weathered oil can be skimmed from the water surface, but it can not be further transferred due to the incapability of the pumping installation.

Skimmers could be classified according to their basic principle of operation into 5 groups:

- weir type devices;
- sorbent surface devices;
- suction devices;
- centrifugal devices;
- submersion devices.

Each of these types has different collection ability and distinct advantages and disadvantages.

3.2.2 *Weir Skimmers*

These devices take advantage of gravity to drain oil off the water surface by means of a weir or a dam aligned with the oil-water interface and a holding tank from which oil is removed by a pump (fig.3.17.). The oil and water plunge into the tank over the edge, which should be carefully adjusted with respect to the interface in order to minimise the amount of water collected. The skimmer is supported by floats to maintain the proper operating depth and this attachement should be capable of lowering and raising the level of the weir.

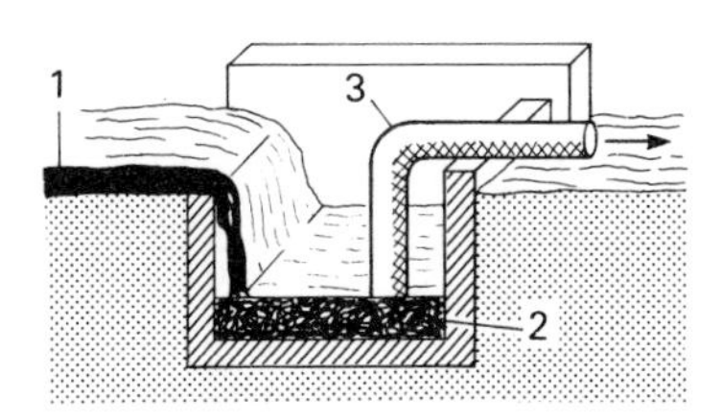

FIG.3.17 Operating principle of weir-type skimmer; (1) oil slick; (2) weir-type skimmer; (3) suction pipe.

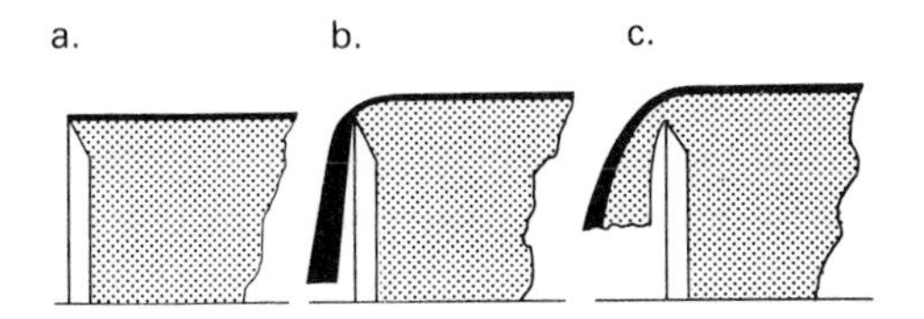

FIG.3.18 Flow of oil over the edge of the weir; (a) no flow: (b) limited flow of oil only; (c) free flow of oil and water.

The efficiency of weir devices is considerably reduced in cold waters, where the viscosity of oil is increased and some oils are below their pour point and will not flow. More viscous oils may accumulate on the edge and thus prevent free flow. The oil will not flow when the edge of the weir is submerged to the oil/water interface (fig.3.18a.). It will begin to flow when the surface tension oil-air will correspond to $V_{crit.}$ for oil (fig.3.18b.). However, when it reaches the surface tension of water, then the water will flow into the weir carrying the oil slick with it (fig.3.18c.).

The presence of debris also reduces weir efficiency, because debris collecting round the edge of weir holds back the flow of oil and screens or prevents the passage of thicker oils or reduces the thickness of the oil film within the screened area. Debris must be manually removed from the skimmer.

Weir skimmers are very susceptible to wave action and they have the tendency to rock back and forth in waves, alternately sucking air above the slick and water below it. In rough water some weir skimmers have a tendency to turn over. Thus in waves the weir skimmers have a very low efficiency and additional trouble may be caused by the loss of prime of external pumps. They should not therefore be used in the open sea, but only in calm protected waters.

The Toscon weir skimmer has two pontoons securing the flotation. A propeller in the stern draws oil and water through a converging channel, beginning at the bow of the skimmer. Two weirs, one at the bow and the other at the centre, skim the oil from the water surface into the skimmer. Each is a horizontal gate attached to the skimmer at the bottom with hinges. The bow weir can be adjusted to admit a varying depth of fluid by the addition or reduction of ballast water inside the weir.

A hydraulic pump and a transfer pump are required to operate the skimmer. The hydraulic pump supplies power to the propeller motor on the skimmer, and the transfer pump transfers the recovered oil and water from the skimmer to the separator.

The Toscon separator is mounted on a trailer, which also carries the hydraulic pump and a diesel driver, plus a transfer pump with hydraulic driver. The separator is of the gravity differential type and uses the difference in fluid densities to separate the mixture. Oil and water recovered from the skimmer enter the progressive-cavity transfer pump through a 2" connection. This fluid then enters the separator through a 2" pipe connected to the discharge of the transfer pump. If the skimmer is collecting 100% oil, the pumping system can bypass the separator. Outlet connections for the separated oil and water are 4" diameter, but can be reduced to 2". Additional facilities are required for storing the separated oil and water (fig.3.19.).

The "Erno" oil skimming system is intended to combine with a hopper-dredger, because these types of vessels have sufficient pump and tank capacity and are specially designed to operate in offshore areas and river estuaries. They are distinguished by a small draft and they can be quickly unloaded, so they could be made available for oil-skimming within a very short time (about 2 hours). This factor is of great importance.

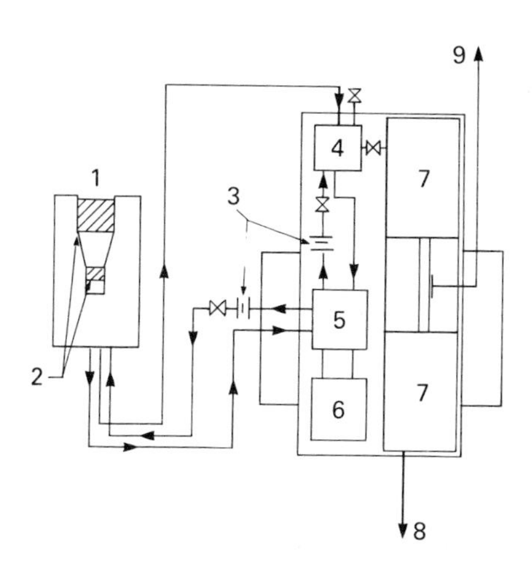

FIG.3.19 Toscon skimmer and separator system; (1) Toscon skimmer; (2) weir; (3) flow metres; (4) motor pump (5) hydraulic pump; (6) diesel engine; (7) separator; (8) water effluent; (9) oil effluent.

The "Erno" system consists of two skimmers, which are permanently stowed on the deck of the dredger. They could be lowered on the water surface by a ship's crane and are easily connected to the ship's side in such a manner that they have freedom to ride the waves. The outer shape of the "Erno" skimmer is similar to a floating dock. The main structure consists of floodable and drainable side walls and the rear crosswall. The draft of the skimmer could be changed by means of buoyancy and trim

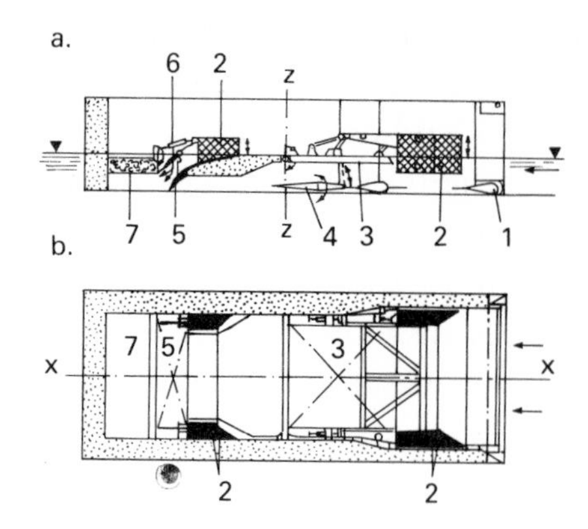

FIG.3.20 "Erno" skimmer; (a) longitudinal section; (b) plan view; (1) stabiliser; (2) float; (3) front separ. flap; (4) stab. flap; (5) aft separ. flap; (6) water outflow; (7) oil sump.

tanks. The skimmer has three functional areas (fig.3.20.):

- a front skimming area with a separation flap fitted with buoyancy tanks and a balanced rudder;
- an aft separation area with a separation flap, containing a buoyancy tank;
- an oil containment area (oil sump).

The skim and separation flaps are adjustable. This feature allows the removal of different types of oil in weathered conditions and in very thick layers.

A balanced skim flap is placed at the front of the system. It is attached to two buoyancy bodies and a balanced wing. It follows the wave and its edge is positioned at the level of the interface between oil and water. Thus it skims the oil layer and leads it to the intake of the main system. The task of the balanced wing is to damp the front separation flap with buoyancy tanks in case of wave breaking. The thickness of the layer, which should be skimmed, could be easily regulated.

After the oil/water layer has passed the intake, it proceeds to the aft separation flap, which safeguards accurate separation. Due to buoyancy tanks it can be adjusted according to the actual water line and it separates the oil layer from water with the front edge of the flap. The overflowing oil proceeds over the flap to the oil sump, from where it can be removed by pumps. The remaining water is conducted via a lower channel back to the sea. The position of the channel outlet is adjustable, depending on the pitch angle of the aft separation flap. The calming area in front of the aft separation flap is divided by a longitudinal bulkhead in order to damp the arising motions.

3.2.3 Sorbent Surface (Oleophilic) Skimmers

This type of skimmer incorporates a surface of a different form, to which oil can adhere while the surface is moving through the oil. This form could be one of the following (fig.3.21.):

- disc;
- drum;
- belt;
- inverted belt;
- rope.

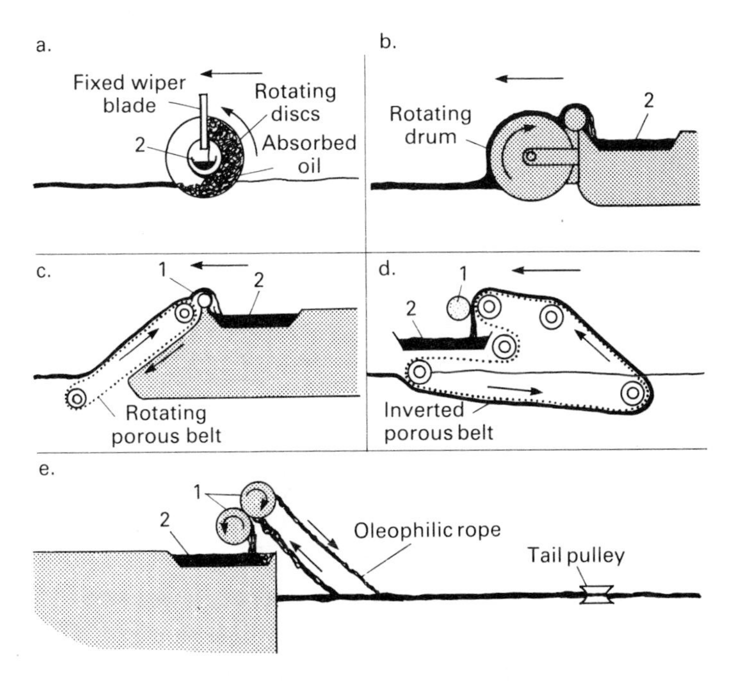

FIG.3.21 Operating principles of various sorbent surface skimmers; (a) oleophilic disc; (b) oleophilic drum; (c) oleophilic belt; (d) inverted oleophilic belt; (e) oleophilic rope; (1) wringer rollers; (2) collection trough; (←) vessel movement.

Oil collected on these surfaces is removed by a wiper blade or wringer roller and is collected into the collection trough, from which it is pumped into the storage facilities on board a ship or on a barge. The sorbent surface varies from metal discs and drums to plastic or fibre belts and ropes. Fabrics and plastics do not pick up large quantities of water and they operate efficiently with relatively thin slicks of a wide range of oil viscosities. Aluminium discs and drums achieve the highest recovery rates with medium viscosity oils, while fabric surfaces work well with a wider range of oils. The speed of rotation of skimming component and the velocity of recovery vessel affect the efficiency of oil recovery. Due to pressure waves there is a critical speed, above which oil is pushed away from the sorbent surface and more water than oil is picked up by that surface.

Debris may interfere with the recovery efficiency and in some cases may cause damage. This type of skimmer is least affected by waves, because the wave action may enhance the oil recovery of certain models by increasing the surface area of the sorbent component, which is exposed to the oil. They are very well suited to work within the containment booms and adjacent to the docks.

In rope design, the rope must be oil sorbent and sufficiently resilient to withstand repeated squeezing. Polypropylene fibres in a specially woven fabric have proved suitable for lighter oils, which can be sorbed into the cell structure. For recovery of heavier oils an oleophilic belt would be more suitable, but moving at a relatively slow rate.

Floating belt mop skimmers have flotation provided by two cylindrical aluminium pontoons of 8" diameter and 60" length. The distance between them is 24" (centre to centre) and the weight of this skimmer is 106 pounds (fig.3.22.). Two continuous 4" wide polypropylene belts are supported by a trough between the pontoons. Oil adheres to the belts on the water surface. The belts are then guided over the trough to wringers, which squeeze the oil out into a collection pan. The collected oil is then pumped to a portable stowage bladder. The oil recovery rate (ORR) of the

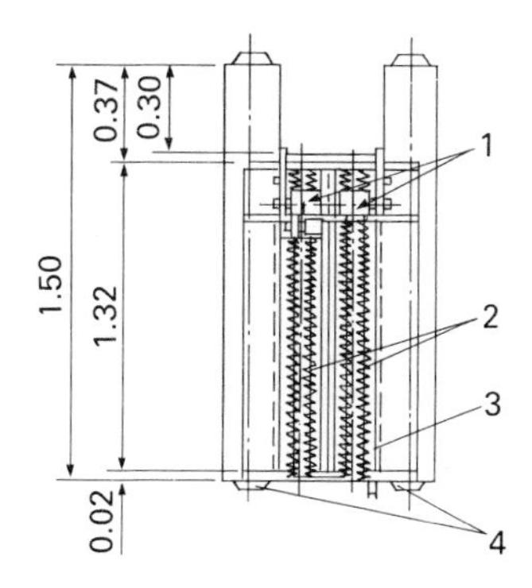

FIG.3.22 Floating belt mop; (1) wringer rollers; (2) two 4" belt mops; (3) oil pan; (4) flotation pontoons.

skimmer is between 7.5 and 11.4 l/min. Oil recovery efficiency (ORE) in calm waters is in excess of 85% and in waves 65 to 70%. The maximum pumping rate is 18 to 21 gallons/minute. The system is driven by a single stage compressor, which is powered by an 8 BHP gasoline engine providing 12 ft^3/min at 100 psi. Compressed air is used to drive both the wringer rollers on the skimmer and the transfer pump, which is of a double-diaphragm type with an intake-discharge pipe size of 1". A polyurethane oil storage bladder of 300 gallons capacity is provided. The skimmer system includes two 1" oil hoses and two 0.25" air hoses. Fig. 3.23. shows a schematic arrangement of the system.

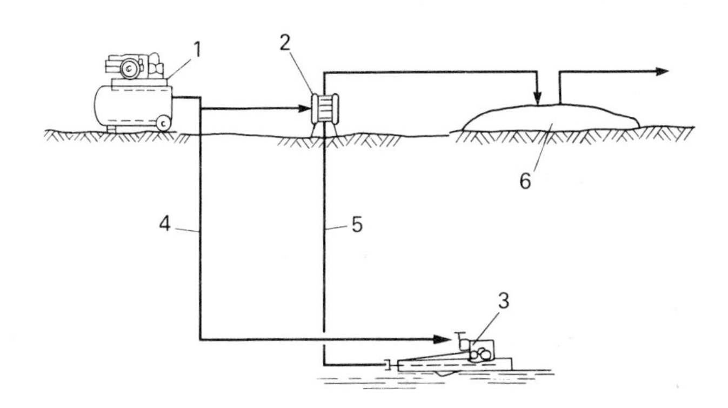

FIG.3.23 Floating belt mop skimmer system; (1) air compressor; (2) 1" diaphragm pump; (3) mop skimmer; (4) 1/4" hose; (5) 1" hose; (6) storage bladder; (7) outlet to vacuum tank truck or a tanker.

Tests of rope mop skimmer have been conducted in ice-infested waters. The skimmer consists of three oleophilic rope mops mounted between the aluminium hulls of a catamaran (4.6 m long and 2.4 m wide). The channel between the two hulls diverges from 813 mm at the bow to 876 mm at the stern to ensure free passage of ice through the skimmer. The three rope mops were each 304 mm in diameter and 9.6 m long, mounted in a loop along the length of the catamaran. At the water surface the mops travelled from the bow to the stern, from where they were pulled up. All three mops were driven and squeezed by a wringer assembly located at the bow. The wringer speed was adjustable to enable the skimmer to operate in a zero relative velocity (ZRV) mode.

The skimmer was equipped with a rotary-type positive displacement pump to off-load recovered oil. A single diesel engine-driven hydraulic pump powered the wringer, the oil discharge pump and other accessories, which included two outboard motors with steering controls and an oil herding water pump. After starting the engine, the skimming operations were remotely controlled from a radio control unit.

Tests have shown, that:

- the rope mop skimmer was able to pick up oil in broken ice with ice concentrations less than 50%. In tests, where ice concentration exceeded 50%, ice jammed at the skimmer inlet and prevented the oil slick from entering the skimmer;
- the ORR (oil recovery rate) varied inversely with ice concentration within the effective range of the skimmer (below 50% ice concentration);
- the ORR and ORE (recovery efficiency) increased with increasing slick thickness. The TE (throughput efficiency - the ratio of oil recovery rate to oil distribution rate, in percent) was lower at the higher slick thickness.

A rotating disc skimmer works on the principle of oil sticking to the surface of a vertically mounted flat disc, rotating about a horizontal axis and being partially immersed through an oil film which floats on the water. The oil can be scraped off from

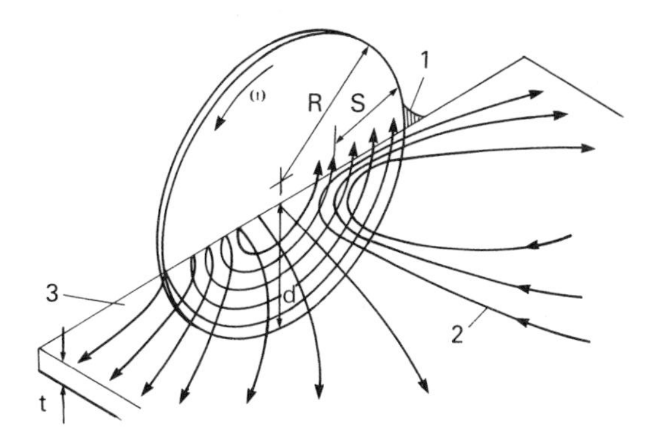

FIG.3.24 Oil film movement produced by a T-disc; (1) tail; (2) stream lines; (3) free surface line; (R) disc radius; (d) depth of immersion; (s) scraped width; (t) oil film thickness; (ω) angular velocity.

the disc surface and transferred to a collection point. A typical range would be from 280 mm to 800 mm, dependent upon the application. The oil collection capacity of a skimmer depends upon the number and size of the discs. A disc skimmer having 80 discs of 820 mm diameter can recover up to 100 tons/hour, whereas a skimmer having 32 discs of 280 mm diameter has a capacity of about 10 tons/hour. The experiments performed to examine the movement of oil films across the free surface in the vicinity of a single rotating disc have shown that collection occurs symmetrically on either side of the downward moving part of the disc. The attached oil film, which is drawn below the surface, is carried through the water region as the disc rotates. Some oil particles will move radially outward and will be thrown off the edge of the disc, dependent upon the rotational speed (3.24). No additional oil is collected by the disc as it emerges through the oil film above the free surface. Experiments to establish the oil recovery rate of a disc as a function of rotational speed, depth of immersion and diameter were carried out with the disc rotating entirely in oils of different viscosity. These experiments have shown that:

- the oil recovery rate increased in proportion to the depth of immersion up to a maximum, when the disc was half immersed in the oil;

- for high depths of immersion (0.84 to 1.0 times the disc radius) the rate of oil recovery varies in direct proportion to the rotational speed;
- for lesser depths of immersion (0.3 to 0.5 times the disc radius) the rate of oil recovery of the disc increased with rotational speed until a limit was reached, after which the recovery rate remained constant;
- small depths of immersion impose a limit on the recovery rate of the disc even when the film is thick relative to the disc radius;
- some oil emerging above the free surface is thrown off the disc to form a "tail".

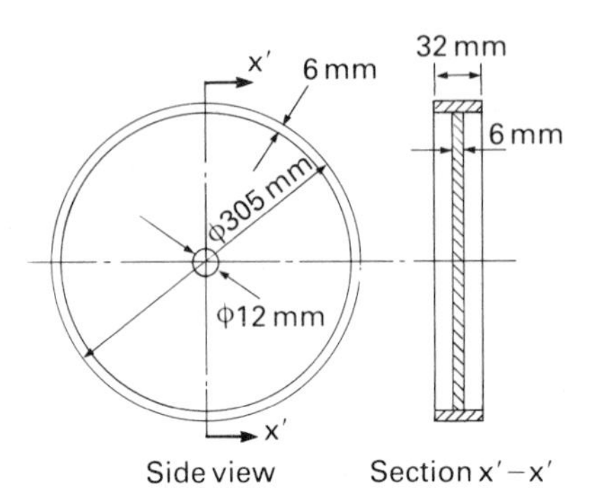

FIG.3.25 Dimensions of a T-disc.

A similar set of experiments were conducted with discs immersed in thin oil films, which have shown that the rate of recovery was identical to that for the oil-only situation until a limiting value was reached. This limit was found to be a function of the oil film thickness, being higher for thicker films. These experiments formed the basis for several design improvements of the geometry of the disc, making it capable of recovering oil at considerably higher rates than the original plain disc. Several improvements were made:

- the T-shaped disc (fig.3.25.) reduced the amount of oil lost at the tail, the cylindrical rim round the outer edge of the disc acted as a barrier reducing the oil loss. An additional benefit is the increased surface area without any corresponding increase in the diameter of the disc. The improvement in

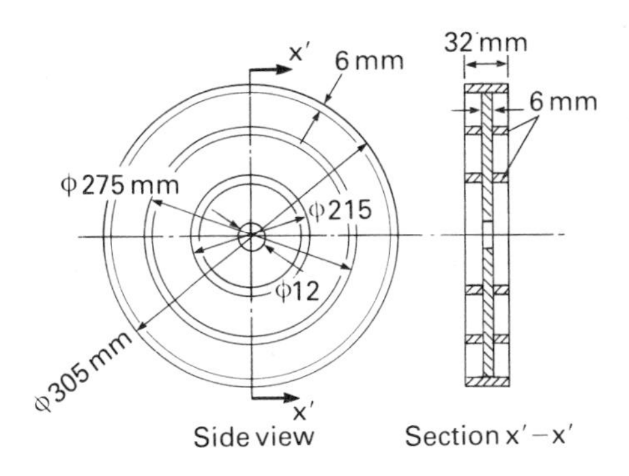

FIG.3.26 Dimensions of ribbed disc.

performance was strongly dependent upon the depth of immersion and rotational speed. When the T-shaped disc was used in a thin 5 mm oil film it was found to recover 20% more oil than a plain disc, but in a relatively thick oil film (11 mm) the improvement amounted to 130%. Thus the T-shaped disc gives the most significant improvement for relatively thick oil films;

- ribbed disc (fig.3.26.) identical in overall dimensions to the T-shaped disc was fitted with two additional concentric circumferential ribs. The ribbed disc was specially developed to improve further the performance of this device at depths of immersion between 60% and 90% of the disc radius. At higher rotational speeds (exceeding 90 rpm) a gain of 134% was measured, compared to 42% of the T-shaped disc;
- recovery of oil in the tail was developed as a rim scoop to collect the oil thrown off the outer edge of the disc (fig.3.27.). This is a noncontacting partially submerged device, which must be suitably positioned close to the emerging side of the disc, where the tail is produced.The weir like geometry of the rim-scoop captures the oil thrown off in the tail, causing it to flow into a collection tray. For a plain disc the improvement in performance at small immersion (20% of disc radius) was about 120% at 30 rpm rising to 240% at 105 rpm. For greater depths of immersion (80% of disc radius) the improvement was generally at a lower level, being 20% at 30 rpm and rising

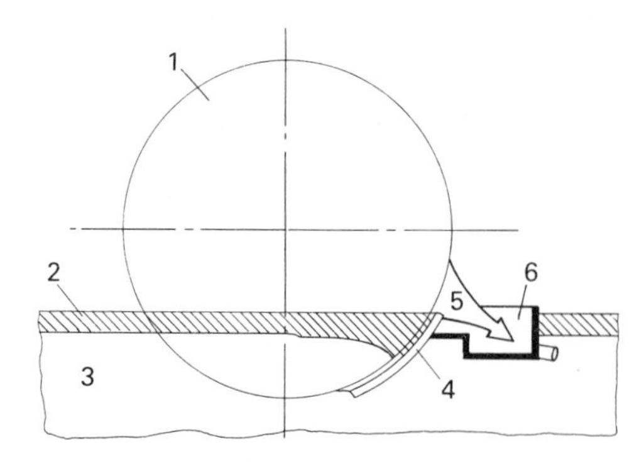

FIG.3.27 Schematic of ring scoop; (1) disc; (2) oil film; (3) water; (4) rim scoop; (5) tail; (6) collection tray.

to 72% at 105 rpm. In oil films 5.5 mm thick the improvement was 34% compared to the plain disc, whereas for an oil film 13 mm thick the improvement was higher still at 85%. A T-shaped disc with a rim scoop has shown for small depths of immersion (20% of disc radius) an improvement of 640%. For larger depths of immersion (66% of disc radius) the observed improvement was considerably lower (about 200%). Practical applications of a rim-scoop would be in locations where there are no waves, such as industrial settling tanks and effluent control systems;

- multiple disc skimmers have revealed that for oil-only situations there is no interaction between two adjoining surfaces and the performance of individual disc inner and outer surfaces remained the same. But when they rotated within an oil film, then the interactions produced a deterioration in the perfomance of the inner surfaces. This interaction depended upon the spacing of the discs, so that the closer the spacing between the discs, the poorer was their individual performance. This is assumed to be the result of the velocity limit for an oil film flowing on the surface of the water. This limit will generally occur near the front edge of the disc.

3.2.4 Other Types of Skimmers

Submersion skimmers work on the principle of forcing oil beneath the water surface by a moving belt inclined at an angle to the surface (fig.3.28.). At the downward end of the belt the oil is removed by a mechanical scraping device, located at the collection well opening. The oil rises to the surface due to its buoyancy and is collected in a well, from which it is pumped to an outboard or adjacent collection facility. The water continues its flow horizontally past the opening. To prevent oil from missing the collection well, the skimmer has to be advanced slowly over the water surface. It is most efficient with low specific gravity oils, which are more buoyant and therefore rise up more quickly into the collection well.

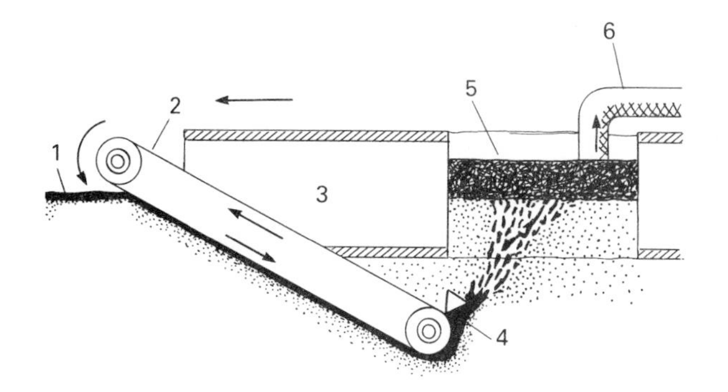

FIG.3.28 Operating principle of submersion skimmer; (1) oil slick; (2) rotating belt; (3) ship; (4) scraper; (5) collection well; (6) pump to storage; (←) vessel movement.

Submersion skimmers do not decrease their collection efficiency in waves to a similar degree to other types of skimmers, or by debris, although many models have screens to minimise the amount of debris passed with oil into the collection well. They could also be used to collect floating sorbent materials. Some models are equipped with "chipper discs" to macerate solid debris. The disc consists of a series of rotating cutters, similar to those used to chop up timber in paper manufacturing.

As the submersion skimmers are normally large, they are usually mounted on or incorporated in a powered vessel. This system, especially if fitted with wide-spread "lobster claw" arms may be used to recover an uncontained slick. The effectiveness in waves is affected by the motion response of the entire system.

A belt skimmer conveys oil from the water surface by adhesion.

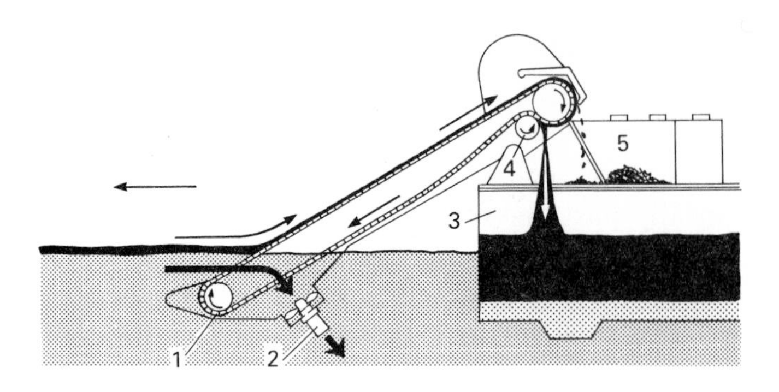

FIG.3.29 Operating principle of belt skimmer; (1) belt; (2) pump; (3) ship; (4) wringer rollers; (5) debris; (←) vessel movement.

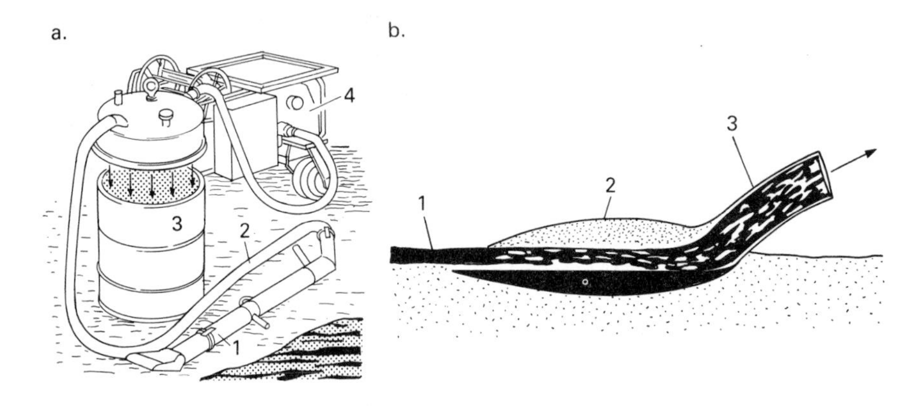

FIG.3.30 Suction skimmer; (a) complete installation; (1) orifice; (2) large diameter pipe; (3) vacuum tank; (4) vacuum pump; (b) operating principle; (1) oil slick; (2) suction orifice; (3) pipe to vacuum tank.

Upward rotating belts carry the oil to their top limit, where it is scraped or squeezed off into a storage tank (fig.3.29.). These skimmers could be used for medium viscosity oils and they also tolerate heavier materials.

Suction skimmers draw oil surface layers through a hand-held or floating suction orifice into a manifold or tank, where there is a vacuum generated by an external pump (fig.3.30.a and b.). The height of the orifice and its vertical position in relation to the surface must be strictly controlled. This is very important

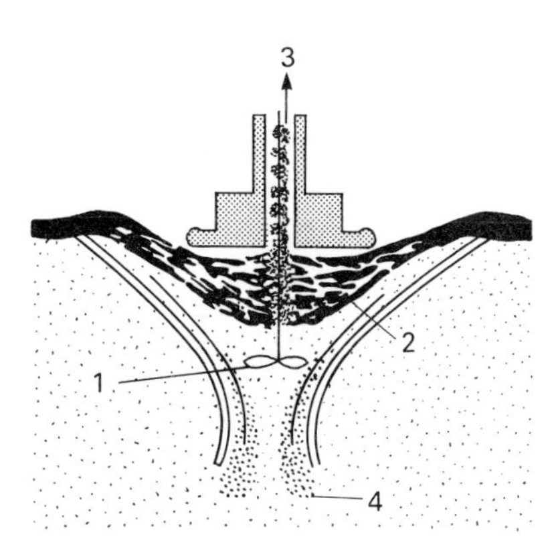

FIG.3.31 Centrifugal (vortex) skimmer; (1) water pump; (2) sinkage of water level due to vortex; (3) oil to separator; (4) water.

when the surface contains waves and elevation varies with time at any point. The overall capacity of most suction skimmers is determined by the orifice, the size of hose and the capacity of the vacuum pump itself. Suction skimmers are extremely susceptible to clogging with debris. They are most effective in calm waters, but recent research has been directed toward designs capable of operating in rough open seas.

Centrifugal (vortex) skimmers work on the principle of creating a water vortex, which draws the oil into a collection area. The vortex is created by means of a propeller forcing the vertical flow of water and sinkage of the water surface (fig.3.31.). The oil is collected in a special collecting box placed in the centre of the vortex, from which it is pumped away by an independent pump. These skimmers are not susceptible to clogging with debris thank to screens, which do not adversely affect oil recovery. But they have current and wave limitations and can not operate in waves higher than 60 cm or currents exceeding 0.5 m/s.

3.2.5 Requirements and Selection Criteria for an Oil Skimming System

There are two classes of criteria for evaluating the proposed skimmer system. First, each system has to meet five "GO/NO GO" requirements. A system failing to meet any one of the five would be rejected. The "GO/NO GO" criteria are:

- the skimmer must be capable of recovering both marine diesel fuel and JP-5;
- the system must be selfsufficient, containing all the components necessary to deploy, power and retrieve the skimmer and to collect, pump and store the oil-water mixture;
- the skimmer must be capable of functioning unattended for up to 4 hours;
- the skimmer must use a positive displacement pump to minimise oil emulsification;
- the skimmer must have a maximum draft of 1 ft.

The second set of criteria are the ranking criteria as given in table 3.1. Weighting factors and utility factors are multiplied by each other and all the scores are added. All technical proposals with total scores greater than 75 points are considered acceptable.

The new skimmer has to undergo two sets of tests: functional tests and confirmation of the skimmer's ability to perform according to established operational criteria.

The functional tests consist of seven specific tests that the skimmer must pass. Any deficiencies are corrected before the first article testing begins:

- the system has to be checked to confirm that it includes all of the components, spare parts, tools and manuals specified in the technical proposal;
- the basic system components are to be disassembled and reassembled to confirm that all the parts fit together correctly without undue force;
- each component of as well as the unit as a whole, is to be checked to see that it is operating correctly. The system is to be operated continuously for 4 hours without any failure;
- the thoroughness of the operation and maintenance manual is to be tested. The personnel should carry out each operation and maintenance step including cleaning;
- a timed test run should be conducted for removing the skimmer from a half-ton pickup truck and bringing it to fully

TABLE 3.1 Ranking criteria for oil skimmer selection.

Weight. factor	Criteria in ranking order	Standard	Utility factor
15	1. Recovery rate	120 gal of oil per hour	<40 gal/h = 0 40 - 79 gal/h = 0.33 80 - 120 gal/h = 0.66 >120 gal/h = 1.00
15	2. Deployment time	20 min from reaching site to operational status	>60 min = 0 40 - 60 min = 0.25 31 - 39 min = 0.50 21 - 30 min = 0.75 20 min or less = 1.00
13	3. Wave impact	Design performance is achieved in waves 1 ft high with harbour chop	calm conditions = 0 0 - 3 in. waves = 0.25 4 - 7 in. waves = 0.50 8 - 11 in. waves = 0.75 >12 in. waves = 1.00
11	4. Equipment validation	Manufacturer's claim validated by actual spill use and by credible laboratory test	unvalidated = 0 validated by user = 0.50 validated by credible laboratory = 1.00
10	5. Oil recovery efficiency	Effluent from recovery device is 80% oil (DFM 5 mm thick, 1 ft wave conditions, 60 °F water)	0 - 20% = 0 21 - 40% = 0.25 41 - 60% = 0.50 61 - 80% = 0.75 81 - 100% = 1.00
10	6. Portability	Transported in bed of 1/2 ton pick up; handled by two people	Meets requirement = 1.00 Does not meet the requirement = 0.00
9	7. Maintainability	Detailed cleaning and repair procedures; no special tools; 8 h training; detail. manual	Meets requirement = 1.00 Does not meet the requirement = 0.00
9	8. Setup manpower	One person can physically handle each component (does not include retrieval)	3 or more people = 0 2 people = 0.80 1 person = 1.00
5	9. Debris impact	Debris of 0.5 in. dia. accepted; larger debris does not interfere with operation	Meets requirement = 1.00 Does not meet the requirement = 0
3	10. Storage capacity	250 gal in a separate unit	<50 or >500 gal = 0 50 - 100 gal = 0.33 100 - 199 gal or 301 - 500 gal = 0.66 200 - 300 gal = 1.00

operational status. Data on length of time and number of personnel required for setup are to be compared to the figures in the technical proposal;

- the portability of the system tested by placing it in the bed of a half-ton pickup truck;
- all maintenance steps must be performed to ensure that no special tools are required; the operator training class should confirm this to be accurate and to require no more than 8 hours for implementation; spare parts are to be confirmed to be in stock at the CES facility.

After successfully passing the functional tests, the system should undergo the following series of tests:

- measurement of the oil recovery rate (ORR) and oil recovery efficiency (ORE) for the following conditions: Navy standard conditions (DFM 5 mm thick in 1 ft waves with harbour chop); calm to 1 ft waves with harbour chop; and 1 to 10 mm oil slick thickness. Results should agree with the reported values;
- ORR and ORE measurements for recovering JP-5 (1 mm thick) under calm conditions;
- confirmation that the unit can perform effectively without an operator in attendance for periods up to 4 hours, i.e. the unit's ORR and ORE should not radically decrease if the operator is not adjusting the controls;
- confirmation that the skimmer is a selfsufficient system by showing that everything necessary for effective operation is provided. The pumping unit especially should be tested for acceptable performance at the maximum mixture recovery rate to ensure that excessive emulsification of the oil-water mixture does not occur. The results should match the reported values;
- confirmation that the storage system can contain the reported volume of recovered mixture. Under fully loaded conditions confirm that the draft is 1 ft;
- confirmation of the debris impact criteria by testing the debris protection system with long grasses, wood, aluminium cans and plastics. The result should be that these items do not

radically reduce skimmer performance. Also, check the pump debris capability with 0.5" pieces of wood and styrofoam.

3.2.6 Selfcontained Oil Recovery System for Use in Protected Waters

This system consists of two 30 ft tow/workboats and the 65 ft skimming barrier assembly. The workboats have identical hull forms, engines and pilot house arrangements. The deck area forward of the pilot house is arranged differently on each: one workboat carries the containment boom, while the other (pump boat) carries the oil recovery pump, oil/water separator, floating oil storage bag, 65 ft skimming barrier assembly and associated suction and discharge hoses. A containment boom could be attached to the skimming barrier and makes up a large, 295 ft skimming configuration.

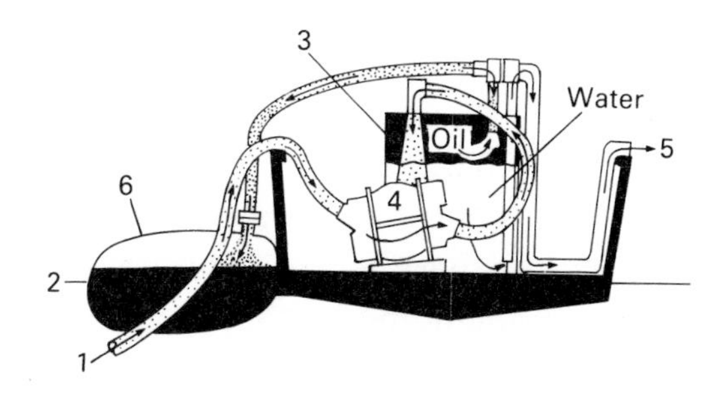

FIG.3.32 Section of oil collection system on a SCOOP vessel; (1) hose for oil/water mixture from weirs: (2) separated oil; (3) 350 gallon separator tank; (4) pump; (5) separated water outflow; (6) 750 gallon oil storage bag.

The oil/water mixture is pumped from the oil recovery weirs by the outboard oil recovery pump and discharged into the separator, where most of the water is extracted from the recovered fluid before being pumped out overboard (fig.3.32.). The skimming barrier weir assembly (fig.3.33.) is two feet high and floats with approximately one foot of freeboard. Two weir sections at the appex are adjustable to reduce the quantity of water flowing into the weir. Oil and water passing over the weir drains into a small tank, which is connected to the 2" hose. An adjustable lead collar ballast weight was arranged to control the weir depth.

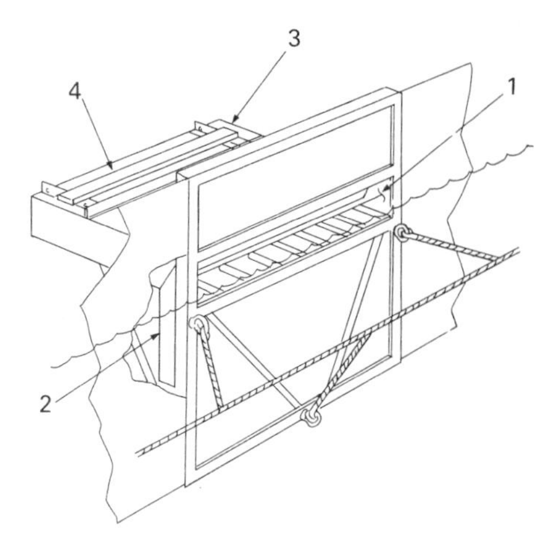

FIG.3.33 Skimming strut; (1) weir inlet; (2) tank; (3) outrigger flotation; (4) adjustable ballast.

The containment boom is designed for ease of storage on board the workboat. The flotation body folds into a stowed position and opens automatically, when deployed (fig.3.34.). A 100 ft section of the boom occupies less than 40 ft^3. Rigid aluminium strut frames support the flotation and the galvanised steel ballast, which is arranged to ensure the proper floating position of the boom. The combined barrier and containment boom were tested over several days including 25 to 30 knot winds and estimated 2 to 3 ft breaking seas.
The 30 ft workboats could be trailered without the need for special overland permits. They are designed to carry all equipment to the spill site in a ready-to-deploy position. Each boat is powered with a 170 BHP engine, giving a speed of up to 25 knots. A high freeboard forward and selfbailing work deck allow personnel to work safely.
The hydraulic pump is a variable displacement pump, pressure and flow compensated. It is directly coupled to the main engine through the clutch. It operates at 1,800 rpm and the engine operates at 2,250 rpm. The hydraulic system includes an oil reservoir, oil cooler, alarm circuits for low oil level and high oil temperature, and a one-quart bladder type accumulator to smooth out the flow surges inherent in the operation of an oil

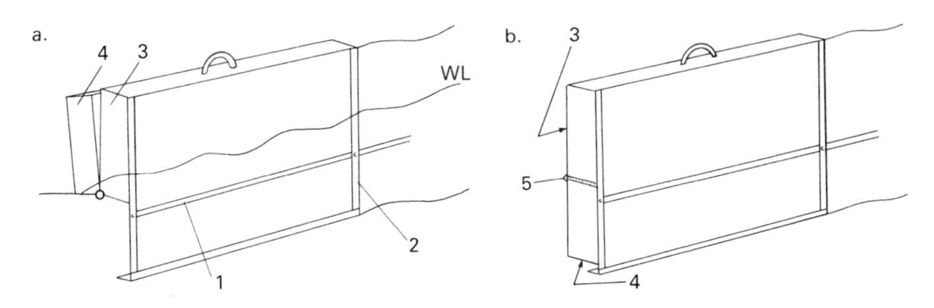

FIG.3.34 Containment barrier; (a) deployed position; (b) stowed position; (1) tension member; (2) rigid strut; (3) upper flotation; (4) lower flotation; (5) fabric hinge.

recovery pump.

The oil recovery pump is a diaphragm pump with a swing-type flapper valve. The pump is double acting, with pumping chambers on each side of the diaphragm.

The oil/water separator contains 12 compartments designed to extend the flow path and maintain the low average velocity required for gravity separation. At 50 gal/min, which is the designed flow rate, more than 5 minutes of residence time is provided. The average velocity is about 0.04 ft/s. In tests the unit operated successfully at flow rates up to 200 gal/min. An 8 ft stand pipe/vent is attached to the separator to allow air to escape. It protects the separator and stowage tank from excessive pressures. The stand pipe is made of clear plastic to allow the fluid height to be monitored. The separator is controlled by valves located on the oil discharge port and the water discharge port. The valves are balanced to maintain the oil/water interface in the separator at approximately half-full level. The level is observed through a clear window on the side of the separator. If both valves are closed down too far, the stand pipe will overflow. A flexible hose attached to the top of the stand pipe directs the discharge overboard.

Four 750 gallon collapsible utility tanks are provided for recovered oil storage. They are compatible with oil, gasoline, and diesel fuel and are serviceable at temperatures from -40 °F to +140 °F. A 4" diameter hose connects the separator with the tanks, which are towed alongside the workboat. A ring and hook arrangement on the workboat and grommets located on the sides of the tank allow the tank to be easily connected, disconnected and towed. Once filled, the tanks are released from the boat and are recovered and towed by some other vessel to a barge or larger storage tank ashore.

3.2.7 Large Sweep Width Skimming System

In skimming the oil slick, the area covered is a function of the skimmer's swathe width and tow speed. Since most oil recovery devices are effective at tow speed of 1.5 knots (0.8 m/s) or less, the only way to increase coverage rate is to increase the device's sweep width. A 300 ft (92 m) sweep system consisting of a skimmer and concentrating boom, towed through a 1 mm slick at 1.25 knots (0.6 m/s) encounters 932 gal/min (212 m^3/h) of oil. A large sweep system is advantageous on large, unified slicks, whereas a vessel-of-convenience skimmer can be deployed more quickly, is maneouvrable (for skimming windrows) and only requires one vessel.

The skimming system described below uses two vessels to tow 660 ft (200 m) of boom in a nozzled catenary with a mouth opening of 300 ft (92 m) (fig.3.35.). The nozzle is formed by a set of two bridles. A weir skimmer at the end of the nozzle skims the concentrated oil. Trailing behind the skimmer is a float-mounted pump, which transfers the recovered oil to the main tow vessel.

The main tow vessel is ideally an offshore supply boat of over 150 ft (46 m) in length. The secondary tow vessel may be a tug of about 1,000 BHP (745 kW), which will have enough power to tow the boom and yet not so much power that low speeds are difficult. The offshore supply boats must resort to clutching their propellers in and out of gear in order to maintain speeds of less than 3 to 4 knots (1.5 to 2 m/s) unless they have variable pitch propellers. Their large aft working deck and low freeboard make offshore supply vessels ideal for deployment of equipment. A 90 bbl (14 m^3) storage/separation tank and hydraulic power supply used to power the pump float are carried on the aft deck. The main tow vessel is linked to the pump float by two hydraulic lines and two discharge hoses.

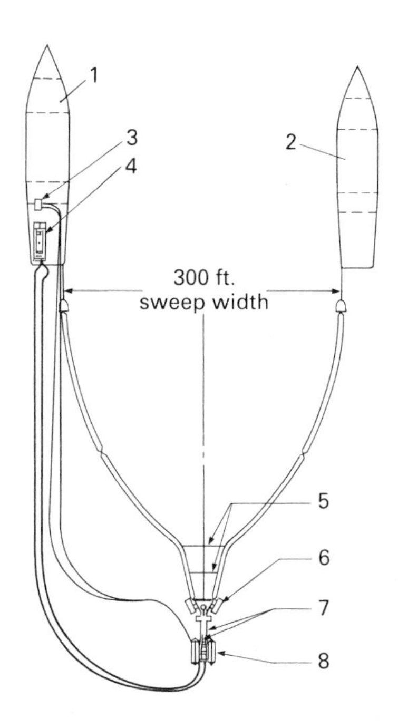

FIG.3.35 Large sweep skimming system; (1) main tow vessel; (2) secondary tow vessel; (3) hydraulic power supply; (4) 90 bbl storage separation tank; (5) bridle; (6) weir skimmer; (7) suction hoses; (8) pump float.

The following benefits arise from using the bridles:

- towing the boom in a catenary creates a reflected wave zone at the back of the catenary, which is the desired location of the skimmer. Placing the skimmer at the back of the nozzle isolates the skimmer from the reflected wave zone and also reduces the size of the zone;
- the main boom towing tension is carried by the bridles. Thus, the nozzle boom and skimmer are subjected to relatively little tension and are better able to conform to waves.

However, bridles must be used carefully to avoid forming cusps in the boom, which create turbulence causing greater underflow failure. Waves do not change the overall boom shape, but they will magnify the tensions in the boom and bridles.

The force increase due to wave action on a boom anchored in a current has been estimated by multiplying the steady state tension by the factor F:

$$F = \left(1 + \frac{2.26c}{V} \sqrt{H_{\frac{1}{3}}}\right)^2 \tag{3.2}$$

where: $H_{\frac{1}{3}}$ – significant wave height;
c – constant with values between 0.1 and 0.2;
V – current velocity.

For example, a 6 ft (1.8 m) sea would increase the tension 1.5 times over the tension in a boom in calm water with a current of 3 knots (1.5 m/s). In sizing the bridles and other components, wave induced tension must be allowed for.

With one bridle the resulting cusp in the boom was 75 °, but with two bridles it was only 50 °. Thus the addition of a second bridle smooths the transition and therefore significantly reduces turbulence. Adding a third bridle would bring the cusps down to about 35 °, but this would increase the complexity of the third bridle, which is not justified in every case. The tension in the main boom at 2 knots towing speed was 2,213 pounds (9.7 kilonewtons), while the tension in the nozzle was 240 pounds (1.056 kilonewtons). This illustrates the extent to which the bridles isolate the nozzle and skimmer from tension, improving the wave conformance.

Adding a large external load to simulate towing the barge causes the boom shape to straighten as expected, but the tensions generated were increased by less than the amount of the added load. The boom has a steeper angle of attack resulting in a smaller normal velocity component and thus less drag. The decrease in drag offsets the added loss.

There are three main options for storing the fluid recovered from a large sweep skimming system: use of the boom to tow a barge, pump over a short distance to a third trailing vessel and pump the fluid a long distance greater than 500 ft (150 m) to one of the tow vessels. For the system developed the third option was chosen for the following reasons:

- boats are generally in short supply at a spill site;
- the addition of a third vessel makes the system too difficult to coordinate. Using a 200 ft (60 m) offshore supply vessel to tow a boom catenary, it can take over 30 minutes to execute a 1800 turn;
- towing a barge generates higher tensions in the boom, reducing the wave conformance;
- with fluid pumped to the main tow vessel, water can be separated from the recovered fluid to maximise oil storage . Also the crew can monitor the fluid stream to determine how well the system is operating.

The viscosity of recovered oil magnifies the problems associated with pumping over such a large distance. Due to nozzle effects, the skimmer will recover oil from a thick pool of oil, which results in a high recovery efficiency, but requires the system to cope with high viscosities. If free water is recovered with viscous oil, the effective pumping viscosity can be reduced, but larger pumping rates and subsequent separation will be required.

In this arrangement the pressure and return hydraulic lines (1" dia) from a 65 BHP (48.5 kW) hydraulic power supply run 500 ft (150 m) to the pump float. The length of the return line generates significant back pressure on the hydraulic motor. In order to avoid running a third hose as a hydraulic motor case drain, a motor was specified which could withstand a high continuous back pressure. For viscous pumping, the hoses may be shortened to 400 ft (122 m) for this system by securing the hoses to the boom and shortening the boom tow lines.

The recovery pump is a positive displacement with displacement 1.48 gal/rev (5,600 cm^3/rev), which allows the pump to run at a lower speed. The larger displacement and resulting lower speed minimise emulsification of recovered oil.

The performance of this pumping system is difficult to determine. More power is required due to the long discharge hoses, but there is less power available due to the pressure drop in the hydraulic lines. It is to be noted that the efficiency of the hydraulic motor also varies. Weathering can increase the viscosity by more than an order of magnitude. Forming an emulsion can increase the viscosity several orders of magnitude, occasionally to the point at which the spill may better be treated as a solid. For the system described the limiting factor is the horsepower available at the recovery pump. The discharge working pressure (150 psi $\approx$ 1.0 megapascal) is not exceeded until after the horsepower limit is reached. Pump cavitation occurs shortly after the discharge hose working pressure is exceeded. Thus, for this system's capacity to be substantially increased, enlarged suction lines, stronger discharge hose and more horsepower would all be needed, but this might be not feasible.

3.2.8 Netting of Viscous Oil

Many oil spills involve heavy crude oil and heavy fuel oil (35%), whereas a further 50% involve medium crude and fuel oils that would be likely to form viscous water-in-oil emulsions. Nets are claimed to be useful in oil pollution control activities for several functions:

- as trawl funnels for collecting and recovering viscous oils, oil lumps and tar balls (fig.3.36.);
- as permeable floating barriers for protecting specific areas against floating oil;
- to encircle floating oil slicks with a permeable floating barrier;
- for surface to bottom protection of specific construction such as water inlets against floating or submerged oil;
- for use with recovery equipment.

Nets are used instead of impermeable barriers, because of the relatively low weight, high strength, low resistance to flowing water and wind and easy manageability. The danger of oil leakage through the net is inherent to its use. The leaking rate will depend strongly on oil viscosity V, mesh area A, the relative netwater velocity U, net thread thickness d and wave conditions. Viscosity and density depend on the fresh oil composition, weathering state, temperature, and on artificial intervention brought

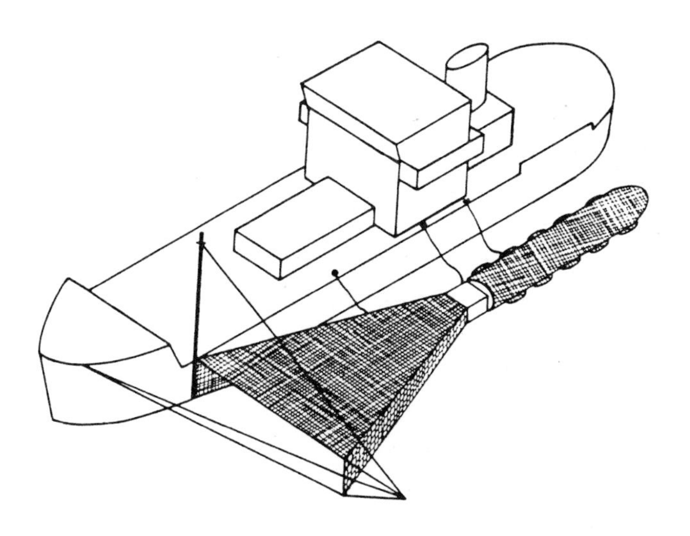

FIG.3.36 Towing a trawl net to collect oil.

about by adding sorbents and gelatinisation agents to the oil. Viscosity measurements on Ekofisk crude samples from the Bravo (North Sea) blow-out indicated a viscosity of 8,000 cS after some hours and of 70,000 cS after two days. The emulsified oil was broken into oil lumps by the mechanical action of the waves.

Tests have been performed in the laboratory with seven types of net to encompass a suitable range of mesh area A from 0.35 to 4.2×10^{-6} m^2 and thread thickness d from 0.25 to 2.00×10^{-3} m, which gave an open area fraction S from 0.10 to 0.74.

Current velocities are generally below 2 m/s in sea conditions and therefore velocities U ranging from 0.1 to 1.0 m/s are most important for assessing applicability, which is directly related to the leakage of oil through the mesh. The loss of oil should be considered relative to the amount of oil that can be recovered by the net and to the possibility of applying other means. A rough indication of the applicability of nets is given below:

- oil slicks can be usefully corralled with an unmoored vertical floating net, if the oil viscosity is >10,000 cS;

- moored vertical nets can be used to protect specific areas, if the oil viscosity is >40,000 cS. However, if the relative water-net velocity is >0.15 m/s, the net can become covered with oil, which can considerably decrease the permeability of the net;
- a trawl net can be used for oil with viscosity >40,000 cS and if the towing speed U is <0.5 m/s. However, the permeability of the net can be decreased considerably and consequently, the inflow of water and oil in the trawl funnel can be reduced.

3.2.9 An Acoustical Method of Collecting and Burning Oil Spills on Cold Open Water Surfaces.

This method yields a clean and controlled burn leaving no residual sludge and enables simultaneous collection of a portion of the oil, if so desired.

The two main phenomena associated with the acoustical method are radiation pressure and atomisation. A force due to radiation pressure may be generated at the interface between two media which differ greatly in acoustic impedance. Thus, when high intensity acoustic energy is focused from below the water surface (and the floating oil), the oil above is projected upwards. This is so because the acoustic impedance mismatch is between the air and the oil on the surface and not between the water and the oil. Because the sound wave is reflected at the interface of air and oil, the radiation pressure is equal to the energy density of the sound wave in the fluid. To project the oil to a height of 1.2 m, for example requires an intensity of 750 watts per cm^3.

To achieve this high density, a focusing ceramic transducer may be used. A parabolic trough shaped to the transducer yields a focal volume like a cylinder. This will levitate the oil in the form of a sheet as wide as the transducer, which is a desirable shape for collection. If the acoustic generator face presents a large angle to the surface, the oil will be projected upwardly, whereas if a small angle or grazing angle is used, it will move horizontally along the surface ("herding"). The ceramic material used is US Navy Code 5400 at frequencies of from 0.5 to 1.0 megahertz for levitating oil. Lower frequencies are used for atomising and burning.

Atomisation of fuel before combustion is gradually becoming a widely applied technique in the prevention of pollution as well as in more economical use of energy. As in levitating of oil, high intensity acoustic energy is delivered from below the oil surface in the form of a cylindrical focal volume. Focusing parabolic trough, single element ceramic transducers may be used (or a sandwich type). This latter has lower frequency capabilities without increasing the thickness of the ceramic material.

High frequency electric energy causes the ceramic material to vibrate at this frequency (piezoelectric effect) producing sound energy. The two basic pieces of equipment are transducers and high frequency power generators (R. F. amplifiers)

The transducer material is a ceramic made from lead titanate zirconate. One type of emitting profile is a machined spherical surface, which produces a spherical focal volume at distances of 7 cm to 30 cm. Another type is a flat rectangular surface, to which a specially modified zone lens is bonded. This gives focal distances of 5 cm to 20 cm.

The input power is supplied through a 3 m coaxial cable with a 100 picofarad per metre capacitance and a characteristic impedance of 50 ohms. The material of the transducer is capable of radiating 60 watts/cm^2 and has a Curie point above 300 °C. The output of amplifiers is directly proportional to the input and the latter is used to control power.

A single operation device is a combination weir and "herding" floating device (fig.3.37a.). It may be operated in a stationary or mobile mode. The transducers for this purpose should be a parabolic trough type or a flat type (8 cm x 15 cm) with a focusing lens. Several of these can be placed end to end with spaces of 30 cm between. Then they are in a line ahead of the weir and as wide as the mouth. These single element transducers would probably be operated at 200 watts on average, depending on the viscosity of the oil. Transducers force only the oil into the mouth of the weir. Without the transducers the weir would draw mostly water. A suction hose removes the oil continuously.

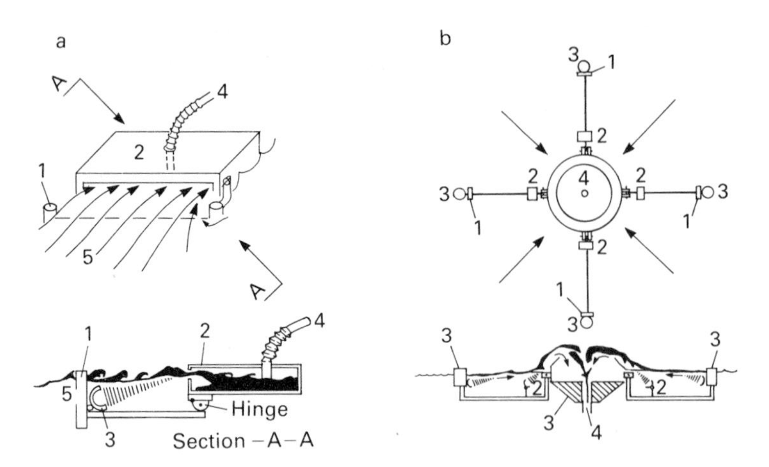

FIG.3.37 Application of acoustical method; (a) combination weir and herding device; (1) float; (2) weir; (3) herding transducer; (4) hose to suction device; (5) oil slick; (b) combination oil herding and levitation device; (1) herding transducers; (2) levitating transducers; (3) floats; (4) drain; (←) direction of oil flow.

A double operation device consists of a combined oil herding and oil levitation device (fig.3.37b.). The transducers are the same as in the single operation device, although operated at higher power. One group of transducers is placed at a grazing angle to the surface, but aimed toward the main body of the device. Another group is placed closer to the body and has a large angle to the surface, but is aimed at the collection pan above the surface. The oil is levitated into the central collection pan above the surface. Below, the pan is connected to the flexible hose and suction device, which continuously removes the oil.

Combination oil herding and atomising devices (fig.3.38.) are very useful in extremely cold climates and for all varieties of oil. Oil is herded to a small boom just above an atomising transducer and below a small stack. A propane torch is ignited by high voltage discharge and its flame aimed so as to follow the splash guard at the base of the stack. The atomising transducer is then turned to higher power and a clean burn results. Some of the heat radiates to the surrounding oil, making the atomisation of very viscous oil very efficient. By means of this device burning of a 0.5 mm thick weathered slick was possible. A 50 watt input to the atomising transducer was found to be sufficient to maintain a clear burn even though the transducer was capable of 2,000 watts of acoustic energy.

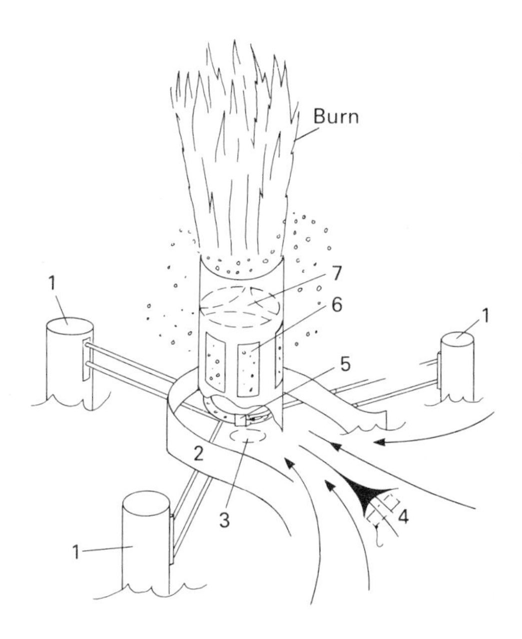

FIG.3.38 Combination oil herding and atomising device; (1) float; (2) weir; (3) atomising transducer; (4) herding transducer; (5) propane torch; (6) burning oil; (7) heat deflector.

If the crude oil is fresh and the temperature warm, a propane torch is unnecessary for lighting the oil. The atomising transducer directly below is turned on. This results in ignition of the slick in this area. It is preferable not to allow the oil to build up to a great thickness as this could result in an uncontrolled burn with contamination of the atmosphere and deposition of heavy sludge. Heat is radiated to the surrounding area and the remainder of the atomising transducers (without the stack) may be turned on. The power is increased until a very clean burn results.

If it has been decided not to collect any oil, the levitating transducers are turned so as to exert a resulting force pushing the oil away and upwards from the apex of the boom. Thus, even though the water may drain down, the oil will not. This enables good control of the burn. If some oil is to be collected, the transducers at the apex of the boom are directed so as to levitate the oil into the collection weir of the fireproof boom. Since the oil has been heated, its viscosity is sufficiently low to be easily suctioned off and collected.

3.3 Oil Recovery Vessels

3.3.1 Classification of Oil Recovery Vessels

There is a great variety of vessels specially designed and built for oil spill combating, as well as many ships which could be fairly easily adapted for pollution combating. There are several criteria to classify these ships such as region of action, method employed, etc. There are ships specially suited for:

- harbours, rivers and narrow estuaries;
- sheltered waters such as sheltered bays and wide estuaries;
- open sea.

As regards the methods employed a ship could have a passive or active recovery system or could carry on board any type of skimmer, which is laid on the water and only powered by the ship, having reception and storage facilities.

3.3.2 Vessels with Passive Recovery Systems

A passive recovery system is illustrated by a barge "Oil Crab", designed and built in the Netherlands (fig.3.39.). It consists of a barge, which is open to the sea through an inlet weir at the front. Oil sweeping arms have been attached to the port and starboard side of the barge's front bulkhead. These sweeping arms extend below the water surface and are folded together to form the stem, when not in use. In an open position and with the barge moving forward, the sweeping arms have the shape of a funnel narrowing towards the inlet weir. Surface oil and a relatively small amount of water enter the barge over the weir, while the side tanks provide the necessary buoyancy. The interior of the barge is subdivided into three separation zones and

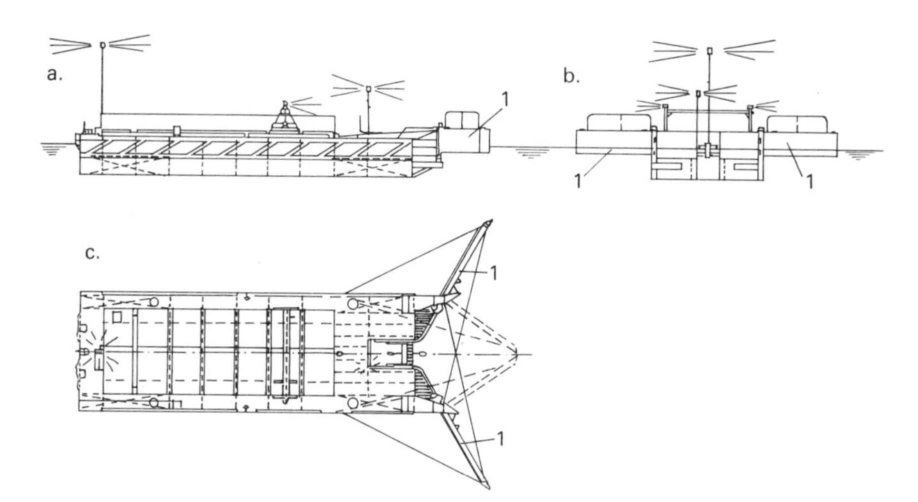

FIG.3.39 Barge "Oil Crab" arrangements; (a) side view; (b) front view; (c) plan view; (1) oil sweeping arms.

TABLE 3.2 Stowage skimmer combination barges.

Type SSC	Length [m]	Width [m]	Height [m]	Sweeping path [m]	Max.throughput [m^3/h]	Stowage capacity [m^3]
125	18	6	2.0	15	350	125
250	30	8	2.0	20	500	250
500	35	10	2.5	25	1,000	500
1000	45	15	3.0	32	2,000	1,000
1500	50	16	3.5	35	2,500	1,500
3000	60	19	4.5	40	5,000	3,000

a clear water discharge section. This is done by means of three transverse bulkheads, which extend from the underside of hatch covers to some distance above the barge bottom plating. Each one of the separation zones acts as a slop tank, where oil is gravity separated from the water. The clear water flows underneath the bulkhead to the following separation zone and finally from the clear water discharge section back to the sea. When the oil/water interface reaches the underside of the bulkhead in the first separation zone, oil and water will flow into the second separation zone, where the slop tank process will continue. Usually oil layers will form in all three separation zones, the thickest layer forming itself in the forward section.

The "Oil Crab" is an SSC 125 type recovery system, SSC stand-

ing for Stowage Skimmer Combination. The following types have been developed as shown in Table 3.2.
The "Oil Crab" could be used in combination with a tug and pushed or towed at a speed up to 3 knots through an oil slick. When the storage/separation zone has been filled to 80%, the recovered oil must be taken ashore. In a water current such as in a river, the "Oil Crab" can be used while stationary, either in combination with an oil boom or with a number of other SSCs moored abreast to form an oil barrier downstream of the slick. In combination with a number of oil booms the SSC can be used to form a U-shaped trap for any oil leak from a spill source, thus keeping pollution under control.
The "Oil Crab" has no crew, it carries no power units and it is ready for use as soon as its motive power is connected. The oil skimmer not only removes the oil spills, but also catches all floating matter. Discharging the barge is by means of a bell pump or grab, depending on the solidity of the matter removed from the water surface.
The "Oil Crab" while not used in cleaning-up operations could be used in maintenance operations of locks, quay walls, bridges and the like in order to reduce the overheads.

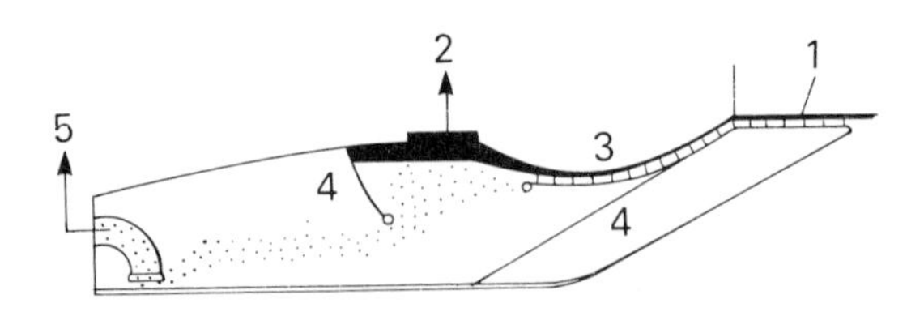

FIG.3.40 The Eimbcke system; (1) oil film; (2) oil outlet to storage; (3) shaped bottom; (4) oil shield; (5) water outlet to sea.

The Eimbcke system features a specially shaped collection and separation tank, that ensures a smooth turbulent free passage of the "oil-free" water out of the back of the tank (fig.3.40.). A 30 m ship was built based on this principle (fig.3.41.) to remain on the station in a gale 8 °B. Its storage capacity for water-free oil is 180 tons. Twin Caterpillar D 353 marine diesels developing 500 BHP at 1300 rpm will give the vessel a top speed of 10 knots.

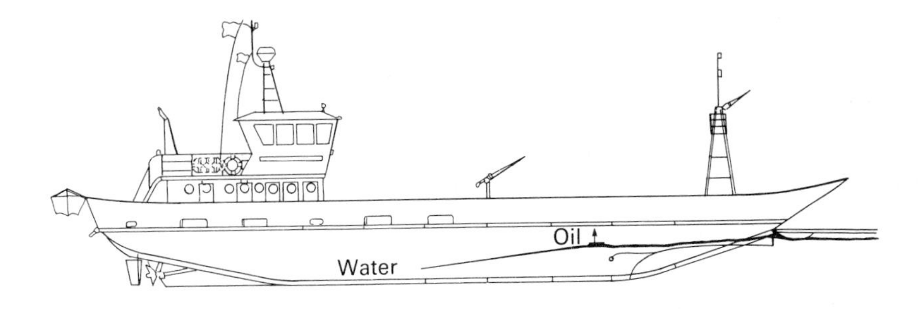

FIG. 3.41 Impala Marine's multi-purpose oil skimmer with Eimbcke skimming system.

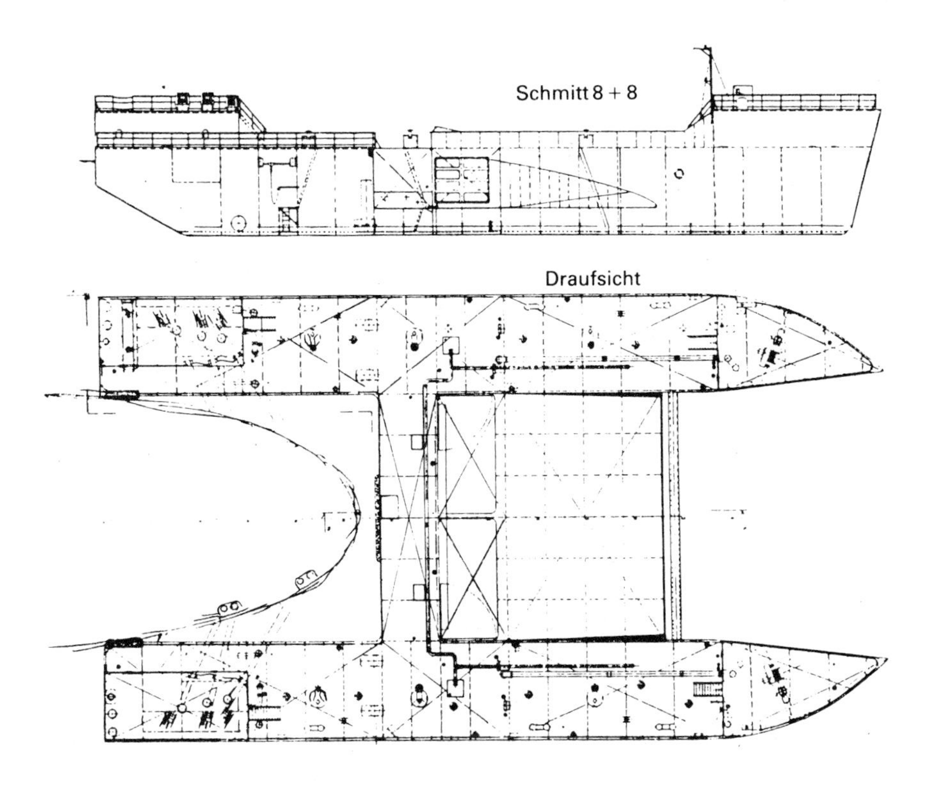

FIG.3.42 Oil skimming catamaran "Mobile Oil Dike" (MOD).

An oil skimming catamaran called Mobile Oil Dike (MOD) built by the Nobiskrug Yard is simple in both construction and handling and is capable of achieving a high degree of oil collection in waves of up 2.40 m. The oil recovery unit comprising two floats, a box transverse girder and sloping ramp has a skimming range of 15 m (fig.3.42.). This unit is non selfpropelled and therefore needs no maintenance as long as it is not in operation. When in operation the unit is pushed by a tug or another suitable vessel. During transit the unit will float on a light draught with an emerged ramp.

During operation in oil polluted areas the unit will be brought to the skimming draught by flooding the oil settling tanks and oil collecting tanks and by taking on ballast. For this purpose openings capable of being locked by ring gate valves will be provided in the aft end which also enable the water taken on during the skimming procedure to be drained off. The oil/water mixture is led over the sloping ramp acting as a preseparator into the wing settling tanks, where the oil is then separated from the water by means of gravity and then pumped into the oil collecting tanks. The amount of oil/water mixture taken on can be controlled by the thrust speed and by trimming and submerging the ramp according to the sea motion. After completion of the operation the oil will be pumped into a barge for transportation.

The principal dimensions of the MOD are as follows:

length over all	about 48.70 m
length on the waterline	48.00 m
breadth over all	27.11 m
breadth of floats	6.00 m
depth to the main deck	5.20 m
draught when empty	about 0.90 m
draught during operation	3.20/4.40 m
tank capacities at 4.40 m draught	1,960 m^3
out of which: oil collecting tanks	2 x 345 m^3
oil settling tanks	2 x 560 m^3
collection trough	150 m^3

3.3.3 Vessels with Active Recovery Systems

The JBF oil recovery system uses the natural propensity of oil to float on water, forcing mechanical separation of surface oil from the water. A dynamic inclined plane (DIP) installed in the vessel's bow forces the oil under the water surface and into the collecting tank. This is achieved by a moving, inclined conveyor belt, which entrains the oil on its surface and then sheds it into a tank, where oil and water separate out (fig.3.43.). The recovered oil is continuously pumped from the collection tank to onboard holding tanks for storage or by flexible hoses to a trailing oil barge. The oil free water is passing out of the vessel.

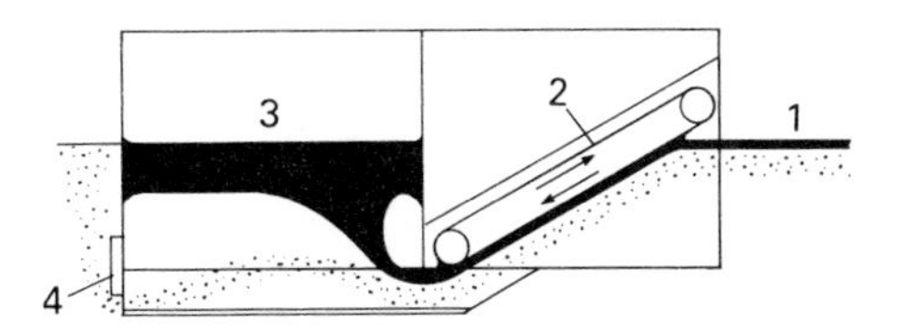

FIG.3.43 JBF Corporation's dynamic inclined plane, (1) oil film; (2) inclined plane; (3) oil collecting tank; (4) water outflow.

Several ships have been built, based on this principle as shown in table 3.3. There are projects for 200, 500 and 750 ton ships. The ships could perform a variety of functions in addition to oil skimming. They could be used as fire boats, tenders and diving support vessels.

TABLE 3.3 Skimming ships fitted with a dynamic inclined plane.

Name of ship	Length [m]	Width [m]	Draft [m]	Speed [knots]	Engines [BHP]	Loaded displ.[t]
Bay Skimmer	20.7	5.2	1.8	10.0	2 x 250	100
North Sounder	22.26	6.1	1.9	10.6	2 x 365	130

The "Bay Skimmer" is additionally fitted out with twin fire monitors with a total output of 250 gallons/min and a 2.000 lbs crane. When fully loaded with a cargo of 10,000 gallons of oil, she displaces 100 tons and is capable of speeds up to 10 knots with bow doors opened or closed. Skimming speed is up to 3 knots with a maximum recovery rate of 500 gal/min, this being the capacity of a hydraulic cargo pump. The effective breadth of the bow with the bow doors opened is 5.2 m. Water jets can be installed on the sweeps to enhance collection towards the

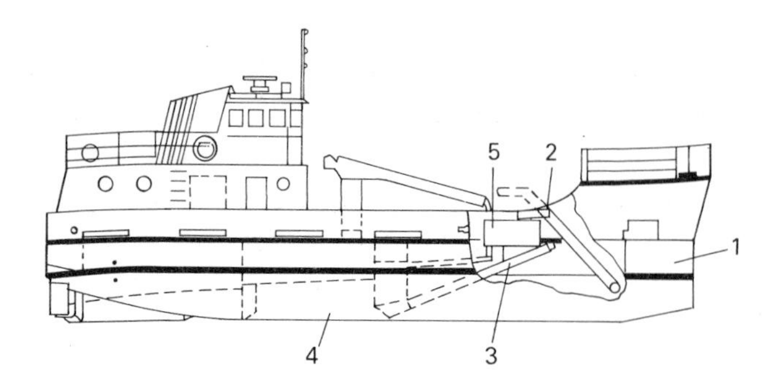

FIG.3.44 The "North Sounder"; (1) bow doors; (2) debris conveyor; (3) DIP conveyor; (4) oil collecting tanks; (5) debris reception tank.

mouth of the vessel. The main propulsion is by twin Caterpillar D-3406, 250 BHP diesel engines with reverse reduction gearing. Power for other services (including electric air compressors, when being used as a diving support ship) is provided by two 25 kW generators. The hydraulic and fire pumps are driven by the PTO shafts on the forward end of the main engine. There is a small cabin for crew, which contains a galley, mess table and two berths aft of the workshop and pump room, above which is a wheelhouse. The "North Sounder" (fig.3.44.) is fitted out with an auxiliary control station at the bow above the oil recovery machinery. From that location the vessel can be operated. A telephone links this forecastle station with the wheel house and other board locations. The spill recovery system actually consists of two conveyor units:

- the debris collection unit, which is a stainless steel conveyor that leads up to debris collection equipment;
- the oil handling unit with a 1.83 m wide belt, which loads oil into the vessel's cargo tank.

The DIP conveyor is located behind the debris conveyor. The 100 mm thick doors are opened and folded alongside the hull by hydraulic power. They could be extended by free-floating sweeps on each side of the hull to enlarge the sweeping width. The main propulsion is by twin caterpillar D-3408, 365 BHP diesel engines, each driving a 1.07 m four-bladed propeller through a 3:1 reduction gearing. Crew quarters for two (3 to 4 while skimming) are located below and aft of the wheelhouse. The vessel may serve as a "command post" for other crew members operating auxiliary boats during an oil spill recovery operation.

In the "Marco" system, the belt lifts oil and debris out of the water. It is able to do this and subsequently to separate oil from debris using a porous filter belt. This allows oil to pass between the strands of the belt rather like a sponge whilst debris rides on the top and, being porous, waves can pass right through, thus preventing turbulence problems. The debris is scraped off the top of the belt and the oil is squeezed out of it. Behind the lower end of the belt under water there is an induction pump, which sucks oil and water towards the belt. This feature is very useful while working around inaccessible piers and jetties.

This system is designed to work between the hulls of a catamaran. For rapid response in bays, harbours and isolated waterways, Marco produces aluminium catamarans 7.9 m and 10.4 m in length powered by twin outboard engines. These vessels are readily transportable by road. The recovery rates are claimed to be up to 100 gal/min in a heavy spill. A larger 17.7 m long catamaran with two filter belts is built to recover 600 gal/min in a thick spill. A pair of water spray booms angled forward from the bow of each hull increase the effective width of sweep to about 13.7 m. A still larger 37.2m long catamaran is planned with a theoretical maximum recovery rate of 140 tons/hour, specially suited for work offshore for long periods.

The "Bodan" system works on the principle of oil adhesion to metal surfaces and utilises a very large revolving drum. The lower half of the drum is in contact with the water. The oil is lifted out of the water by this drum and then scraped off on the down turn.

An interesting development in oil pollution vessels has been made by McGregor-Navire Oil Recovery. The oil is collected and concentrated from the water surface by booms at the sides (fig.3.45). The booms guide the oil/water mixture into open ducts, where rotating bristle chains are located. The oil adheres to the bristles, whereas clean water flows through and out from the for-

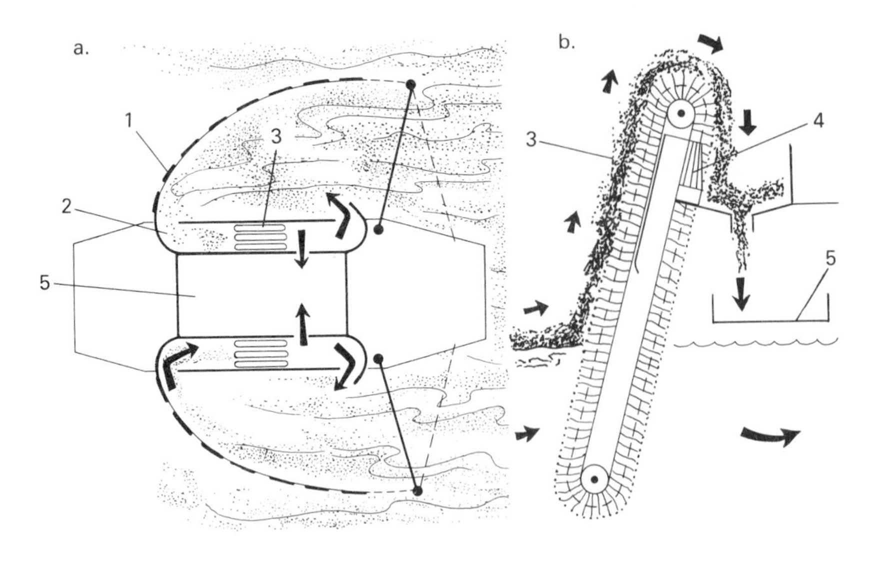

FIG.3.45 McGregor-Navire Oil Recovery vessel; (1) boom hydraulic spreading; (2) oil/water mixture; (3) rotating bristle chains; (4) scraper; (5) oil tank.

ward gates. The oil is removed from the bristles mechanically by means of a cleaning device and flows down to the storage tank by gravity - no pumps are required. The storage tank could be heated to facilitate transfer of recovered oil to a tanker or to large floating rubber sacks. The whole operation could be managed by 2 to 3 men on board. The unit is fitted with a diesel engine giving a speed of 10 knots, whereas the skimming speed does not exceed 2 knots (1,0 m/s). For extending the booms, two hydraulic cranes are provided. Vessels of several sizes are produced in lengths from 10.5 to 17.5 m.

3.3.4 The "Scissor" Coastal Tankers

The Luehring shipyard has designed and built two twin-hulled "Scissor" tankers, which in the open position could serve as very large oil skimmers (fig. 3.46.). Both parts of the hull are connected aft by a hinge and engaged forward by a special locking device. Each hull is equipped with one main propulsion unit (rudder-propeller) and an auxiliary retractable rudder-propeller forward. Both systems have controllable pitch propellers and a slewing range of 360 °. The main superstructure and accommodation block is fitted on the starboard hull, whilst machinery and pump rooms, air conditioning and exhaust ducts are arranged in both hulls.

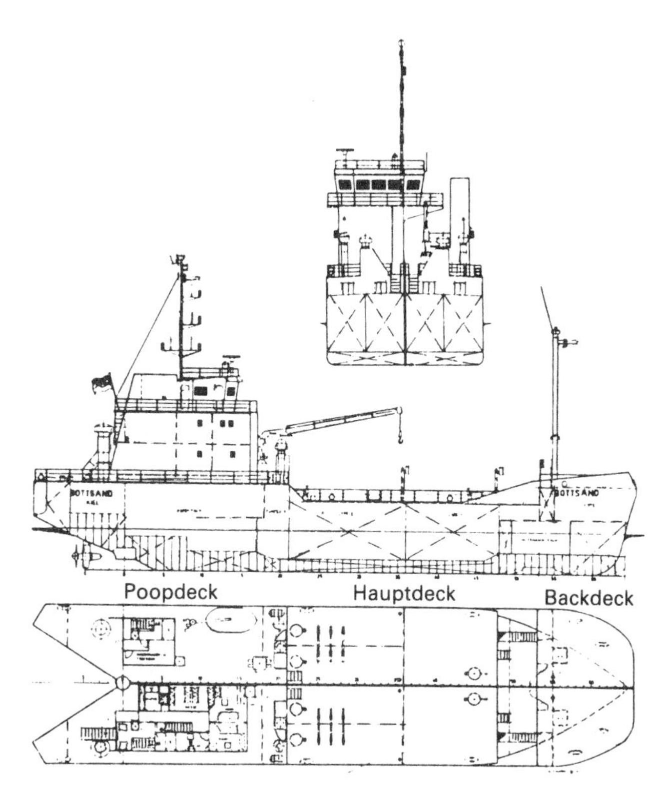

FIG.3.46 The principles of "SCISSOR" coastal tanker.

On arrival at a spill the hulls are driven apart by the two forward propellers and then locked aft. The stresses in the hinge are within the allowable limits when the ship is moving with the hulls open against the waves 1.8 m high. An optical-acoustic alarm warning system is fitted in order to safeguard against too high stresses in the hinge due to weather conditions. For skimming in high seas the ship has to travel with the waves, thus creating a sheltered pocket between the hulls. In the aft part of the open triangle there is a concentration of oil film and the waves are damped by the thick oil film, allowing it to flow quietly through the vortex into the reception tank. The four independent rudder-propellers give excellent manoeuvrability and enable the vessel to move into the slick. This gives the possibility of positioning the ship so as to concentrate a thick film of oil aft in the vee formed by the two hulls.

TABLE 3.4 Particulars of "Scissor" tankers.

Items		Name of the ship	
		THOR	BOTTSAND
Year of completion		1981	1984
length over all	[m]	34.50	46.30
width	[m]	8.20	12.00
height	[m]	3.40	5.20
draft	[m]	2.50	3.10
deadweight	[tdw]	n.a.	650
tank capacity	[m^3]	281	790
sweeping width	[m]	31.00	42.00
main engines	[kW]	2 x 220	2 x 600
auxiliary engines	[kW]	2 x 44	2 x 150
speed	[knots]	9	10
wave height	[m]	1.5	1.8

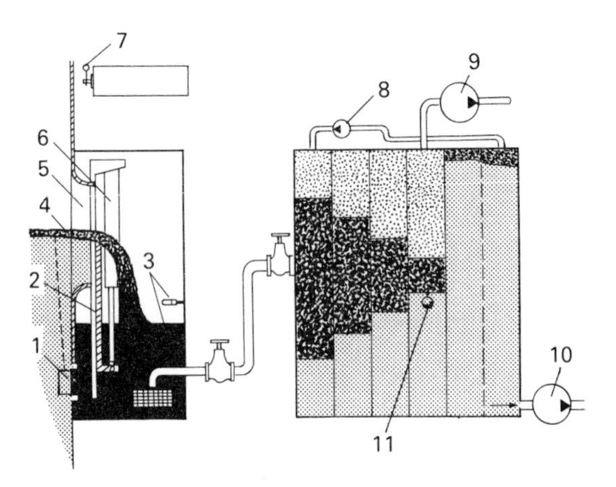

FIG. 3.47 Water/oil inlet and separation installation by Jastram; (1) depth sounder; (2) gate valva; (3) float cut-off; (4) oil film; (5) inlet opening; (6) hydraulic actuator; (7) hand control; (8) rest oil pump; (9) oil pump; (10) water pump; (11) oil sounder.

On the "Bottsand" a special installation made by Jastram Werke was fitted in each part of the ship consisting of a reception tank

and a separator (fig.3.47.). At the apex of the triangle formed by the two ship parts a large opening is arranged with a sliding cover which is hydraulically operated with electronic control. The signal is given by a level indicator arranged below the opening and which measures the level of oil-water mixture in the reception tank. This valve could be controlled by hand as well as from the position on deck, just above the inlet opening, where the actual situation could be easily observed. The upper edge of the sliding valve should be at such a level that only a thin layer can enter the tank irrespective of the waves. Some water must always enter the tank over the vortex, especially when high viscosity or partly emulsified oil is being skimmed. The water plays the role of a carrier, enabling thick oil to enter the tank quickly, from which it is transferred into the separator. The separator works on the principles of gravity and coalescence. The oil/water mixture flows from one section to the next one without any turbulence and thus the oil separates from the water. Finally the remaining water goes to the final section of the separator where the remaining oil is separated from the water.
An additional installation for adding emulsion dispersing agents is fitted with several gauges, sending signals to minicomputer, which controls all the system. Thus, the whole installation can work automatically and unattended. Water containing about 100 p.p.m. of oil is pumped out from the ship in the region between the two hulls. Oil containing about 3 to 5% of water is pumped into a storage tank.

3.3.5 Devices Enabling the Adaptation of Non-purpose Built Ships

It is important, that in the case of a spill as many non-purpose built ships could, within the shortest possible time, be adapted for oil spill combating. Therefore the single-ship sweeping system is chosen, in which the ship not only tows the sweeping boom, but also handles the oil skimmer and the skimming operation itself, since the collected oil is brought right alongside the ship by the boundary layers. The oil collected in the pocket is hardly affected by the forward movement of the ship. Debris, which often severely reduces the efficiency of a skimmer system, can be readily removed using ordinary boat hooks from the deck. Operators can easily keep visual control of the skimmers, ensuring that they are used to maximum efficiency.

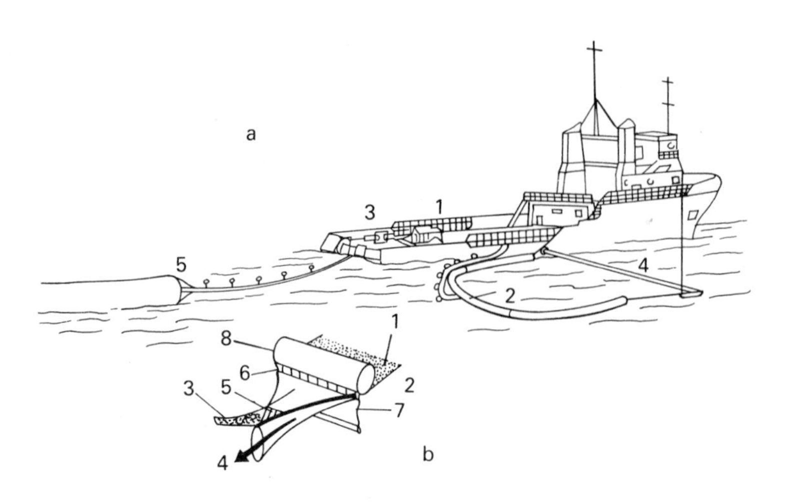

FIG.3.48 The "SIRENE 20"; (a) general layout; (1) air compressor; (2) skimmer; (3) pumping unit; (4) jib; (5) flexible floating tank; (b) skimmer; (1) oil film; (2) water; (3) oil suction hose; (4) water outlet; (5) pump; (6) strainer; (7) skirt; (8) float.

There are many different concepts, some of which will be described below:

The SIRENE 20 is a French direct suction skimmer integrated with an inflatable boom (fig.3.48.). The device comprises a small flexible boom, whose central part opens in a slot and whose lower lip limits the entry of water into a pocket, that varies in form to ensure thickening of the slick. While water drains freely to the bottom of the pocket, the upper oil-rich layer is transferred to a storage container towed by or integrated into the vessel carrying the pumps. The two double acting diaphragm pumps are driven by a 70 BHP compressor. The Sirene has a 20 m swathe and can travel at 1 to 3 knots. It travels optimally at 2 knots, so that its encounter rate, i.e. the volume of pollutant it encounters, is about 70 m^3/h for a 1 mm thick slick. This skimmer could be towed by two ships in a "U" configuration, but its use by a single ship through a jib is most appropriate. This system could also be used while moored in a river or estuary facing incoming slicks.

The ESCA offshore skimmer is designed to be used with a small tanker. It includes a weir skimmer on a catamaran-type support connected directly to the tanks of the ship by a large diameter hose (fig.3.49.). A valve allows direct flow from the weir toward the tanks of the ship through a hole in the hull a few metres under the waterline. The oil-water mixture enters the tanks by gravitation alone, which limits the formation of emulsion and allows rapid settling and immediate extraction of water, which is discharged to the sea.

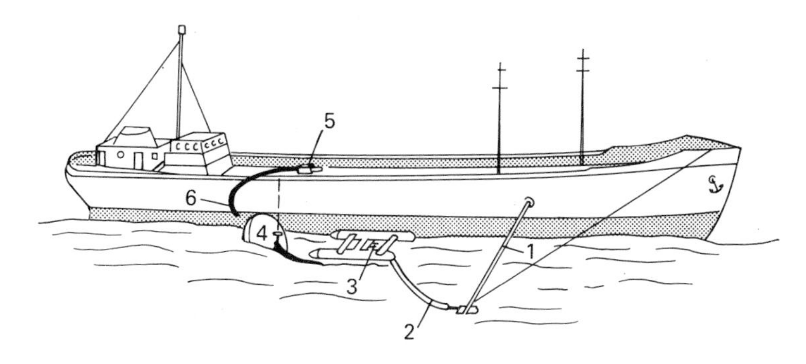

FIG.3.49 The ESCA offshore skimmer; (1) jib; (2) deflecting boom; (3) weir; (4) valve; (5) pump; (6) water discharge.

A deflecting boom enlarges the swathe of the skimmer head, so that the total skimming width of the recovery device is 12 m and can be 20 m using a longer jib. The boom is 14.5 m long, consisting of beam floats and a vertical screen made up of a semirigid plastic skirt and fastened to the top and bottom of a vertical frame. The skimmer is a catamaran (3 m wide and 4 m long) made of steel. The semiflexible hose is 400 mm in diameter and equipped with a connecting flange. Projecting slides on the ship's hull guide the flange down to its position in front of the opening in the hull - in this case 1.5 m below the waterline. Watertightness of this connection is obtained by eccentric clamps operated from the deck. The opening in the hull is closed by a valve fitted inside the tank and operated from

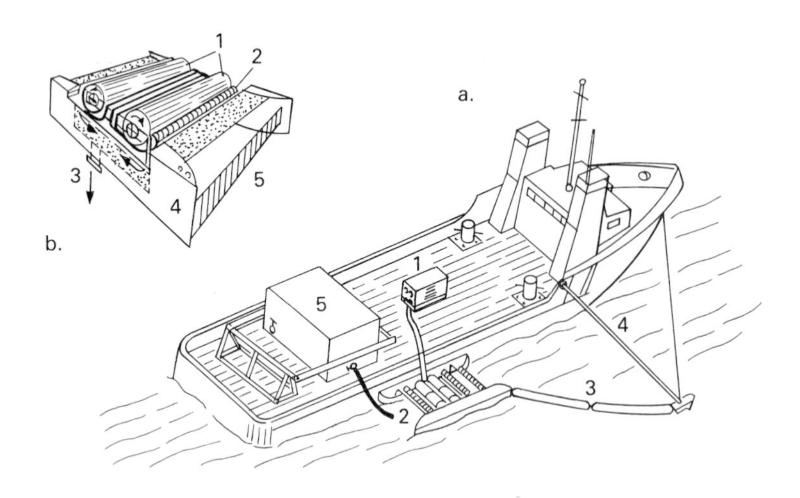

FIG.3.50 The STOPOL offshore skimmer; (a) general layout; (1) hydraulic power pack; (2) skimmer; (3) deflecting boom; (4) jib; (5) tank; (b) skimmer; (1) oleophilic drums; (2) Archimedean screw; (3) double screw pump; (4) floats; (5) trash screens.

the deck.

The STOPOL 3P system comprises a CAROL skimmer, consisting of two contrarotating oleophilic drums on a catamaran shaped floating frame and its power pack, a deflecting boom and a jib (fig.3.50.). The two drums consist of a cylinder (600 mm diameter and 1.2 m long) covered with an oleophilic skim. The drums rest on a frame of four floats held still by crossmembers. This floating frame holds the hydraulic motors and supports the double screw volumetric. The transfer pump is located on one side of the parallel paired floats. Its maximum capacity is 50 m^3/h of oil. The CAROL skimmer is 3.9 m long and 2.3 m wide and weights 950 kg.

All three systems have been experimented with and the results are given in table 3.5. The average thickness of a slick is the ratio between the rate of oil encountered and products of the swathe by the speed of advance. In fact, the real thickness of the slick encountered was two to three times the ratio since the width of the slick was two to three times less than the swathe of the skimmers.

TABLE 3.5 Skimmer performance test results.

Device tested	For'd speed [knots]	Sea state [°B]	Oil Volume [m^3]	Distr. rate [m^3/h]	Oil slick thickness [mm]	Oil slick encount. [m^3]	Oil Perfomance % TE	RE	ORR
SIRENE 20	2	3 - 4	14	22	0.3	14	72	40	10.0
ESCA	2	3 - 4	7	17	0.4	6	50	99	8.5
STOPOL 1	1	1 - 2	7	21	0.4	7	75	75	9.0
STOPOL 2	1	1 - 2	7	10	0.2	7	80	80	6.0

TE - throughput efficiency; RE - recovery efficiency; ORR - recovery rate.

It is possible to establish a final comparison of the three skimmers according to the test results. Test conditions were very different in many respects. However the following conclusions could be made:

- The SIRENE 20 is the least affected by waves and would work in seas to force 4 °B. Its recovery efficiency is good enough to consider working with flexible floating tanks. On a given thickness of oil its encounter rate is higher than that of the other two devices. On the other hand it is limited to less viscous products due to its suction length.
- The ESCA is most robust. It is also the least affected by variations in viscosity and thickness of encountered oil. Since it recovers large amounts of water with the oil, it is necessarily associated with a vessel or barge with large storage capacity.
- The STOPOL is more selective and can be recommended for recovering small spills using vessels without a large storage capacity. Its oil recovery is low, but it is expected to increase with increasing viscosity of oil, as is generally seen in large spills. It is certainly the most sophisticated device.

The VOSS system (Vessel of Opportunity Skimming System) consists of a skimming barrier towed in a catenary alongside a single vessel with an outrigger. The skimming weirs are connected by suction hoses to a pumping subsystem on deck. A diesel-hydraulic power pack drives the pumping subsystem and collected oil is pumped to storage through a gravity-type oil-water separator (fig.3.51).

The VOSS barrier is based on the centre section of the high seas skimming barrier. A 65 ft length barrier was selected for easy

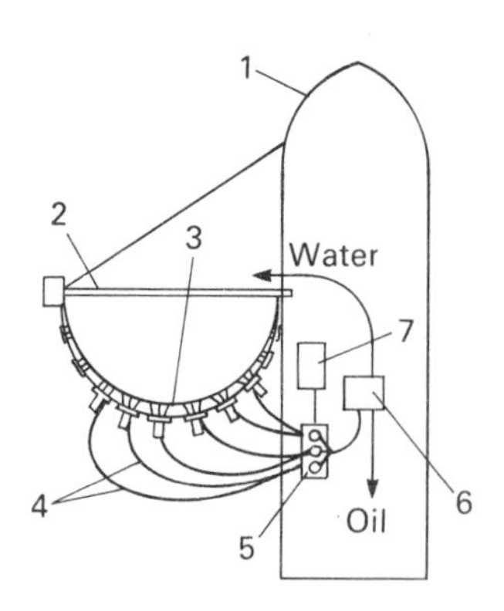

FIG.3.51 Single VOSS configuration (port side); (1) support vessel; (2) outrigger; (3) skim barrier; (4) suction hoses; (5) pumping subsystem; (6) oil/water separator; (7) power pack.

handling and to preserve the manoeuvrability of the support vessel. This gives an effective sweep width of 40 ft plus half the vessel's beam. Each vertical strut supporting the barrier curtain is individually attached to an external tension line to avoid tensile loading of the curtain itself. A slack line runs along the barrier below the waterline to maintain the curtain slack between the struts.

The six centre struts of the barrier skim the contained oil. Each skimming weir has an open weir slot located at the waterline on its oil-side face. Relatively high weight and large waterplane area and the external tension line give the barrier good seakeeping characteristics in waves. This wave response is necessary so that the weirs remain near the waterline for efficient skimming. The barrier is towed alongside by an outrigger. A foam-filled float supports the outriggers outboard end and incorporates a barrier towpoint below the waterline. A universal joint attaches the outrigger to a mounting on the vessel rail. Guy lines maintain the outrigger position during skimming operations. The VOSS pumping system consists of three selfpriming pumps and a hydraulic control system. Flexible suction hoses link the pumps to the skimming weirs. The pumping subsystem is mounted on a skid and operated on the vessel deck.

Hydraulically driven, double acting diaphragm pumps are used to minimise emulsion formation by maintaining low fluid velocities. Their design allows them to pump debris up to 3" dia. The subsystem has a capacity of 750 gallons per minute, which is the maximum expected flow over the six skimming weirs. The hydraulic control system senses the stop of flow at the end of each pump stroke. Flow is reversed when either the true stroke end is reached or a large debris is encountered, providing protection against pump clogging.
A portable diesel-hydraulic, skid-mounted power-pack drives the pumping subsystem. An air cooled diesel engine was selected as the prime mover for reliability and safe operation in explosive atmospheres.
A rectangular, gravity oil-water separator of flow rates about 200 gal/min completes the system. Incoming flow passes through an inlet diffuser cone and 12 consecutive separator compartments divided by vertical baffles. Open spaces over the baffles allow debris and ice to pass. Gravity separation of the oil and water is promoted by the separator's low fluid velocities and extended flow path. Valves on the oil and water outlets control the separator oil-water interface level. The highest oil recovery rate was 530 gal/min at a weir recovery efficiency of 76%. This occured with heavy oil at a low speed of 1 knot.
For skimming thin slicks a diversionary wing of light weight contain- ment boom could be used (fig.3.52.).

3.3.6 Adaptation of a Hopper Dredger

A trailing suction hopper dredger has a large hopper, in which oil can be stored. As the oil has a lower specific gravity than the dredged material, this capacity can be supplemented by using some of the air casings as storage space.
The following considerations form the basis for the choice of such an oil recovery system:

- the system is required to be operative on a recovery vessel within a few hours, thus all the necessary equipment must be permanently and readily available on board. Any adaptations

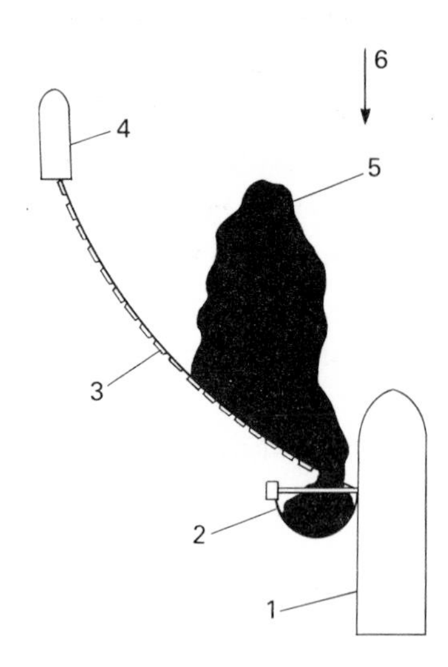

FIG.3.52 VOSS in operation with diversionary containment boom; 1 - support vessel; 2 - VOSS skimming barrier; 3 - containment boom; 4 - auxiliary tow vessel; 5 - oil slick; 6 - direction of current.

to enable the vessel to perform this role must be effected by the crew in a few hours;
- the system is also required to function reasonably effectively in bad weather and in conditions where the significant wave height is about 1.5 m coinciding with wind force 5 °B;
- it must be an active system, i.e. the vessel's speed in its recovery role must be utilised;
- the capacity of the recovery system must be as large as possible in relation to the weight and required power;
- the system is required to work just as effectively with thin layers of oil, approximately 0.1 mm;
- the system requires to be relatively insensitive to floating debris, seaweed, etc.;
- the discharge of cargo of high viscosity from the hopper must be accomplished in as short a time as possible;
- the vessel must be most manoeuvrable at low speeds in order to catch the slick patches dispersed over the surface.

For containment and collection of slick oil, "Hydrovac" sweeping arms could be used with a trailing suction hopper dredger. This system consists of two sweeping arms, one on each side of the vessel, which project at an angle of about 60 ° to the direction of movement of the vessel. Each arm (fig.3.53.) consists of one or more partitions, which extend into the water and are attached to a structure, which in turn is connected to the float. The oil floating on the water is guided to a skimmer in the armpit of the sweeping arm, which operates on the "cream-off" principle. The oil gathered in the collecting chamber of the skimmer is transported to the hopper by a submersible pump. The speed of the sweeping operation can be from 2 to 3.5 knots.

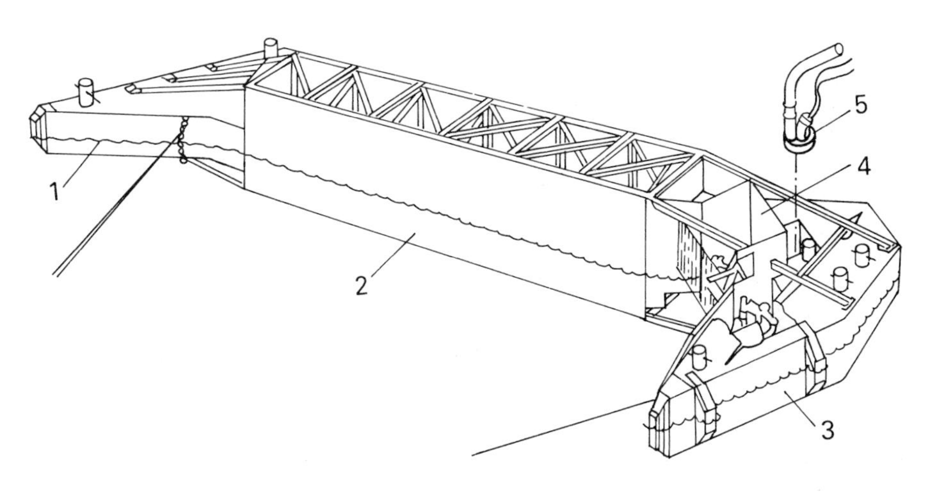

FIG.3.53 Sweeping arm; (1) external float; (2) partitions; (3) armpit float; (4) collecting float; (5) submersible pump.

The length of the arm could be up to 20 m and each arm is equipped with two special hydraulically driven submersible pumps. These pumps could be independently controlled and they have a capacity of 360 m^3/h at a pressure of 3.5 bars in a liquid of viscosity 900 cS.

The open hopper is used for containment of oil of flash point lower than 61 °C, which is an inflammable liquid. Ships transporting such a cargo are subject to the regulations, which perscribe the use of closed tanks. Light fractions evaporate rather quickly (within 6 to 24 hours) from spilled crude oil and, as result, the flash point of most crude oils spilled at sea rises above 61 °C. There is then no question of an inflammable liquid and the regulations laid down for tankers in this respect no longer apply, but some relevant requirements must be satisfied.

The following modifications are required:

(a) – subdivision of the vessel:

- cofferdams around the hopper - the following spaces may also be regarded as cofferdams for this purpose: existing dredge pump room forward and extra cargo pump room aft (for transfer of oil);
- extra cargo pump room for transfer of oil - the electric motors for the pumps must be separated from the pumps by means of a gastight bulkhead;
- engine room must be situated aft;
- no deckhouse is allowed above or in front of the hopper.

(b) – electrical installations:

- electrical installation in cofferdams must be explosion proof;
- installation above the deck to a certain height above the hopper must be explosion proof;
- all deck machinery is required to be explosion proof (i.e. of hydraulically operable design) or is required to be switched off during oil recovery operations.

(c) – fire prevention:

- the panelling and insulation in accommodation spaces must meet certain fire-resistant standards;
- a fire extinguisher must be installed in each space;
- emergency exit from each cabin (kick-out doors);
- entrances to the engine room must be provided with selfclosing doors;
- foam fire extinguishing monitors are to be located on a platform above and near the aft bulkhead of the hopper;
- in the aft cargo pump room a permanent fire extinguishing appliance is to be installed.

(d) – various supplementary requirements:

- supplementary ventilation facilities to be provided in: cofferdams, pump room aft, accommodation space and engine room;
- two motor lifeboats with gravity davits;
- portable gas detectors to be available on board to measure gas concentrations in various spaces after recovery operation or to measure gas concentrations on deck and in spaces during oil recovery;
- permanent gas detection to be fitted in the fore and aft ends of the vessel, above the hopper, in the accommodation space and on the sweeping arms. Alarm devices to be provided;
- a foam producing system for blanketing the hopper with foam in a few minutes in case any gas appears above the oil hopper;
- a helicopter deck should be provided to bring officers for oil recovery operations;
- an operation control room in the deckhouse and a laboratory is required;
- for removing the oil from the deck, sweeping arms etc. a detergent cleaning system should be installed. The hopper should be cleaned at a land based cleaning station;
- a demulsifying system for injecting a chemical product at the suction side of the pumps in the sweeping arms as well as inside the hopper is required.

The conversion of existing trailing suction hopper dredgers into ships for oil combating is very complicated and expensive (up to 50% of the new building price). In some cases conversion is virtually impossible, e.g. ships with the deckhouse over the hopper and suction pipes on the aft deck.

When the dredger is required for combating an oil slick, the dredger has to stop its work and set course for the scene of the accident, discharging on its way any dredging material from the hopper. The crew has to get the dredger ready for its new work. This entails:

- closing the gastight doors and switching on the associated apparatus;
- switching off all the non-explosion proof electrical equipment

and switching on all the safety equipment;
- preparing all gas detection apparatus for operation and checking for proper working;
- in the event of high-viscosity oil being involved, fitting provisions in the hopper in order to facilitate unloading of the recovered oil;
- checking the special filters for the air inlets to the engine rooms and accommodation spaces;
- checking the foam producing installation for blanketing the hopper surface.

On arrival at the spillage site, the permanently fitted sweeping arms will be put overboard by means of two hydraulically operated cranes. The forward speed of about 2 to 3.5 knots causes the arms to adopt a particular position, which can be regulated by altering the length of the retaining wires. The optimum angle with respect to the heading of the vessel is assumed to be 60 °. The clean swept path has a width of about 50 m, depending of the breadth of the ship and the length of the arms.

The oil is guided to the armpit of the sweeping arms, where hydraulically driven, submersible pumps are situated, delivering the oily mixture to the hopper (fig.3.53.).

At the start of operations the hopper should be filled with seawater and the oil-water mixture gradually introduced into the hopper, just beneath the water surface to minimise turbulence and to promote separation. As much water should be pumped out as oil-water mixture is pumped in. The oil stored in the upper part of the hopper can be pumped to an accompanying tanker or discharged in the harbour.

3.3.7 Pollution Combating Ships for Open Waters

Vessels for oil recovery so far described were relatively small ships with a very limited capability to stay at sea in rough weather. Under the regional conventions, the member countries are obliged to have ships and equipment serving not only for containment and combating oil pollution, but for unloading and towing damaged tankers to safety. In order to limit expenses these ships should be capable of rendering additional services when they are not in use for spill combating. Employing these ships for alternative services such as hydrographic, ice breaking, fire fighting, etc. is the best guarantee that these ships will be ready at any time for oil pollution combating. For economic reasons any such ship must be designed as a multipurpose ship. Very good examples are the oil recovery ships "Mellum" and "Fasgadair". M/S "Mellum" was specially designed and built for oil recovery. Additional tasks were given such as buoy tending, ice breaking and fire fighting. Thus the following additional reqirements have to be met:

(a) – as buoy tender:

- sufficient deck area to tend the buoys;
- the height of the deck above the water surface should not exceed 1.0 to 1.8 m, depending on the contents of the tanks;
- a crane capable of lifting a 12 tons load at 1.5 m distance from the ship side should be fitted;
- the stability should be such that at lifting of 12 tons load from the water the heel angle should be relatively small;
- the manoeuvrability of the ship should be very good.

(b) – as an ice breaker:

- the hull should be reinforced for navigation in ice to the E3 class according to the requirements of a Classification Society;
- the machinery output should be sufficient for an ice breaker;
- the form of the fore end should be optimal for ice breaking;
- indirect cooling of the machinery installation through the outside shell heat exchanger.

(c) – as an oil pollution combating ship:

- sufficiently high power of the main propulsion for the rescue services and free running speed;
- a large deck area in the after part of the ship with the possibility of working over the stern;
- the possibility of carrying a great number and variety of skimmers, sweeping arms, etc. and sufficient length of booms packed in containers;
- high capacity reception tanks for the oil/water mixture;

TABLE 3.6 Principal characteristics of M/S "Mellum" and M/S "Fasgadair".

Item		M/S "Mellum"	M/S "Fasgadair"
National flag		F.R.G.	U.K
Year of construction		1984	1981
length overall	[m]	71.5	75.9
length between perpendiculars	[m]	63.8	70.0
breadth moulded	[m]	15.0	13.0
depth moulded	[m]	6.5	5.5
draft	[m]	5.25	4.0
deadweight	[t]	1,500	1,860
recovered oil capacity	[m^3]	990	1,850
machinery	[kW]	4 x 1.650	2 x 875
speed	[knots]	16	11.5
bollard pull	[kN]	1.100	n.a.

- auxiliary equipment for supporting rescue ships assisting damaged ships such as high sea fenders, pumping sets and piping connections etc.;
- a travelling crane of sufficient lifting capacity and outreach to enable work along the ship sides P and S and over the stern;
- excellent manoeuvrability with the possibility of continuous regulation of speed, especially at low speeds;
- a steering position on the bridge as well as forward and aft;
- an explosion proof electrical outfit;
- a fire extinguishing outfit enabling work close to a burning tanker.

All the above named requirements had to be fully satisfied in order to make this ship a very workable tool for all three very varied services. The crew comprises 14 men and in case of emergency an additional six men could be accommodated. Two additional berths are in the hospital. The ship is so designed that she can work in a heavy oil spill without any danger of causing an explosion of air/carbon-hydrogen mixture. All the accommodation is accessible through gas proof entrance locks and at several points inside the enclosed spaces; measuring gadgets are fitted with a central observation position on the bridge. The ship is fitted out with two 15 m long sweeping arms, each containing one 350 m^3/h pump.

For laying out the sweeping arms, there are two sets of special davits installed with control of all the activities required to place the arms in working position in the water. The electro-hydraulic power unit for servicing the davits is fitted in the engine room. The six reception tanks have a total capacity of 990 m^3. They are fitted out with all the necessary instruments preventing overflow and facilitating observation of the oil separation in tanks.

The ship carries booms of 600 m total length and all the equipment necessary for off-loading cargo and fuel from damaged ships packed in suitably sized containers, which could be easily handled by ship's crane or carried on board the damaged ship by a helicopter.

The towing winch with drums accommodating 1,000 m of tow line of 62 mm dia is placed on the main deck. The travelling crane has a lifting capacity of 12 tons at a radius of 13 m and 8 tons at 15 m. The maximum lifting speed to follow the movement of the ship in waves is 120 m/min. This arrangement guarantees that the lifting wire will never become slack and thus a sudden pull on the load is prevented. The ship is fitted out with a foam fire fighting installation. Two fire fighting cannons are fitted on the beam joining the funnels and another two in front of the bridge. Each cannon has an output of 4,000 l/min. Two of them are distance controlled. Foam could be sprayed in the same quantity, when desired. Spraying nozzles are fitted on the deck houses and on the decks to protect the ship from catching fire from outside.

The engine room is fitted out as a temporary unattended machinery installation. To improve the manoeuvrability of the ship a 700 kW bow thruster and a twin Hinze rudder at the stern is fitted. Joy-stick steering is installed in places to facilitate steering from any position which might be required.

The M/S "Fasgadair" is a conversion from a heavy lift Ro-Ro ship M/S "Brunneck". She is intended to work mainly in coastal and inshore waters in waves up to 3 m and she is used as a training ship. She is essentially a vehicle ship, from which specialist equipment can be deployed. Oil skimming equipment has not been built into the structure of the vessel, but will be handled over the side with cranes, while an aluminium bow ramp is pro-

vided for landing vehicles straight onto the beach.
The capacity of the storage tanks is 1,850 m^3 of recovered oil. They are equipped with steam heating coils and facilities for hot water tank washing.
Provision is made for a basic crew of 12 men to man the ship on a day to day basis, but this would increase to about 25, when dealing with an emergency or when running training courses. There are 10 double and five single berth cabins. The port deckhouse is fitted with a changing room and laundry, by which all personnel enter the accommodation, when the ship is operational. The deckhouse with living quarters and the bridge is placed on the starboard side. A lane along the ship's centre-line on the upper deck is available for vehicle and ramp storage. Sited to port of this lane is most of the cargo pipe work and associated equipment. Two 12" mains run along the side of the ship, one for loading and one for discharging, though they are interlinked to allow the redistribution of cargo and slops between the nine tanks. Two discharge points are provided on the port side and it is intended that oil will be transferred to an attendant tanker while at sea. Over this pipe work is a nitrogen tank and an evaporator for tank purging and a 12 m^3/h - 15 p.p.m. oily water separator for treating the tank washing slops.
All the boats and skimmers would be deployed either over the bow or the starboard side of the ship. For this purpose M/S "Fasgadair" has been fitted with two 10 ton cranes, each of 10 m outreach. An 8 ton gantry fitted to the M/S "Brunneck" has been retained for general duties, though this is not suitable for use at sea.
On the uppermost deck there is a pollution control room, which is essentially an office with VHF radio facilities and a commanding 360 ° view. The deck below comprises the combined bridge and cargo control room. Controls for the engines, rudder and bow thruster are duplicated in an aft conning position, from where there is a good view of the weather deck and stern.
Adjacent to the aft control position is the cargo distribution panel from where pumping operations can be controlled. The

three cargo pumps draw from the centre tanks. Remote controls for butterfly valves are provided, allowing the tanks to drain into the centre tank. Gas monitoring alarms are also fitted here.

Operationally the ship is capable of containing the slicks in inflatable booms, recovering the oil from the sea and spraying dispersant chemicals. Two skimmers by Vikoma have a total capability of 150 m^3/h of oil, which is fed to the cargo main from where it could be routed either to the No.2 centre tank (buffer tank), to a discharge manifold or to any of the cargo tanks. Provision has been made for the injection of chemicals at various points in the cargo pipe work.

The ship is equipped for a containment role with three floating water and air filled booms, which can be deployed from the work boats. One of these boats is an 11 m dedicated boom laying vessel with a speed of over 20 knots capable of deploying its 400 m boom in 20 minutes. It is powered by two 250 BHP turbocharged diesel engines, which drive water jets. It is fitted with special boom handling equipment and requires a crew of only two. The ship carries other booms, which are the Vikoma Sea Pack 400 m and the Vikoma Coastal Pack 250 m. The Sea Pack boom is stored in a dumb barge and is inflated with water and air by a small Volvo engine. The Coastal Pack boom is a lighter unit, which comprises a GRP hopper mounted on the stern of a workboat, from where the boom is payed out.

Dispersant can be sprayed either from the ship or from the workboats. There are two 13.75 m stainless steel arms located on the back end of the forward deckhouses.These light-weight units are stowed longitudinally and can be deployed under remote control from the cargo control area. Three smaller selfcontained units using 6.0 m arms have been supplied for use with work boats. Provision was originally made for the storage of 36 tons of dispersants, but now an additional 170 tons are carried in the No.1 wing cargo tank.

Additionally two workboats are provided for boom handling and two small rigid-floored inflatables - one for rescue purposes and the other one for use in association with the Coastal Pack.

3.3.8 Design Requirements for an Oil Pollution Combating Ship

As there are many different sorts of crude oils and their products, the ship must be capable of easily employing several methods of pollution combating. Thus at least one mechanical oil removal principle must be used, which could be deployed with the vessel stationary or in motion, and facilities to combat dispersed oil by means of dispersants, if the sea is too rough for mechanical removal. Such an outfit allows the ship to be used in the best manner regardless of the sea conditions. The ship must have sufficient deck area for working this gear and tank capacity for collected oil-water mixture and dispersants. If the ship is used for other purposes, i.e. surveying, diving support, etc., these tanks would be used for ballast.

The size of the ship depends on many factors such as:

- the size of the expected spill;
- the distance of the spill from the coast;
- the required deck area;
- the tank capacity;
- the draught restrictions;
- the navigational characteristics.

At present there are no specific design requirements and regulations concerning oil pollution combating ships. But these ships should be ready to tackle a disaster immediately and therefore the design must be based on handling and storing oil with a flash point <61 °C, although the flash point of spilled crude oils changes very quickly, especially during the first few hours through evaporation of the volatile components. Experience shows that a flash point of 61 °C and above is reached not sooner than after 24 to 28 hours.

The ship should reach the scene within 3 to 4 hours after the disaster, as the oil spreads very quickly. Booms should be laid or dispersant spraying must commence as quickly as possible to limit the spreading of oil or to prevent the formation of "chocolate mousse". Thus, when the home port and the location of the disaster are far apart, the ship must be capable of high speeds irrespective of weather conditions (up to sea state 5 °B). For oil recovery low speed is required, generally not exceeding 1 knot, but manoeuvrability must be excellent at that speed. For the above reasons a bulbous bow and bow thruster as well as a stern rudder of special design may be required.

The ship should have good trim under all working conditions. This requirement will have a direct impact upon the location of tanks and equipment. The freeboard must be sufficient in order to reduce the probability of deck wetness in irregular seas. Of great importance are vertical accelerations due to heave and pitch at service speeds.

Special attention must be given to acoustics through appropriate insulation, floating platforms, sound absorbing bulkheads, etc. The noise level in accommodation should be less than 55 dB(A). In order to improve the stability at low speed, antiheeling tanks might be desirable.

Cargo tanks should be equipped with steam heating coils and with Butterworth facilities for hot water tank washing. There must exist the possibility of adding chemicals to the tanks in order to facilitate the natural separation of the oil/water mixture. There should be two mains of large diameter running along the side of the ship - one for loading and one for discharging. They should be interlinked to allow the redistribution of cargo and slops between individual tanks and unloading the ship, while pumping the spilled oil. Two discharge points are to be provided at one side of the ship in order to transfer oil to an attendant tanker while at sea. A 15 p.p.m. oily water separator is to be provided for treating the tank washing slops. Provision should be made for the injection of chemicals at various points in the cargo pipe work.

The basic crew should be limited to a minimum, keeping in mind that in case of action a specialised crew will have to be taken on board. This is normally done in such a way that basic crew lives in single cabins, which have additional beds fitted. Each single cabin is very easily converted into a double cabin.

As the cleaning up of crude oil is a filthy procedure, special consideration must be given to changing rooms and laundry, through

which all personnel would enter the accommodation, when the ship is operational.
The pollution control room should be on the uppermost deck with a commanding 360 ° view. It must be fitted with VHF radio facilities and with a repetitive display of airplane side scanning radar.
Controls for the engines, rudder and bow thruster should be duplicated in an aft conning position, from where there is a good view of the stern.
Cargo distribution panels could be placed either in the proximity of the bridge or adjacent to the aft control position. Remote control for valves should be provided there as well as control for butterfly valves. Gas monitoring alarms should also be fitted there.
Fire detection systems give audible and visual warning of fire in at least 12 positions around the ship.
For this type of ship, the medium speed propulsion machinery with a reduction gearing and controllable pitch propeller is best. In order to keep the machinery compartment as short as possible because of the restricted length of the vessel and a very limited requirement of full power the most economic solution is a two engine installation with couplings and one generator driven from the main engine, covering all requirements during the normal work of the ship. Two auxiliary diesel generators should be provided of a total output equal to the shaft generator, so in case of failure these two generators can cover all the power requirements. An emergency generator is to be provided to cover all the requirements in case of emergency. Auxiliary generators, oil combating machinery and oil fired marine boilers should be fitted in an auxiliary engine room.

CHAPTER 4

Oil Spill Combating on Shores

4.1 Impact of Oil on Shores

4.1.1 Shore Processes

The shore zone includes the beach, the surf zone and nearshore waters, where wave action moves the bottom sediments (fig.4.1.). The extent and configuration of the shore zone depends on the parent geology of the land, the waves and winds, the range of tides and the wave deposited structures. Beaches are dynamic systems undergoing perpetual changes. Continuous dissipation of energy can with time wear away coastal rocks (erosion). Yet at different locations waves can transport sand and build up beaches (accretion).

Waves in deep and shallow water behave differently. As the wave moves closer to the shore, the depth of water will decrease and so the speed of the wave will decrease. When the crests of waves are running parallel to the shoreline, then the slowing down will be accompanied by a gradual decrease in wavelength between consecutive crests. When the crests of waves are running at an angle to the shoreline, then the direction of their travel will be changing with decreasing depth of water with a tendency to parallel the depth contours (refraction). In straight deepening coasts the wave crests will bend parallel to the shoreline (fig 4.2). The portions of waves which are in shallow water will travel more slowly than the portions in deeper water. Thus no matter what the direction of the incoming swell, the breakers will break in lines parallel to the beach.

Irregular bottom topography refracts waves in a complex way

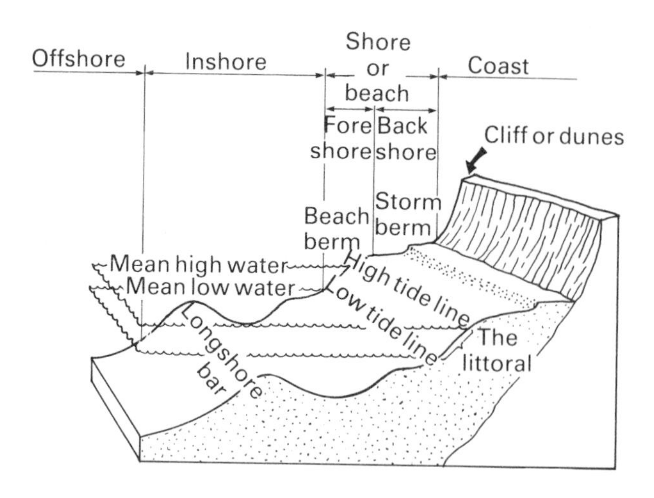

FIG.4.1 Beach profile illustrating the terminology.

and produces variations in wave height and energy along the coast. Waves over a canyon are reduced in height due to divergence, when those at either side, where the waves converge, are higher (fig 4.3a). An offshore shoal will cause the wave to increase due to convergence and be smaller on both sides (fig.4.3b). This appearance has direct impact upon the formulation of the coastline.

As the wave moves into shallow water its speed will depend on

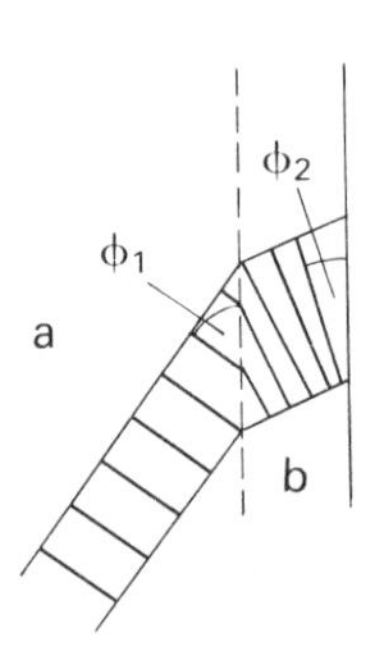

FIG.4.2 Wave refraction; (a) deep water; (b) shallow water; (Φ_1) angle of approach of the wave; (Φ_2) angle of refraction.

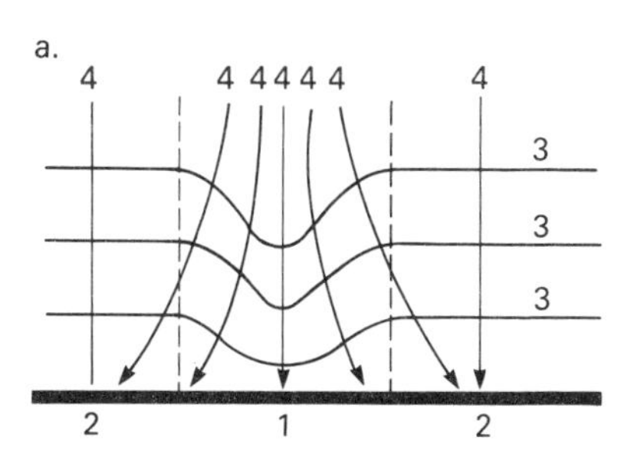

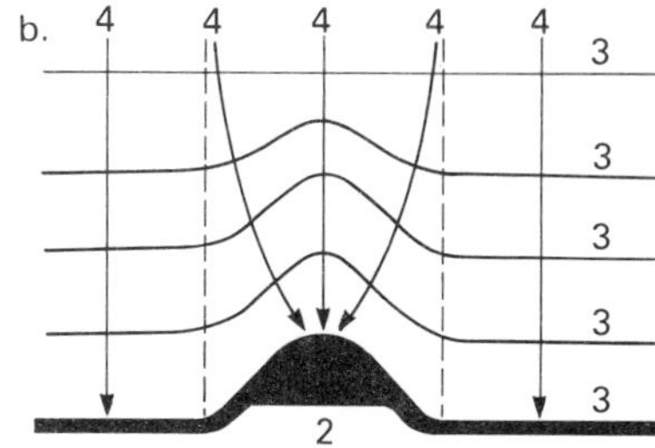

FIG.4.3 Irregular bottom topography; (a) with a canyon; (b) with an offshore shoal; (1) region of divergence; (2) region of convergence; (3) depth contours; (4) wave rays.

the depth of water H and the formula becomes:

$$C = \sqrt{g * H} \tag{4.1}$$

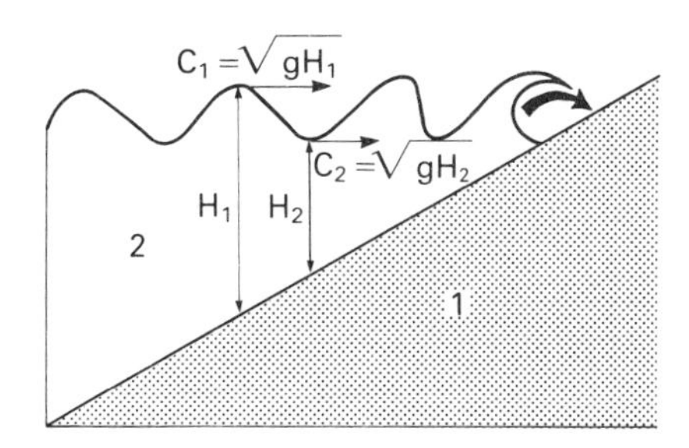

FIG.4.4 Occurrence of breakers over sloping beach; (1) beach; (2) seawater.

But as it runs up to the shoreline the value of H will differ for consecutive wave crests and troughs (fig.4.4). At the crests, H will be greater than at the troughs ($H_1 > H_2$) and the wave will be moving faster at the crests than at the troughs ($C_1 > C_2$). Each crest will steadily overtake the trough in front until there is an unsupported overhaging wall of water (crest) and at this point the crest topples over to produce a breaker.

There are several types of breakers (fig 4.5.):

- spilling breakers - over a nearly horizontal beach for a long distance out, with incoming waves gently breaking at their tops;

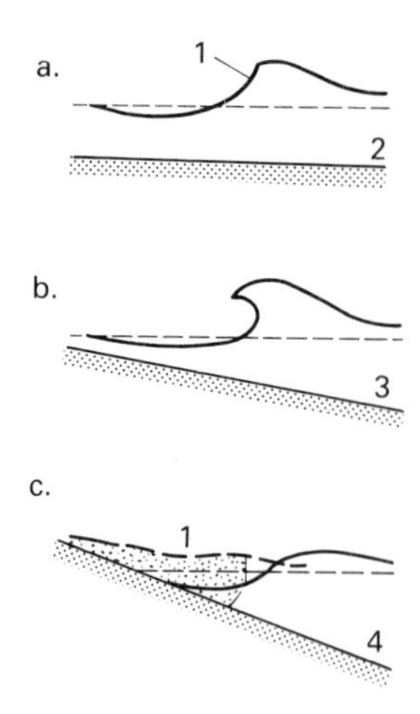

FIG.4.5 Types of breaking waves; (a) spilling breaker; (b) plunging breaker; (c) surging breaker; (1) foam; (2) nearly horizontal beach; (3) steep beach; (4) very steep beach.

- plunging breakers - over a steep beach;
- surging breakers - over a very steep beach, where they do not break, but simply collapse, sometimes not even foaming.

The steepness of the bottom in the run up to the beach is the most important factor in determining the type of wave formed (fig.4.6.). The most orderly swell arises from waves produced by storms thousands of miles away.
The height of the tide and the wave action have a direct impact upon the shape and the nature of the beach. The larger the waves the more energy is pounding the beach and thus the erosion of beaches occurs during heavy storms. Large waves remove the flat sandy part of the beach, which is carried out to sea by a back-wash and collects offshore, forming a ridge of high ground (a bar) (fig.4.7). The incoming waves break over the bar, dissipate their energy and cease to erode the beach. Thus if there is sufficient sand in the foredunes for building an offshore bar, then erosion will be temporary. The most common reasons for the disappearance of foredunes are:
- mining the sand from the beaches;
- levelling of the dunes to build beach front houses;

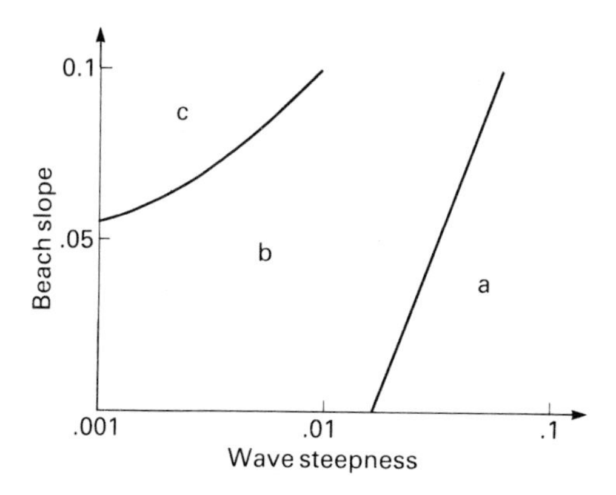

FIG.4.6 Dependence of type of breaker from the beach slope and the wave steepness; (a) spilling breaker; (b) plunging breaker; (c) surging breaker.

- destruction of the dune vegetation by cattle or humans. The vegetation holds the dunes together and once it disappears the wind will blow the dunes away.

FIG.4.7 Erosion of sand dunes; (1) erosion breaker; (2) backwash; (3) bar; (4) drift.

Beaches could be gradually restored during calm periods, when a drift shorewards exists, carrying with it sand from the offshore bar and depositing it on the beach. The bar gradually disappears and a wide berm is formed. This process is called accretion (progradation). The waves must have sufficient energy to accrete - they must raise the sediments from the sea floor, carry them to the beach and deposit them in their swash. Small and fine particles are favoured, because they are light enough to be lifted and propelled forward. The relationship between the velocity needed to move a particle (the erosion velocity) and the particle size is illustrated in fig. 4.8. High velocities are needed to pick up large particles because of their weight. Surprisingly, high

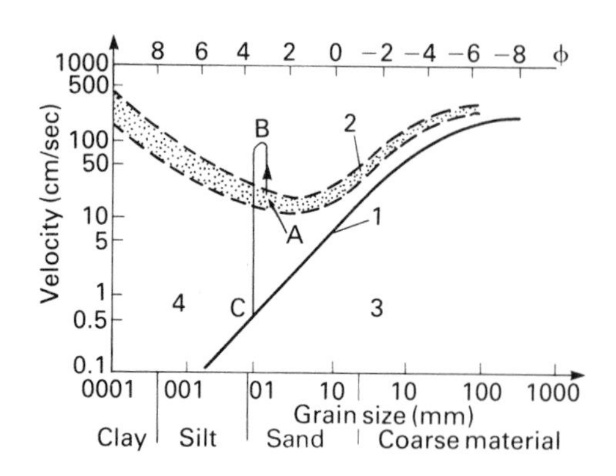

FIG. 4.8 Hjulstrom curve of erosion and deposition for uniform material; (A) lifting velocity of particles; (B) peak velocity; (C) fall velocity of particles; (1) fall velocity; (2) erosion velocity (lifting velocity of particles); (3) deposition (below fall velocity); (4) transportation (above fall velocity).

velocities are also needed to lift small particles. They form a colloid, in which they are kept together by electrical forces. But when the erosion velocity has been reached, they will tend to move as a suspended load and they will remain as suspended sediment until the flow velocity drops below the fall velocity, when they will settle out.

In the nearshore zone two wave induced current systems exist (fig.4.9.). These are:

- longshore currents produced by an oblique wave approach to the straight shoreline;
- a cell circulation system of rip currents and associated longshore currents, which are narrow strong jets of seaward-flowing waters.

A longshore current flows parallel to the coastline in the nearshore zone and it is responsible for nett transport of sand and other beach material along the shore. The velocity of this current decreases quickly to zero outside the breaker zone and thus the sand settles forming sand bars across river mouths or a beach tied to the coast at one end, free at the other.

Breaking waves approaching the beach produce a rise in the

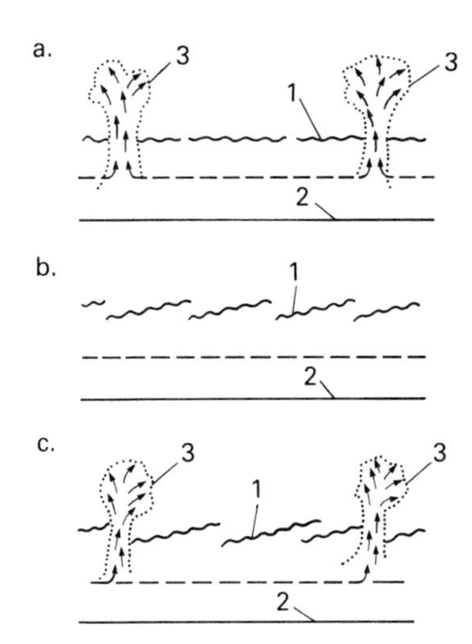

FIG.4.9 Wave induced current systems; (a) cell circulation; (b) longshore current; (c) resultant nearshore current pattern; (1) breaker zone; (2) shoreline; (3) rip currents.

mean water level - the higher the waves the greater is the wave set-up; therefore the currents flow from the positions of highest breaker heights and turn seaward as rip currents at positions of lowest wave height.

4.1.2 Classification of shores

The world's shorelines can be divided into several groups varying in wave energy and tidal range. The size of beach sediments can range from very coarse-grained (gravels) to very fine-grained (silt and clays). Along all coasts, several partitioning components remain relatively constant for medium to light oils, e.g. evaporation (30 to 50%) and biodegradation/photo-oxidation (0 to 5%). Others may vary substantially.

The physical environment of the coastal systems could be divided into three categories:

- high to moderate energy rocky shores;
- coarse-grained beaches;
- fine-grained beaches.

For this consideration it is assumed that the oil is spilled fresh in the nearshore zone, but it may be spilled offshore and weather as it travels toward the surf zone and the shoreline, losing its lighter components through evaporation.

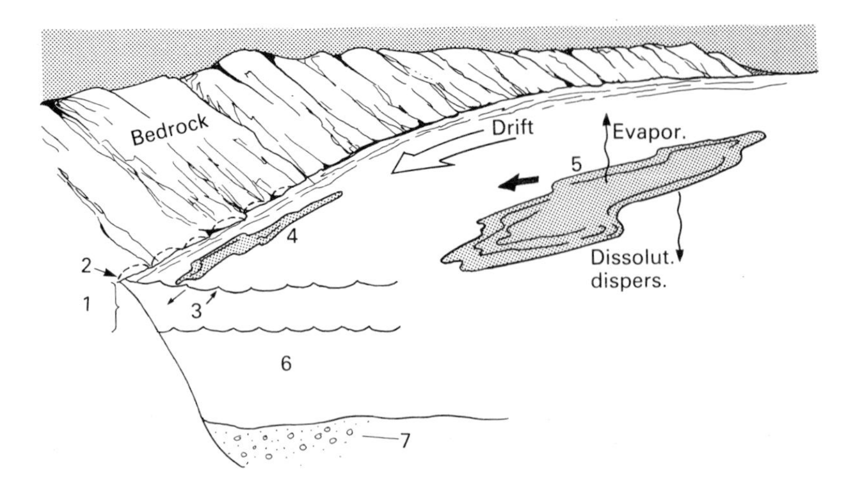

FIG.4.10 Oil partitioning and interaction along a moderate to high energy rocky shoreline; (1) tide range; (2) wave run-up/reflection; (3) region of dispersion; (4) patch of oil held offshore by wave reflection; (5) incoming oil; (6) little to no suspended material; (7) bottom sediments.

High to moderate energy rocky shores encompass rocky headlands, steep rock-dominated shores and wave-cut rock platforms. As wave energy is decreased, oil coating of the rocky shore increases. Depending on wave energy, oil will be held approximately 10 to 30 m off the coast by wave reflection and therefore be influenced primarily by longshore transport and chemical processes (fig.4.10.). Under very low wave conditions, a small amount may adhere temporarily to the vertical slope depending on the slope, biological encrustations and tidal range or area of exposure. With increased wave energy oil will again enter the marine environment and will be subjected to chemical and mechanical processes. Evaporation will be a major factor and can be up to 50%. Emulsification in the surf zone would increase the general mass of oiled material. Under calm conditions the maximum extent of shoreline stranding would be the rocky, intertidal zone. Since only low concentrations of suspended particles are present along these coasts, oil agglomeration and sinking are very minor. Bottom sediments are commonly well-washed boulders or bedrock not appropriate for oil incorporation.

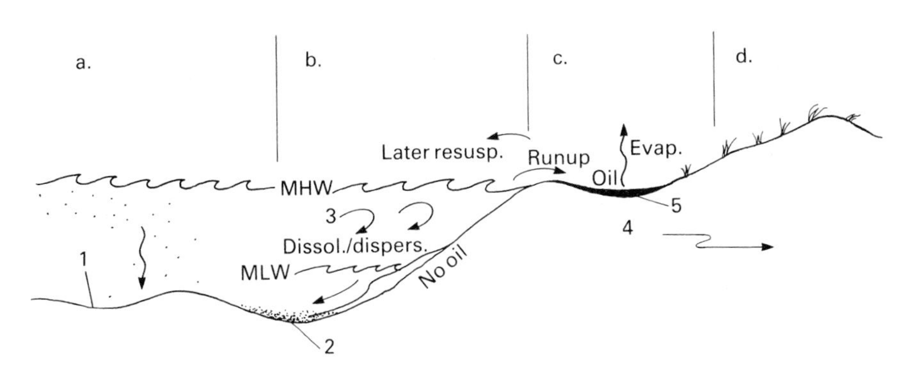

FIG.4.11 Oil interaction along a coarse-grained beach; (a) offshore; (b) beach face; (c) backshore; (d) dunes; (1) oil in water column; (2) oily sediment; (3) region of dispersion/dissolution; (4) primary zone of deposition; (5) oil; (MHW) mean high water; (MLW) mean low water.

Coarse to fine-grained beaches include gravel and sand beaches with silt and clay fractions below 5%. Wave energy may vary from high to low. The incoming tide will deposit oil along the upper portion of the beach (fig.4.11). If tide levels are high, oil can reach the backberm and backshore areas. The lower beach face will remain free of oil. On an outgoing tide, oil may temporarily coat the beach face if oil quantities are great. This oil will lift off and be transported toward the backshore or along shore by currents on the next incoming tide. Oil will reside along the backshore until sufficient wave activity reaches these areas to erode the oiled sediment or mechanically disperse the oil through abrasion of oiled sediment by grains rubbing against each other. The oil on each grain may asphaltise and then oil and sediment may be transported offshore. Once in the active surf zone again, mechanical dispersion and chemical dissolution will continue, although longshore and wind generated currents may transport the remaining oil to another shoreline type. Here the major factor will be the grain size (particle size distribution) of the beach. Oil can penetrate into the beach and be buried as an oiled sediment layer. The depth of oil penetration and burial increases as grain size increases. A moderately to heavily oiled, fine-grained sand beach will show 2 to 5 cm of oil penetration and up to 15 cm buried within one month and will have about two months of oil persistence because of little variation in the sweep of topographic profiles. On mixed sand and gravel beaches, oil penetration and burial can exceed 65 cm and residence can approach a decade, if oil is deposited above major wave activity and allowed to become asphaltised.

Most oils are less dense than water and can penetrate sediments only within the nonsaturated zone located above the water table. The velocity of oil penetration and flow through the soil pore spaces is dependent on the permeability of the sediment and the viscosity of oil.

A wide range of permeability exists for naturally occuring sediments. High permeabilities (greater than 0.2 ft/min) allow for rapid adjustment of the water table in response to tidal changes. The velocity of fluid flow through sediment is inversely proportional to the dynamic viscosity of the fluid. Assuming a ground and water temperature of 50 °F and the viscosity of the crude oil to be approximately 30 times higher than that of seawater at the same temperature, the velocity of fluid flow through soil pore spaces will be 30 times slower for this crude oil than for seawater.

As oil migrates through the pore spaces of "clean" sediment, a portion of the oil becomes trapped in the soil and remains behind. The migrating oil is eventually immobilised through progressive entrapment of the oil by the soil. The immobilised oil is present in the soil in two forms:

- thin films that are adsorbed to the surfaces of the soil particles;
- discrete micropockets of oil held in the soil pore spaces by capillary tension.

The maximum depth of oil penetration into the beach sediment is limited by the volume of oil on the beach and by the position of the water table. In the lowest portions of the beach, only a limited amount of oil penetration is possible, because the water table at low tide drops only slightly below the sediment surface in that area. On the upper beach, oil penetration may be limited by the available oil volume and not by the water table position.

The presence of burrow structures of all sizes within sandy sediments may enhance the penetration of oil into subsurface layers. Having penetrated into sediments, the oil is likely to degrade at a slower rate than on the surface. If the degree of oiling is light and the toxic properties low, reworking of oiled sediment by large infauna such as *A. marina* may return the oil to the surface, permitting further degradation. As penetration of the oil into the burrow structure can occur within 24 hours of application, in the event of a spill efforts should be made to clean up as soon as possible after the oil has beached. Oils less viscous than mousse enter and penetrate the burrow system more rapidly.

Fine-grained beaches include mud-dominated tidal flats and marshes and such areas as the organic rich fine-grained shores fronting major depositional deltas. Tidal currents and wave activity are commonly very low, while concentrations of suspended particles are very high. Estuaries are the most typical environment of these areas. They are very important since they:

- have high sedimentation rates, which tend to incorporate oil within shoreline deposits;
- are sheltered, thereby inhibiting wave-, current- and storm-related processes and increasing oil persistence;
- contain high quantities of suspended matter, which increases oil and sediment transport to the bottom;
- contain fine-grained offshore sediment, thereby increasing potential oil incorporation into bottom sediments.

Oncoming spilled oil, in turn, acts as sediment and will be deposited in the same areas as the sediment on the surface of the marsh or sheltered tidal flat. Some fresh oil will refloat on the following high tide and may exit on the subsequent ebb tide, so that oil accumulation in these sites may be 50 to 100% of total incoming oil. Onshore winds will tend to increase the final value. On the surface of the flat or marsh, oil will be more prone

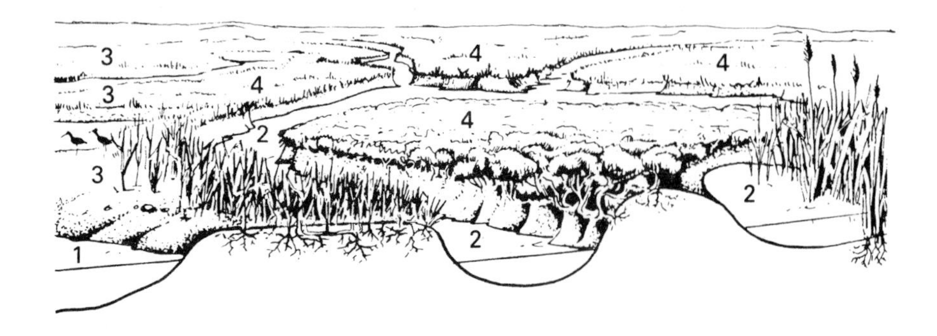

FIG.4.12 Salt marsh; (1) sea shore; (2) channels; (3) muddy tidal flats; (4) salty marsh.

to evaporation and chemical weathering. Mechanical weathering is negligible. While in or on the surface of the water column, oil will interact with suspended particles, whereupon it will probably sink to the bottom and be incorporated into the bottom sediments. Oil can be incorporated into bottom sediments directly through dissolution as well as dispersion. In these interactions aliphatic components are much more likely to interact with suspended sediments than are aromatic compounds. The persistence of oil in sediments is generally great. Three years after the "Amoco Cadiz" accident approximately 50% reduction in oil content in the estuarine sediments was found, whereas in offshore sediments this reduction amounted to 90%. The estuarine model (fig.4.12.) includes oil interaction in a channel, adjacent mud flats and on upper marshes. The channel contains high concentrations of suspended particulate matter, causing oil sedimentation and incorporation into fine-grained bottom sediments. The mud flat portion indicates oil stranding along the upper intertidal zone with leaching or oil penetration into flat sediments and transport toward the channel through the ground water. In marsh regions, oil will be deposited on the marsh surface with same leaching and penetration into the marsh sediments. Some minor oil export occurs during each successive tidal cycle.

TABLE 4.1 Summary of spilled oil mass balance for different shoreline type components (in percentage of total spilled).

Partitioning components	Moderate to high energy rocky shores	Coarse-grained beaches	Fine-grained beaches
Evaporation	30-50	30-50	30-50
Dissolution	10-15	10-15	5-10
Sedimentation	0-2	2-5	10-20
Shoreline stranding	0-1	20-35[1]	40-60
Biodegradation and photo-oxidation	0-5	0-5	1-2
Particle formation and transport	Remainder	20-50	Remainder

([1]) - Subject to refloating, additional transport and weathering

A summary of three major shoreline types and the relative breakdown of spilled oil components is given in table 4.1.

4.1.3 Shoreline Sensitivity

The shoreline of a potential spill impact area can be divided into units, each with a specific morphology. As oil enters each unit, it will (to varying extents) evaporate, dissolve, interact with suspended particles and sink, biodegrade, photo-oxidize, be transported to the next unit or strand on the shoreline. In the last case, oil will reenter the aquatic system after a given time and again be exposed to the same processes.

The ultimate goal for an oil pollution combating strategy is to minimise the potential for environmental damage. This can be overcome by using the right type of equipment at the right place and at the proper time in an operation. To accomplish this in practice, it is necessary to identify the environmental resources, evaluate their sensitivity to oil pollution and oil pollution combating methods and to recommend appropriate countermeasures to be taken in oil pollution situations.

Oil pollution might have a huge potential impact on fisheries due to tainting of fish and other types of sea food and effects concerning the most sensitive egg and larval stages. The breeding season and maps of breeding areas for economically important species should be worked out. Generally it is recommended to use mechanical recovery equipment in case of oil pollution.

If it is unavoidable to use dispersants, it is necessary to be aware of tainting problems. Aquaculture ponds are specially sensitive to the use of detergents, whereas mangrove areas could be saved by using appropriate dispersants.
Distribution of coral reefs along the coasts should be indicated as they are inhabited by many species of marine fauna, which could be affected by dispersants. There still exists a lack of sufficient agreement on the use of detergents.
The revenue from tourism is of extreme importance to some regions. Thus, oil pollution in tourist areas should be especially carefully considered. Industrial plants, such as power plants and refineries, require large amounts of cooling water. In some areas there are salt producing plants, which are most sensitive to oil.
The collected information is in its nature dynamic, not static. It needs to be updated at intervals and in particular oil spill situations. Sometimes lengthy research is required concerning aquaculture and coral reefs, as well as wind direction and velocity, current patterns and seawater temperatures.

4.1.4 Environmental Sensitivy Maps

Environmental sensitivity maps should be developed to streamline decision making by the command team. The maps identify priority areas that would require a maximum effort for protection, clean-up and conservation. The system is based on a sensitivity scale of 1 to 10 with respect to the expected persistence of hazardous material spills along the coast. The index is based on the geomorphology of the area, coastal processes and the amount of physical energy on the coastline (Table 4.2.)
On the maps of the coastline the following information should be identified: water depth, current velocities and distances across the inlets for deployment of containment booms. Areas where loading and unloading of heavy equipment is possible should be indicated, as well as stockpiles of booms, sorbents, clean-up equipment, etc. The information regarding the areas of special biological importance should indicate localities of oil-senstive, protected or commercial species and communities. In order to

TABLE 4.2 Sensitivity index of shores.

Sensit. index	Shoreline type	Comments
1	Exposed rocky headlands	Wave reflection keeps most of the oil offshore. No clean-up necessary.
2	Eroding wave-cut platf.	Wave-swept. Most oil removed by natural processes within weeks.
3	Fine-grained sand beaches	Oil does not penetrate into the sediment, facilitating mechanical removal if necessary. Otherwise, oil may persist several months.
4	Coarse-grained beaches	Oil may sink and/or be buried rapidly making clean-up difficult. Under moderate to high energy conditions, oil will be removed naturally from most of the beachface.
5	Exposed compacted tidal flats	Most oil will not adhere to, nor penetrate into, the compacted tidal flat. Cleaning is usually unnecessary.
6	Mixed sand and gravel	Oil may undergo rapid penetration and burial. Under moderate to low energy conditions, oil may persist for years.
7	Gravel beaches	Same as above. Clean-up should concentrate on high-tide/swash areas. A solid asphalt pavement may form under heavy oil spill.
8	Sheltered rocky coasts	Areas of reduced wave action. Oil may persist for many years. Clean-up is not recommended unless the oil concentration is heavy.
9	Sheltered tidal flats	Areas of great biological activity. Oil may persist for years. These areas should receive priority protection by using booms or oil sorbent material. Clean-up avoided.
10	Salt marshes and mangroves	Most productive of aquatic environments. Oil may persist for years. Protection of these environments should receive first priority. Burning or cutting to be avoided.

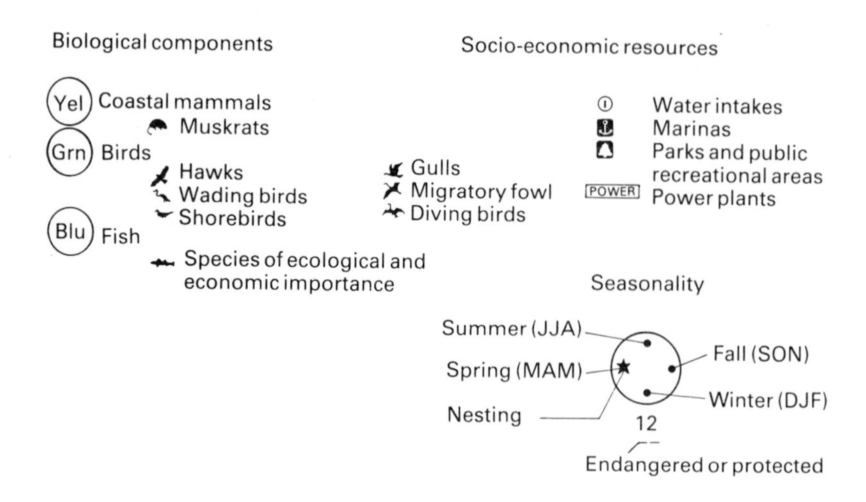

FIG. 4.13 Keys to symbols and information presented on shoreline sensitivity maps.

make this information easily readable, special symbols and coloured circles should be used: yellow – mammals, green – birds, blue – fishes, orange – shellfish (fig.4.13.). The silhouette in the centre of the marker refers to the major ecological group and the number refers to a species or species group listed in the manual. Such maps should be prepared in advance for each season of the year with indications of which particular species are most likely to be present with respect to reproduction, migration and feeding behaviour. In case of a spill it is then known at once which areas of the coast should be especially carefully protected.

Important recreational sites, nature reserves and industrial plants, especially industrial cooling water intakes, have to be indicated on the maps. A list of whom to contact concerning each particular site, in addition to a predesignated clean-up plan is important. Positioning of open-water and harbour booms, open-water skimmers and areas where the inlet is small enough to be infilled to protect the interior sensitive area have to be indicated. A priority ranking of spill protection should be indicated - the first line of defence is to protect the beaches from offshore oil reaching the coast by means of open-water booms and skimmers; the second line of defence is to prevent oil from entering major inlets and the third line of defence is to prevent oil from entering sensitive small creeks and salt marshes.

Appropriate logistical support is essential for any successful action. There must be sufficient and well trained manpower and adequate equipment available. Information about its handling, the best methods of clean-up, treatment and disposal of collected material, access roads and communication devices are essential conditions for a successful oil spill combating operation.

4.2 Shoreline Clean-up

4.2.1 Introduction

Many oil spills on the open sea cause pollution of shorelines despite efforts to combat oil at sea and to save the coastline from any damage. The clean-up is usually straightforward, but it is very labour intensive. Inadequate organisation and resources as well as adverse weather conditions may increase the damage caused by oil. Fog may severely restrict skimming operations and at times prevent overflights to locate oil concentrations and to direct the necessary equipment.

The clean-up might be complicated by oil lying submerged in the nearshore surf zones adjacent to the areas most heavily affected. New impacts from the submerged oil might become a daily occurrence. Thus repeated beach cleanings are necessary. Submerged oil may come up on the beach at a considerable distance from the location of the spill and some weeks or even months later. The amount of submerged oil should be described and the mode of its transfer onshore, which is in conjunction with naturally occurring sand accretion processes. To accelerate this process, several methods should be deployed to recover the submerged oil, including the use of a high-capacity vacuum truck, cement pump and heavy machinery. It is unavoidable that an undetermined quantity of oil will mix with sand in the surf zone and will be buried.

An effort should be made to minimise the amount of sand removed from the beaches. Oil initially removed from the beaches

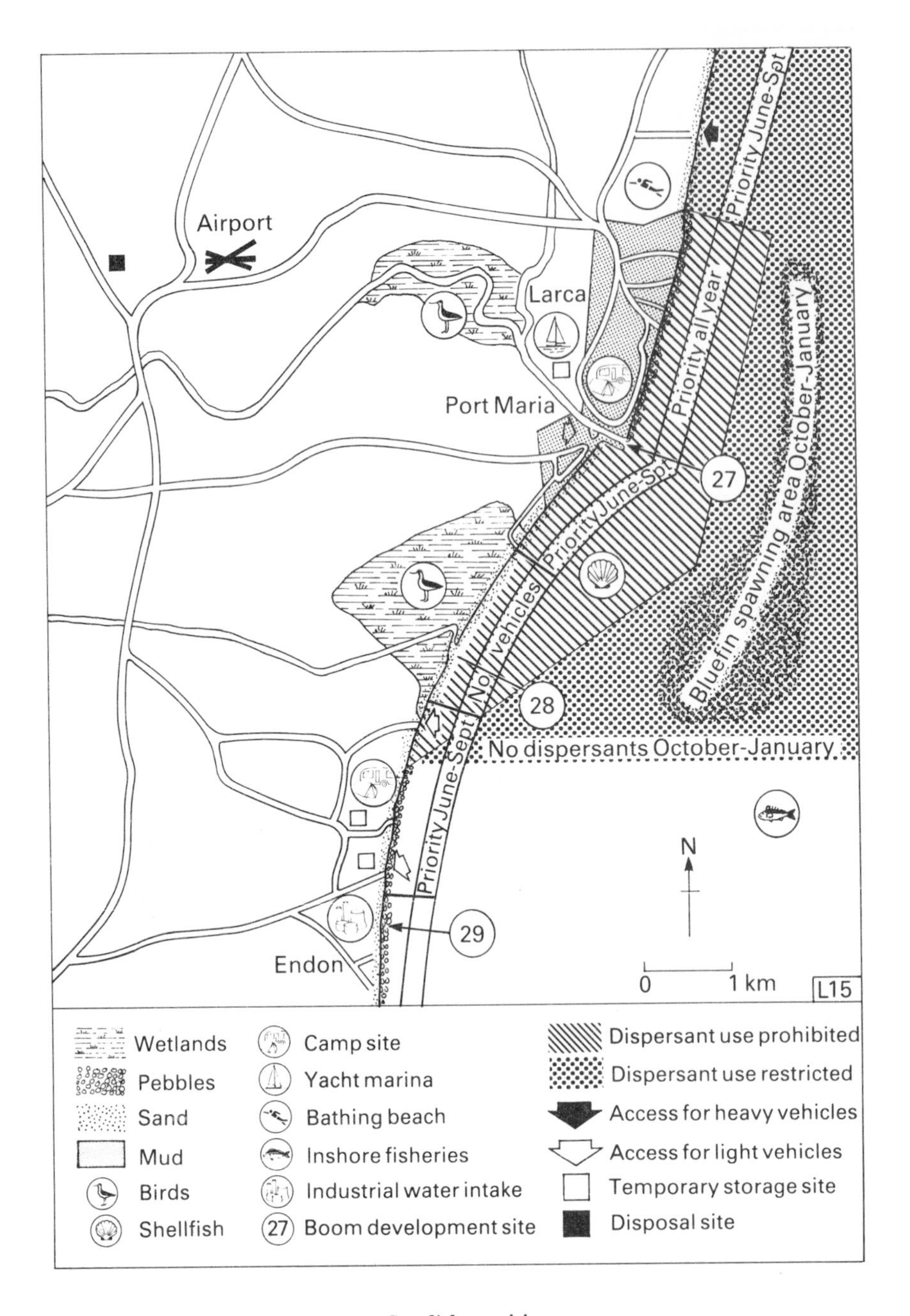

FIG.4.14 Swedish sensitive area map.

contains very little sand. As the clean-up progresses, it becomes necessary to remove sand, which is contaminated with oil from within the tidal zone and to stockpile it above the tidal zone. It is necessary to determine if the stockpiled sand requires classification as a waste (>10% of oil content). Such sand may need to be removed from the entire beach area. Moderately contaminated sand could be relocated and spread in the back beach area adjacent to existing sand dunes. Lightly oiled sand could be spread in the non-tidal, mid-beach area.

The best method of removing large quantities of oil from the impacted beaches is by using road graders to move the newly beached oil above the high tide zone to prevent it from washing back out into the surf zone. In a large spill, the necessary equipment might amount to more than 50 graders and 100 25 cubic-yard dump trucks.

During the beach clean-up a scientific support crew should perform transectional surveys and underwater diving surveys of the beach and surf zone in order to determine the movement and location of submerged oil. In depth chemical and physical analyses of oil should be made prior to its disposal to ensure that placement of oil contaminated sand would present no severe future environmental problems due to toxic components and that water-soluble fractions were no longer present in weathered oil, thus minimising the possibility of toxic leaching from the sites.

During the beach clean-up conflict may arise over who owns the contaminated beach sand. Location of disposal sites might be complicated by ongoing litigation regarding where the state's property ends and the private landowner's property begins. The state may want to remove the sand from the beach front, while the landowners may want the sand to be left where it was moved to the dune line, seaward of the vegetation line. Thus for planning purposes, the state should identify sites that might be used for disposal of large quantities of contaminated materials and sand. The sites should be catalogued by location, type of materials, amount of materials and degree of contamination of the materials.

4.2.2 Manual Clean-up

Manual clean-up is very labour intensive and is used only in special circumstances such as:

- when access to the shore is difficult;
- when the coast profile is unsuitable for employing mechanised equipment;
- when large amounts of debris and boulders are on the beach;
- where the environment is very sensitive, i.e. marshes.

In manual collection only human muscle power is being used. In some cases small hand transportable power sources could be employed. The tools should be simple to use, easy to clean and store, simple to handle during transportation and in the clearance area. Stones and rocky shores take a great deal of time to clean. For this purpose specially selected and designed brushes made of suitable brush material are used, which are easily cleaned by wiping against the scraping edges and extractor on the holder ring. Collection of oil from the surface of the water is effected by means of a scoop. Oil and water are poured into a floating ring made of aluminium section. The ring is so elaborated that oil and water can be raked over the edge and into the ring. Inside the ring is an inner sack of plastic with a hole in the bottom. The oil inside the ring deposits itself on the water and squeezes it out through the hole in the plastic bag. Thus a thick layer of oil is built up in the collection ring. The oil is emptied either by hand with a scoop or with a pump.

Collection of oil and scraping of tools is best done in plastic sacks of 12 to 15 kg volume. They require very little storage space and transport space in relation to the volume of oil that can be collected in them. The sack is mounted in a special, low sack-holding ring without a fixed bottom, resting on its outer edges on the beach. The upper edges of the sacks are folded out over the top edge of the holding ring and held in position by means of the narrow outer ring with two scraper edges, against which the oil can be scraped off.

The problems of storage and transport for all the equipment kept in oil combating stores have been solved by applying containers with ready made modules for a combating crew of five men. To reduce the volume of equipment, a quick-coupling system between the shanks and tools has been developed. Thus the number of shanks has been reduced from 30 to 8 for one work crew. The tools are stored in racks, so that one person can pick up a shank, a sack holder and a set of tools and then work on his own regardless of the type of beach or the type of oil.

When larger amounts of oil reach the shoreline, then portable mechanical equipment developed in Sweden could be used. This system consists of three basic units:

- hose pump;
- hydraulic unit;
- water injection pump.

The hose pump and hydraulic unit are each equipped with support wheels enabling them to be rolled along on the ground, where the terrain so permits. Otherwise the units can be carried by two or four men. The water injection pump is carried like a rucksack and only needs to be taken along if very heavy oil (e.g. grade 5 fuel oil at 0 °C) is to be pumped. When pumping very heavy oils, water is added at the suction nozzle and possibly at some additional point along the 14 m long suction hose so as to be able to maintain a reasonable collection capacity.

The hydraulic drive permits stepless speed control and reversing of the pump direction. This is an advantage if the suction nozzle has to be cleared of large twigs, etc. that may have clogged it. The pump housing is a casting of magnesium alloyed aluminium and in all other respects every effort has been made to minimise its weight. The maximum speed of the pump is set for 42 rpm and its displacement is 3.75 dm^3 per rotation. This means that the maximum capacity at a suction head of 1 m can be set at about 10 m^3/h. With very heavy oil fuel grade E05, the minimum capacity will be 1 m^3/h at an ambient temperature of 0 °C. The wall thickness of the displacement hose and the resilient plastic lining of the pump body enable the pump to cope with sharp-edged solid impurities up to 25 mm in size, when the free flow diameter is 50 mm.

This equipment can be carried by a crew of 5 men for 1.5 km in 1.5 hour in completely roadless terrain. It could be carried on a small caterpillar vehicle capable of travelling on bare ground. Such a vehicle could be driven by a small combustion engine used in a hydraulic unit. This hydraulic unit could be used for driving the hose pump and the fan required to supply air to the combustion furnace.

The pumping system for transfer of intensively emulsified oil collected during an oil clearance operation should be adaptable to various types of oil with a pour point higher than 0 °C, likely to be encountered, and capable of pumping contaminated oil. The capacity should be at least 50 m^3/h at a counterpressure of 5 bar. The equipment should be easy to handle and transport. It should consist of:

- pump equipment satisfying the requirements of selfpriming, crushing normally encountered foreign matter such as bottles, plastics, pieces of timber, etc. to a pumpable size, resistant to chemicals;
- oil suction head with equipment for heating and injection of steam or water (hot/cold);
- a steam injection flange for injection of steam/water into a hose line.

This pumping system could be transported on trucks or boats.

The efficiency of manual clean-up activities decreases with time as oil becomes scarcer. As seen from the table 4.3. the effort measured as the number of man-days was roughly constant, but the oil quantity collected dropped by 99% over the clean-up period and the unit cost ($/t) increased sharply.

TABLE 4.3 Efficiency of manual clean-up.

Clean-up period	Effort (man-days)	Oil quantity collected (t)	Cost (US$)	Unit cost ($/t)
Stage 1	136,800	2,270	1,560,000	748
Stage 2	143,400	200	748,000	4,069
Stage 3	102,150	20	236,000	712,835

4.2.3 Seawater Flushing

Opinion is divided on the use of seawater flushing as a method of cleaning up oil pollution because of its impact on different sediment types and its effects on the biota. The principle is to raise the water table locally so that surface flow is initiated down the beach, thus floating off the oil and perhaps freeing oil that has already penetrated surface layers. The floated oil may then be herded into trenches and recovered mechanically or by hand. But the physical disruption of sediments may cause additional incorporation of oil into sediments as well as mechanical damage to the biological community.

Low pressure flushing uses ambient water streams at any pressure low enough (generally < 7 atm) so that beach material and organisms are not removed. Flushed oil is concentrated by containment booms and is removed by skimmers. It can remove most free oil without removing or damaging organisms. It is particularly effective in sensitive habitats like marshes especially if it is done from boats rather than from land. This method might be most successful due to its high effectiveness and its minimal biological and sedimentological impact on the shore, where:

- the oil is viscous and has not been treated or can not be treated with chemical dispersants;
- the slick is in a thin (< 10 mm) relatively continuous strand line;
- the potential shore is accessible and the response capability is rapid enough to prevent deeper sediment penetration or burial of the oil by mobile sediments;
- the sand and mud are relatively firm and of sufficient thickness so that the limited sediment erosion that will occur does not totally remove the substrate and prevent the recolonisation or burrowing by infauna;
- there is a gently sloping, poorly to moderately draining (< 30 cm depth to water table) shoreline seaward of the stranded slick;
- a local supply of water is available for pumping, which will not result in salinity stress on fauna;
- the pump rate is of the order of 2 l/s and not directed by a hose pipe nozzle of excessive pressure;
- there is an area of quiescent standing water downshore for

containment and retrieval of the flushed oil by a combination of booms and skimmers.

Such conditions may be found with ridge and runnel systems on exposed, sandy beaches, where several bars and ridges may be separated by water saturated troughs. These would permit the flushing and recovery points to be arranged sequentially across the intertidal zone (in severe oil pollution incidents) as well as along the shoreline (for a single oil strandline). These conditions may also be met in lower energy estuarine, fluvial and deltaic environments, where geomorphological features such as flood and braided channels may be used as temporary water receiving bodies. More research is obviously needed, particularly in lower energy environments, where the residence time of the oil will be longer and the need will be greater to remove and to recover the oil without the help of chemicals before its burial or translocation. The technique offers considerable promise in these areas, particularly where the scale of clean-up required lends itself to a "modular" system of relatively small scale teams and hardware. However, flushing techniques should not be used where:

- fluid muds and low shear strength muds and muddy sands would cause problems with erosion and transport or where mixing of the oil and sediment would occur;
- coarse sand and gravel beaches may preclude successful raising of the water table and result in deeper penetration and mixing of oil and sediment during flushing;
- trenching (lined or unlined) in areas of high water table may not combine with flushing techniques;
- the mechanical disturbance of flushing and the additional trampling damage would rule out its use on many salt marsh and mangrove shores, although narrow fringing marshes along microtidal inlets may be workable from boats;
- any shore where either access and/or short to medium term oil storage is difficult would not be suitable.

High pressure flushing (hydro-blasting) using high pressure (>7 atm) water streams is often adopted to remove oil from the substrate. Flushed oil is then led to a recovery area. This method

is used for cleaning jetties, piers, sea walls, pilings and rocky shores. However, it removes organisms and it could be particularly damaging to low energy, high productivity shorelines or other habitats where recovery is slow. Oil penetration into sediments could be increased by high pressure water streams and further impacts could result from transport of equipment over the oiled and adjacent habitats, especially in fresh water marshes.
The use of dispersants can sometimes also assist oil removal by means of high pressure flushing. However, with more viscous oils the dispersant simply acts to release oil from the rock and does not produce dispersion. Undispersed oil should be carefully collected to prevent recontamination.
In tropical climates hot water washing is likely to be less effective than in temperate climates since oil exposed to the sun becomes baked onto the rock. Small areas can be cleaned by sand blasting.
On cobbles and shingles water at high pressure could be used to flush surface oil to the water's edge, but some oil will also be driven into the beach. Some oil will inevitably remain in the body of the beach after the surface stones have been cleaned. This oil will slowly leach out as a sheen over a period of weeks. Removal of all the oily stones will rarely be a practical option, but if it is contemplated, it must be ascertained that the removal of stones will not cause serious erosion of the beach.
Where there are open areas in the mangroves, then high pressure flushing on a large scale may be considered practical. Flushing should be of such magnitude (fig.4.15.), that the currents generated by the water pressure will circulate through the large numbers of mangrove roots. Local flushing with a single nozzle is quickly dampened by the tangle of roots and is generally ineffective. Flushing can be equated to the wave and tidal cleansing action that minimises the oil impact.

4.2.4 Mechanical Beach Cleaning Equipment

Large oil spills inevitably result in massive quantities of oil being washed ashore. The main task for local authorities is to lift off as much as possible of this semi-liquid stranded emulsion

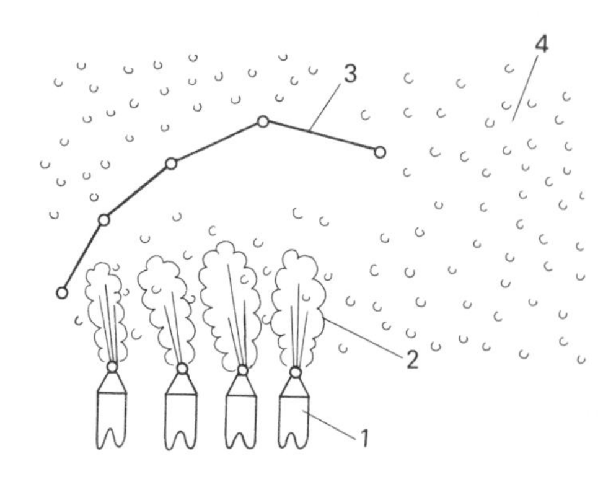

FIG.4.15 Mangrove flushing; (1) nozzles; (2) flushing currents; (3) barrier boom; (4) mangrove roots.

by a variety of techniques and then, when the layer of residual emulsion is less than 6 mm deep, to employ manual methods or to spray dispersant about 0.5 hour before the flood tide reaches the layer of oil, so that the action of the surf can break up the residual oil and carry it away in the form of very small suspended droplets for natural decay and biological attack to completely destroy the oil.

The techniques for lifting off the thick layer of oil will depend very much on the nature of the beach, which is characterised by:

- gradient and firmness of the beach, which determine the possibility of employing mechanised equipment and its size;
- particle size distribution of the beach material;
- natural water table movement with tidal cycle;
- quantity and nature of oil covering the beach.

Of great importance is the possibility of using available agricultural and sludge disposal equipment. Several experiments have been carried out in order to establish the usefulness of this equipment for cleaning up the beaches.

The following vehicles and mechanised equipment could be used on the beaches:

- a standard vehicle mounted snow plough should be used at an optimum speed, because faster speeds will result in the emulsion flowing around the end of the blade as the bow wave spreads out in front of the blade. This results in thicker layers of emulsion being formed on either side of the scraped swathe (approx. 1.5 to 1.8 m wide) made by the advancing blade. Very little sand is included in the emulsion. The overall clearance rate is estimated at about 47 m^3/h;

- a standard tracked bulldozer with blade setting 3 to 5 mm above the sand leaves clean sand patches, where the blade slices through the top layer of sand under emulsion. Consequently, a large amount of sand mixed with emulsion is pushed into the trench and sinks to the bottom. At the same time, emulsion has been pushed by the tracks of the vehicle into thicker furrows on each side of the track marks. The tracks leave depressions even on firm sandy beach, which eventually fill with emulsion. The vehicle tracks also deposit large lumps of emulsion mixed with sand some 0.3 m in diameter and 3 cm deep. Afterwards, hand scrapers have to be used to push the remains of the emulsion in the furrows into the trench. A considerable amount of sand is mixed with the emulsion and this mixture sinks to the bottom of the recovery trench making subsequent oil pick-up difficult. The overall removal of the emulsion could be estimated to be roughly 50 m^3/h excluding extra sand collected. Attention should be drawn to the fact that a lot of emulsion had been sunk by sand and required extended agitation to release its oil onto the surface of the water in the recovery trench;
- a digger fitted with a rubber blade attachment on the back of the bucket, tipped 90 ° to its normal position. An improvised wiping blade, consisting of rubber inserts forms a snow plough stiffened between two rough cut timber planks and secured with nails, was fastened to the bucket by means of four large G clamps. The device cleaned the mousse into the trench with a negligible amount of sand entrained, but some oil flowed round the ends of the wiper blade, which could be easily prevented by two men, one on each end of the blade, equipped with curved hand scrapers holding the mousse within the width of the wiper blade. The overall clearance rate could be estimated

at about 55 m^3/h;

- a farmyard scraper, tractor mounted was fitted with an all-rubber blade supported on a metal bracket as standard equipment. In clearance runs, the tractor is driven backwards because the scraper attachment is fitted to the rear hydraulics of the tractor. The wiper blade is selflevelling when it touches the ground under the weight of the scraper attachment and wiper blade. In practice the wiper blade is always slightly angled, the bottom of the blade nearest to the tractor being just clear of the sand. The ends of the "wings" of the blade cut into the beach slightly, when the main part of the blade (at right angles to the vehicle movement) was wiping across the surface of the beach. This resulted in more sand being collected than the rubber wiper tested on the bucket. The blade was used in an overlapping mode as the mousse passed around the ends of the blade and thicker furrows were formed on each side of cleared swathe. After one pass the oil thickness was reduced to 1-2 mm. The second pass over the same swathe resulted in a considerable amount of sand being pushed into the collection trench. The overall clearance rate was about 20 m^3/h;
- a front loader tractor was first used with the rubber wiper blade to clear the emulsion remaining on the sand after an experiment to float the emulsion into the trench. The emulsion was pushed down the beach into the trench, taking very little sand with the oil. Using the front loader with a metal bucket the thickness of the emulsion layer was reduced from 35 mm to 12 mm. Sand was disturbed from under the oil by the teeth of the leading edge of the bucket and was entrained with oil. A further 10 passes reduced the layer thickness to about 0 to 5 mm, leaving clean sand patches, where the bucket sliced through the top layer of sand, but a large amount of sand was collected in the bucket mixed with the oil. The deep tread tyres resulted in deep depressions being left in the beach, which filled with the mousse. The treads also formed patches of mousse mixed with sand (0.3 m diameter and 3 cm

deep) which remained on the beach. The overall clearance rate was estimated at about 35 m^3/h.

The above presented equipment has overall clearance rates from about 20 m^3/h to 55 m^3/h at a depth of roughly 4 cm of emulsion. At the finish of each successful scraping operation the average depth of residual oil was about 1 to 2 mm, but varied between exposed sand and lumps of emulsion mixed with sand about 6 mm thick. These areas could be successfully cleaned by application of dispersant about 0.5 to 1 hour before the arrival of the flood tide. A reasonable clearance speed was about 3 km/h. At higher speeds more pollutant spilled from the blade edges, at lower speeds sand entrainment became troublesome.

Rubber blades performed well and wiped the emulsions into the trench with a negligible amount of sand being entrained with the emulsion. The metal blade equipment, where the blade rested on firm sand, collected an unacceptable quantity of sand with the recovered oil.

The presence of sand mixed with the emulsion caused the collected oil to sink beneath the water in the trench and created problems with the subsequent recovery of oil from the trench. Oil submerged below the water had first to be released from the sand by agitation or dug out of the trench with the sand and disposed on land as solid waste.

The springsweep air conveyor system consists of a 2 m^3 recovery hopper, a large air transfer pump driven by a water cooled diesel engine and a small air compressor to operate the pneumatic control valves. The system is mounted on a flat bed lorry for ease of movement and it operates by collecting oil and water to fill the hopper and then discharging the contents of the recovery hopper to storage. The air conveyance system is fitted with a fan-tailed recovery nozzle (0.4 m by 0.1 m) at the end of 15 m of 150 mm diameter flexible hose. The rate of recovery, allowing for the time of discharging oil to the reception skip, is estimated as about 12 m^3/h.

A small road vacuum tanker (3.7 m^3/min exhauster) is able to skim oil and water successfully with a 75 mm diameter hose. The overall rate is low in scraping operations because of gross sand entrainment and thus the pollutant is not fully collected. When a large percentage of pollutant is floating, then the emulsion could be collected at a rate of 4.5 m^3/h. For scraping operations this rate is about 10 times lower.

A large road vacuum tanker (11.4 m^3/min exhauster) gives a rate of oil pick-up of about 5 m^3/h. The large tanker is easier to use, because it holds more liquid (11.6 m^3) and the recovery does not need to stop part way, as with small tankers when full, to allow the contents to separate and to run off the bottom layer of water.

A farm slurry tanker towed by a tractor might be very useful, but because of the size of the tank it has to be emptied at frequent intervals.

All the above named suction based recovery devices could transport pollutants at viscosities of 50 to 100,000 cP over relatively short distances provided that water was entrained as lubricant. The mechanical screw pump appeared to work more efficiently in the absence of water. The overall rates of recovery indicate that the maximum pick-up rate of about 5 m^3/h is limited to the rate at which the suction nozzle can be fed by hand.

The use of water filled trenches was only suitable for the recovery of emulsions that remained floating on the surface. Emulsions mixed with sand sank to the bottom and were very difficult to handle except by digging out the mixture and removing it from the beach. The use of plastic sheeting to line the trench slows down the clean-up operation by blocking the fan-tail recovery nozzle and it is not considered to be practical.

4.2.5 Specialised Beach Cleaning Equipment

For collecting oil deposited on the surface of the beach two types of specialised equipment could be used:

- collecting drum;
- beach scraper.

The collecting drum (fig.4.16.) is based on the principle of collecting oil selectively with an oleophilic material. Different types

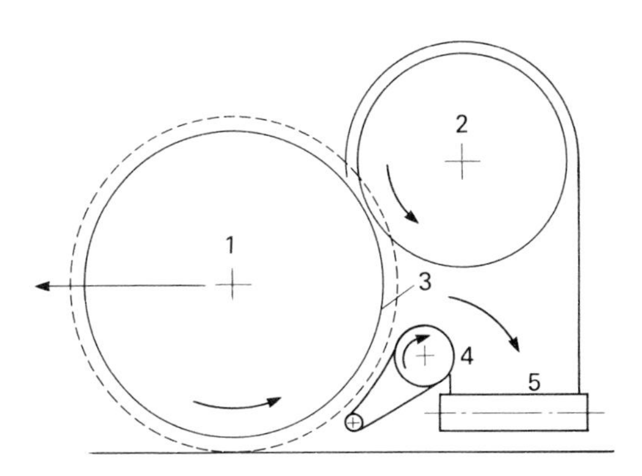

FIG.4.16 Working principle of collecting drum; 1 - collecting drum; 2 - rotating brush; 3 - cleaning blade; 4 - sand collector; 5 - lateral conveyor.

of covering materials can be used for picking up a wide range of oily deposits. On land the oil is often thick and the viscous layers can be agglomerated with sand. Thus the absorption properties may vary considerably. Thus physical adherence or wedging has to be relied upon. A research programme gave full information concerning optimal materials and design of selective drums.

A selective roller consists of a collecting drum which is covered with a suitable material, which can pick up oil of varied thicknesses. The drum is cleaned by a rotating brush and a cleaning blade or a comb, depending on the collecting material. The collected waste is stored in a rear bucket or transported by a high capacity conveyor directly to a wagon, driven separatly alongside the drum.

A beach scraper (fig.4.17.) has a front collecting blade which, in order to increase the efficiency, is vibrating. The waste is transported by a conveyor to a bucket. The conveyor is powered by a hydraulic pump, driven by the tractor power take-off. The front blade is vibrated by an eccentric axle, actuated by the hydraulic motor, which can be disengaged at will. Instead of a bucket, the scraper could be fitted out with two side conveyors, one on the right side and the other on the left. The slope of the conveyors is adjustable and they are entirely retractable for driving on roads.

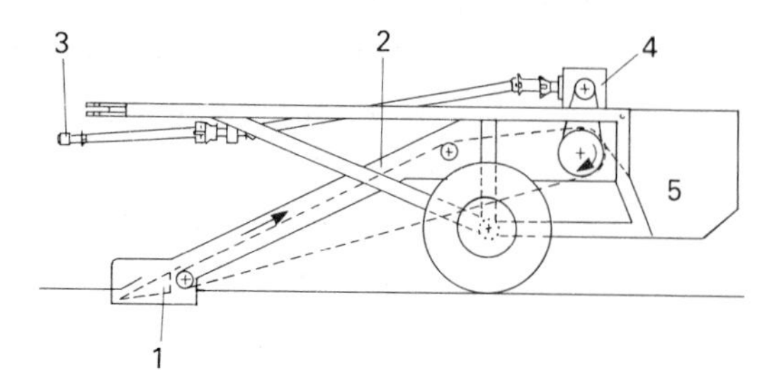

FIG.4.17 Beach scraper; (1) front collecting blade; (2) screening mesh; (3) tractor power take-off; (4) hydraulic drive; (5) bucket.

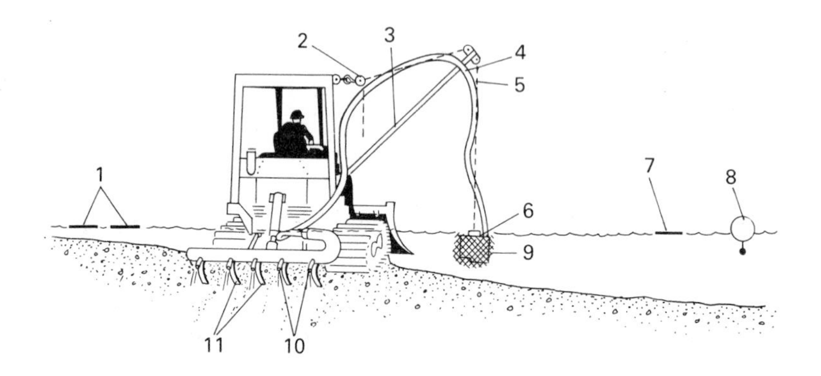

FIG.4.18 Bulldozer with multiple rippers and water jets; (1) sorbent sweeps; (2) chain hoist for adjusting pump level; (3) strut on pivot; (4) water hose; (5) cable; (6) centrifugal pump run from bulldozer hydraulics; (7) sorbent sweeps; (8) oil containment boom; (9) screen around pump; (10) water jets behind ripper teeth; (11) multiple ripper teeth.

When oil gets deeply entrapped in beach sediments, then vigorous mixing of oil contaminated sediments in the presence of water is the only method of freeing the oil. Several types of vehicles were tried out (table 4.4.), but the best type was a bulldozer with ripper assembly and water jets (fig.4.18.). The beach areas were isolated by containment booms and sorbent sweeps to facilitate oil recovery (sorbents were weighed before and after use) and to prevent escape of floating oil or recontamination of the beaches. The bulldozers were operated parallel to the shoreline in water depths 6 to 24 inches. The ripping depth typically ranged from 12 to 15 inches. Portions of the beach that were heavily oiled required more than 40 passes with a bulldozer before the visual rate oil emergence from the beach decreased to a very small amount. These segments of the beach usually required two full days of beach agitation (four full tide cycles) to be completed. Beach segments that contained relatively little oil were completed in one day using one bulldozer. The presence of cobbles up to 10 inches in diameter presented no significant difficulties. However, beach conditions prevented successful beach agitation in such areas as:

- soft sediments at the beach surface;
- proximity to piles or between closely spaced piles;
- presence of steel cables and anchor chains.

Tidal currents or wind usually concentrated oil at one end of the segmented work area. This concentration could easily be removed by rope skimmers or a vacuum truck.

TABLE 4.4 Beach Agitation Equipment.

Equipment description	Effect	EAdvantages / Disadvantages
1. Bottom jetting with barge mounted water nozzle	limited	Can be used in areas not trafficable; requires barge; jet immediate vicin.
2. Beach jetting with presurised water hose	limited	Cumbersome to move hoses; beach erosion; labour intensive
3. Rubber-tyred tractor with cultivator rake	moderate	Manoeuvrable; limited rake depth penetration; suitable for small areas
4. Bulldozer with teeth or plough bits on front blade	effective	Manoeuvrable; typical depth of agitation up to 12 inches
5. Bulldozer with hydraulicly operated ripper teeth	effective	Fast; routine operation; typical depth of agitation 12 to 18 inches
6. Bulldozer with ripper teeth and water jets	very effective	Fast; easy operation; typical depth of agitation 12 to 18 inches

4.2.6 Beach Sand Processing

When the shore is cleaned by a mechanical shovel, the oil is picked up together with a lot of sand. Transporting the oil/sand mixture in large quantities would mean pollution of the dunes and aquatic environment by leaks during transport operations to the disposal depots. Therefore sieving on the beach could be selected. The working depots can be arranged directly on the

beach. The depots should be fenced off with warning notices placed around them. In order to prevent contamination of the ground, plastic foil should be layed on the floor.
The oil/sand mixture contains about 10% of oil and the sand to be returned to the beach should not contain more than 0,5% (by weight) of oil or less if possible.

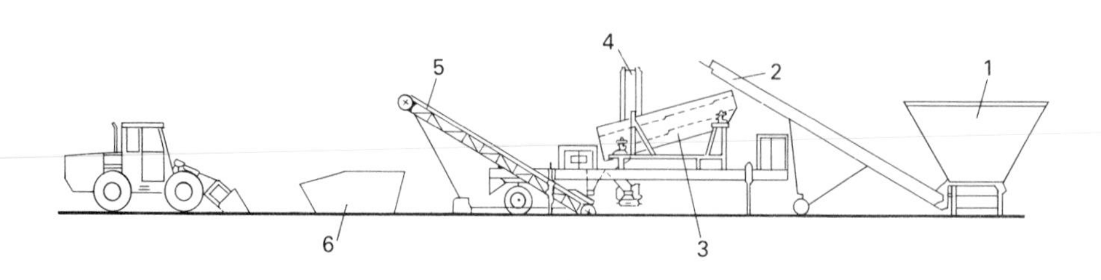

FIG. 4.19 The sieving technique; (1) doser; (2) oil/sand mixture conveyor; (3) trembling sieve; (4) sand conveyor; (5) oil lumps conveyor; (6) container.

The collected mass is poured into a doser (fig.4.19.) and conducted to a trembling sieve, which consists of an upper and lower deck with the necessary sieve holes to enable the oil/sand lumps to fall through to the lower deck. The lower deck consists of two sieves. The first section is a normal sieve with a mass area of 10 mm. A special sieve gauze has been applied to the second section with the mass area of 4 or 6 mm. By varying the amount of flow and by controlling the amplitude and frequency of the sieve sections, it is possible to control the process. The oil/sand lumps are sieved out while the clean sand falls through to the lower deck into the sand collection funnel. Using a vibrating sieve prevents oil lumps from being in permanent contact with the sieve deck. This prevents adherence and blocking of the sieve's mesh. Sieving of the collected mass totalling 30,000 m^3 took place at six different locations.
After sieving on the beach, the sand fraction was dispensed on the beach between the high and low water lines. In a few days

there was no trace of oil, which has been released into the sea. Another method was burying the sand fraction near the high water line on the wet beach. Burial was carried out with the help of dumpers or earth removal wagons, attached to tractors and on excavators. After burial the beach was levelled by bulldozers. A small amount of oil still present in the sand fraction was taken up and disposed of by the sea biologically during ebb and flow movements.

The sieved oil fraction has to be stored in containers which should be transported to a depot or storage area of the thermal rinsing installation where the processing of oil fractions takes place. In order to prevent soiling of the storage area by oil, the ground has to be covered with plastic foil or a sand layer of appropriate thickness.

The basic principle of oil contaminated sand washing techniques is on-site cleaning of sand in mobile equipment, which can be operated either close to the polluted beach or at a centralised location. Washed sand is then returned to the beach or used for other purposes. Such equipment was successfully developed in France (fig.4.20.). The contaminated sand is fed into a loading funnel and from there is taken by a belt conveyor to the horizontal drum scrubber. At the same time, the wash water is heated in a heat exchanger and cleaning agents are added. The mixture of polluted sand, warm water and cleaning products is stirred in the drum scrubber and then passed onto a 5 mm grate to eliminate bulky products such as gravel, stones and seaweed. The recovered sand is then washed again with recycled water before going into a hydrocyclone. The oil-rich liquid phase leaving the overflow, which contains fines, whose maximum diameter is defined by the adjustment of the cyclone, is carried by gravity into settling tanks. Washed sand leaving in the underflow of the cyclone drops onto the vibrating belt for the maximum elimination of interstitial water and then may be stored before being returned to the site from which it was taken.

The washing plant has an output of 18 tons of washed sand per hour. Regular sampling of sand and water is necessary. Wash-

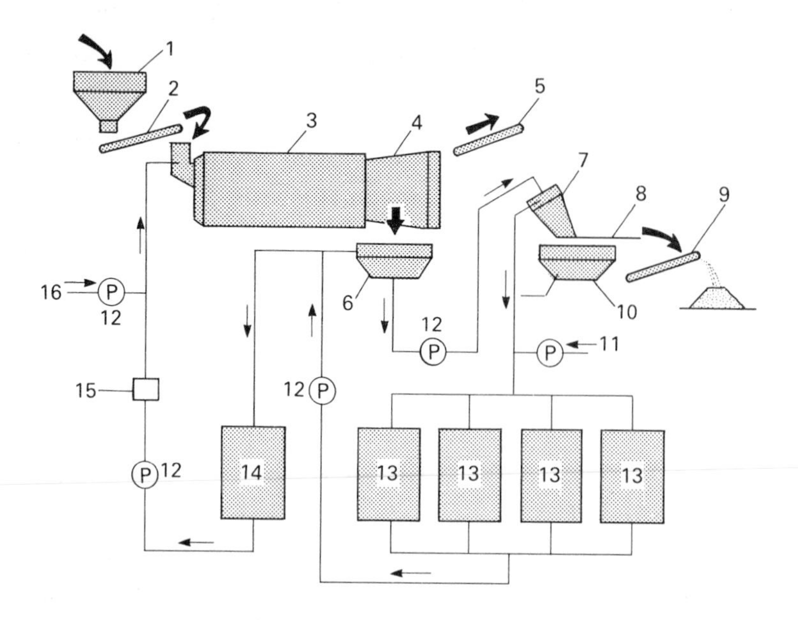

FIG. 4.20 Oil contaminated sand washing installation; (1) loading funnel; (2) belt conveyor; (3) horizontal drum srubber; (4) 5 mm separation grate; (5) gravel transporter; (6) sand washing; (7) hydrocyclone; (8) drainer; (9) sand transporter; (10) draining tank; (11) flocculating agents; (12) pump; (13) settling tanks; (14) washing water tanks; (15) heat exchanger; (16) cleaning products.

ing quality is excellent as the oil content of polluted sand was reduced with water at 30 °C from 20 to 22 g of oil per kg of sand to 1.5 g/kg for recently polluted sand and in the case of highly weathered sand to 6 g/kg. Water recycling reduces water requirement to 5 to 6 m^3/h necessary to compensate for water removed with the washed sand and with the pollutant skimmed on the surface of four settling tanks of 15 m^3 each. Washing water and flocculation products are stored in 500 litre tanks.

Sand washing has as its final purpose the return of the washed sand to its original site. The main factors likely to influence the recolonisation of the washed sand are the residual pollutant content and the particle size of deposited sand. The residual oil and any residual washing products may disturb the recolonisation of sediments. Depositing of treated sand in a pile improves the recolonisation possibilities due to natural rinsing.

4.2.7 Storage and Maintenance of Response Equipment

Response equipment may be stored for many months without being used and when the oil pollution occurs it is required to operate in all conditions for extensive periods continuously and reliably. Thus it should be stored as advised by the manufacturers to guarantee perfect condition. These requirements depend on the type of equipment, and the materials of which it is made in order to prevent corrosion and/or deterioration. Thus a dry, well ventilated storage building should be used with control of humidity, temperature and exposure to UV radiation. Stored equipment should be protected from theft and damage by pests. The store should provide a clear working area for cleaning the equipment to remove oil and salt water after use and for maintenance. Booms folded or reeled for storage should be regularly unfolded or unreeled to prevent the material sticking together or creases forming which lead to points of weakness. There should be good access to the equipment for inspection and maintenance as well as for road vehicles and lifting equipment for rapid loading of the equipment in case of an emergency. Workshop facilities for small repairs are also required.

There are several systems of maintenance:

- calendar system, under which the equipment is inspected on a fixed time schedule, i.e. monthly or annually;
- equipment working hours, where the equipment is inspected when a certain number of working hours have been reached;
- equipment break down, where equipment is inspected and repaired as defects are reported;
- condition monitoring, where readings are taken from the equipment at regular intervals to ensure it remains running within designed criteria;
- a combination of some of the above systems.

Maintenance of response equipment is a very specific undertaking as there are not many real repairs to be done, but it is mainly the control of actual condition and capability to work reliably in all circumstances. The failure of equipment results mainly from lack of use rather than from overuse. Thus inspection and control must be treated most seriously and all procedures must be carried out exactly. The role of personnel carrying out the maintenance and of the operators is of special importance. It is advisable to use single teams to deploy, test and maintain the equipment. With this system, each operator is given the responsibility for an equipment stock pile.

A computerized maintenance system can be of considerable help in planning operations and the workload of maintenance based on the full information on each item of response equipment stored in the computer.

4.3 Oil Pollution Combating Operations

4.3.1 Stages of Combating Operations

Oil pollution combating operations can be divided into four stages:

- stage 1 - protection of coastline against oil spilled at sea in the proximity of the shores;
- stage 2 - removal of heavy contamination and floating oil;
- stage 3 - clean-up of moderate contamination, stranded oil and oiled beach material;
- stage 4 - clean-up of lightly contaminated shorelines and removal of oily stains.

There is no need in every situation to go through all the above stages. In some cases oil may be left on shores to weather and to degrade naturally. The type of treatment is dependent on the nature of the coastline, kind of oil, climatic conditions and several other factors.

4.3.2 Stage 1 - Protection of coastline

After an oil spill at sea, it is necessary to determine its probable trajectory and the potential threat to the coastline and any offshore and land structures. The type of oil, its components and physical properties should be determined. If there is even a small chance that the spilled oil may reach the coast before being totally removed from the sea surface, then all preparatory work should start to prevent the spill from reaching the coastline by getting booms installed in appropriate locations and near-shore equipment brought to the vicinity of possible employment. The combating teams should be alerted and brought into a state of readiness. Certain parts of beaches should be protected by bales of straw or hay or by shallow water booms.

Of special importance is the protection of intakes of any desalination plants or cooling water of power stations or oil refineries. The operators of these plants should be notified without any delay. They have to monitor the quality of water and plant operation. The consequences of closing down these plants should be determined and alternatives should be considered. Oil getting into filters and heat exchangers may upset their normal functioning. Product water should be checked for contamination by aromatic fractions of benzene and by alkyl derivatives of benzene. Benzene has been found to be a human carcinogen. Thus its content in drinking water should not exceed 5 p.p.b. (parts per billion).

The best protection of desalination plant and power station intakes is accomplished by appropriate design and construction of the plant. Closing down the desalination plant should be possible when oil is approaching the plant's water intake filters. The consumers of desalinated water must then be informed about the situation in order to limit the consumption of water and to provide alternative sources of supply.

Intakes should be located well below the water surface. It is the standard practice to have rotating screens in front of the intake to remove trash and debris, which might be drawn in. Some plants have "subsea intakes", which are buried below the seafloor. The water must first pass through the sandy sea bottom, which acts as a natural filter to remove contaminants. The first line of coastlal defence is achieved by laying several rows of oil containment booms strung so as to deflect the oil from the shores and the intakes. For intakes below the surface, fish nets are strung over the openings to catch large tarballs and sunken oil globs. Another technique has been to surround the intake with hoses releasing air, which form a bubble screen. The rising bubbles cause vertical motion in the water column, which lifts tarballs to the surface. The tarballs are then scooped up with the net. Finally, accurate spill trajectory modelling permits the plant to close down the intakes, while the slick passes. Before using dispersants to protect the intakes of a plant, it should be verified that the currents will carry the dispersed oil away from the intakes, because a dispersed slick into the water column may make it much easier for dispersed oil to enter the intakes.

If oil enters a multiflash desalination plant it may be necessary to first use a solvent to loosen the oil particles. Next, the plant should be flushed with soap and water and finally it should be flushed with fresh water. Oil clings to heat transfer surfaces, disrupting the heat transfer processes.

4.3.3 Stage 2 - Removal of Heavy Contamination and Floating Oil

On high to moderate energy rocky shores, the removal of heavy contamination and floating oil should be carried out by means of vacuum trucks or vacuum tank trailers, skimmers and pumps. Employment of vacuum trucks requires access roads to the water edge and good load carrying capacity of the ground. Rocks protruding above the beach can be obstacles preventing the vacuum trucks from reaching the oil pools on the coastline. Many skimmers do not function well in shallow water or in waves and thus their use may be very limited. The water collected with the oil should be allowed to settle and then drained off before the oil is taken away for disposal. On tidal shores, the oil settles on the rocks and must be flushed off, forming pools on the water edge. To prevent this oil from entering the sea, it may be necessary to use floating booms or floating ropes to hold oil onto the shoreline during collection.

Where vehicles are not able to get sufficiently close to the water's edge, the oil has to be picked-up manually. This method of oil recovery is very labour intensive because it employs primitive tools such as buckets, scoops, shovels, dust bins, etc. Viscous oils are more readily removed by manual methods than lighter oils, which will require the use of sorbents for complete recovery. The most effective are synthetic materials such as polyurethane foam and polypropylene fibres. In the absence of these materials, local naturally occuring materials such as straw, hay, peat or chicken feathers could be used. The oil-sorbent mixture can be collected with forks and rakes and carried in plastic bags or small containers from collection point to temporary storage.

A shoreline consisting of cobbles, pebbles and shingle is probably most difficult to clean satisfactorily, because much oil has penetrated into the spaces between the stones, deep into the beach. The first stage of clean-up for this type of shoreline consists of pumping the liquid oil where possible or removing it by hand. The poor load-bearing characteristics of such beaches can inhibit the movement of both vehicles and personnel.

Shorelines consisting of fine-grained sands are generally regarded as a valuable amenity resource and priority is given to cleaning them. Intertidal sand flats are often biologically productive and important for commercial fisheries. Recreational beaches often have good access although on some shorelines temporary roadways have to be constructed to allow heavy equipment onto the beach, which is capable of collecting bulk oil from the beach relatively quickly, but there is danger of removing beach substrate. Excessive removal of sand may result in beach erosion. In such cases manual methods must be used.

Sand beaches are frequently unable to support any vehicle without its wheels or tracks sinking into the sand and causing oil to be pressed deep into the beach. Vehicles driven onto the beach may become immobilised once loaded. Thus demands for amphibious multipurpose vehicles for cleaning intertidal mud flats are ensue. Such vehicles should have:

- capability to swim and to drive in silty parts of the mud flats;
- very low ground pressure;
- couplings for integrated recovery and clean-up modules.

Flat hard packed beaches may support heavy vehicles such as graders and front-end loaders (fig.4.21.). The grader's blade isset

FIG.4.21 Method of using graders and front-end loaders for concentrating and collection of oil on a uniformly polluted beach; (a) front-end loader; (b) grader; (c) scheme of consecutive runs by a grader.

to skim just below the beach surface and the oil and sand are drawn into lines parallel to the shoreline. The grader works down from the top of the beach and the collected oil is picked up by front-end loaders.

An alternative method for tidal beaches is to flush oil into the trenches dug parallel to the water's edge or to the sea fenced off by a floating boom (fig.4.22.). In a similar way a sump may be dug out, the oil flushed down the beach and deflected into the hole by suitably positioned boards or bunds built of

FIG.4.22 Flushing oil off the stones into the sea fenced off by a boom for recovery by a vacuum truck; (1) water flushing; (2) oil; (3) boom; (4) vacuum hose; (5) vacuum truck.

sand. Trenches dug in clean sand just above the high water mark should be lined with plastic sheathing to minimise contamination of the sand. Oil collected in trenches or sumps can be removed using pumps, vacuum trucks or tank trailers.

4.3.4 Stage 3 - Clean-up of Moderate Contamination

Where rocky shores are cleaned from mobile oil, further cleaning can be achieved by washing with hot or cold water under pressure depending upon oil type and equipment availability. Typically water is heated to about 60 °C and sprayed at 10 to 20 litres per minute from a hand lance operating at 80 to 140 bar. On occasions, higher temperatures are needed and even steam might be required to dislodge viscous oils. This method effectively removes oils, but it also removes any living organisms which may take 5 to 10 years to reestablish a new rocky intertidal

FIG.4.23 Team of men working with front-end loader.

community. It can increase penetration of oil into surrounding sediments.

On sandy oily beaches, oily sand is best removed by teams of men using front-end loaders to transport collected material to temporary storage sites at the top of the beach (fig.4.23.). Each man in a combating team collects between 1 and 2 m^3 per day by this method. Front-end loaders can remove directly as much as 100 and 200 m^3/day/machine, but at the expense of at least three times as much clean substrate. As a rule, the oil content of sand collected by machines is between 1 and 2%, whereas that collected manually contains 5 to 10% oil. To make most efficient use of each front-end loader, the men should collect oily sand into piles deposited on plastic foil or fill 200 litre drums placed at intervals along the beach. The loaders should work from the clean side of the beach in order to prevent oil from spreading up the beach. If the oily sand has to be collected into bags, they should be relatively small, similar to the bags used for carrying fertilisers or sugar.

4.3.5 Stage 4 - Clean-up of Lightly Contaminated Shorelines

After pooled oil is removed, a coat of oil may remain strongly adhering to the seawall and rocks. Tests of several methods of cleaning the seawall should be conducted in order to find the most effective one. These would be:

- hydro-blasting with low pressure warmed sea water - this may leave stains on rocks;
- a high pressure "sugar" sand blasting - very effective on the rocks, but not as effective on concrete seawalls;
- a dry "sugar" sand blasting - very effective for remaining oil residue and marginally faster than hydro-sand blasting, but very abrasive to the concrete walls and less effective on the rocks than hydro-sand blasting;
- use of dispersants - negative results as the dispersants are unable to penetrate the thick oil coating to provide the desired dispersion action. No environmental damage was observed as the result of using the dispersant.

Based on the above tests hydro-sand blasting would be selected as the prime clean-up method on the seawall. The use of dispersants is not advisable on cobble and pebble beaches, because it tends to carry the oil further into the substrate.An exception may be oil of very low viscosity. The greasy film remaining on the stones after cleaning with high pressure water could be removed by pushing the top layer of stones into the sea, where the abrasive action of waves will rapidly clean the surface of stones. This could not be done if the bottom layer of stones is also stained by oil. The rebuilding of the beach profile might take several years.

After removing the top layer of sand from the beach, the remaining substrate is likely to have a greasy texture and may be discoloured. This is unacceptable for recreational beaches. In such cases small quantities of dispersant could be applied, allowing it to contact the substrate for 30 minutes before being washed by the incoming tide. In non-tidal regions or in areas without strong surf action, hosing may be necessary to achieve good dispersion.

Tidal beaches could be repeatedly ploughed or harrowed at low water. Thus the oil is mixed with a greater volume of sand and more frequently exposed to weathering processes. The oil remaining after the clean-up of dry sand beaches forms nodules of oily sand up to 50 mm in diameter. These nodules and tarballs are washed up along the high water mark and could be picked up by conventional beach cleaning machines, which remove the top surface of the beach to a preset depth and pass the sand through a series of vibrating or rotating screens. Clean sand is allowed to drop back onto the beach, while nodules and tarballs are retained.

In order to restore the beach to its original use in the shortest possible time, clean sand could be brought in and spread over any lightly oiled sand. This sand should have the same grain size as the original sand so it behaves in a similar way. Finer grained sand would be quickly washed away. When sufficient notice is available before the spill reaches the beach, it may be possible to move some of the sand above the high water mark. This material can be replaced after the beach has been cleaned.

4.4 Oil Combating in Ice

4.4.1 Behaviour of Oil in Ice

Crude oil of specific gravity about 0.88 would fill an open crack in floating sea ice of specific gravity about 0.92 to a depth of about two-thirds the thickness of the ice, after which any additional oil flow would spread out over the upper edges of the ice sheet. A lighter oil would penetrate the crack to an even shallower depth (approximately one-third the ice thickness for a specific gravity of oil of 0.7) before reaching equilibrium and spreading out over adjacent ice floes. Oil could be forced beneath the ice, if the ice near the crack/lead is grounded and sufficient volumes of oil accumulated over the opening in the ice.

In warmer waters crude oils spread to a fairly uniform equilibrium thickness of the order of a few hundredths of a milimetre. In cold waters such equilibrium conditions are commonly achieved in thicknesses of a few milimetres. The thickness of the oil layer in cracks and leads would be increased further by the natural containment of the surrounding ice.

Natural under-ice surface roughness will again provide massive oil accumulation points. These irregularities will strongly depend on irregularities in snow cover of the ice. Oil spilled beneath ice will not be transported by currents until the water movement relative to ice is in the range of 0.15 to 0.25 m/s.

4.4.2 Detection of Oil under Ice

The detection and location of an oil spill are essential in determining the nature and extent of any response necessary to deal with that spill. The detection of oil within or beneath the ice could involve direct sampling with ice augers or corers, the use of underwater inspection devices or the use of surface operated equipment involving impulse radar and various acoustic techniques. A promising method of detecting oil under ice involves the use of induced fluorescence. Crude oil fluoresces strongly in the visible spectrum when it is illuminated with ultraviolet light. A compact, portable pulsed ultraviolet source similar to

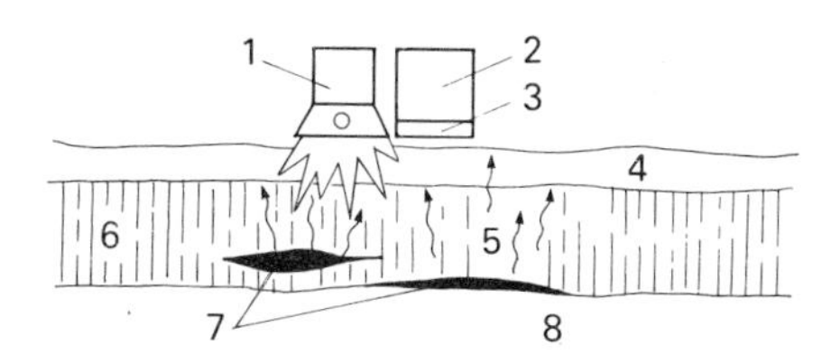

FIG.4.24 Detection of oil by induced fluorescence; (1) UV strobe; (2) photodetector; (3) optical filter; (4) snow; (5) fluorescence; (6) ice; (7) oil; (8) water.

a photographic strobe light might be used to flash an intense beam of utraviolet light onto a small area of subsurface of the sea ice (fig.4.24.). The light would need to be sufficiently intense to penetrate snow cover and ice. Oil excited by a suitable wavelength of ultraviolet light fluoresces strongly at a much longer wavelength in the visible spectrum. A photodetector mounted adjacent to to the ultraviolet light source should detect fluorescence from any oil present.

4.4.3 Behaviour of Oil under Ice

Oil tends to fill the natural subsurface pits and troughs commonly found beneath stable snow drifts associated with prevailing winds over an open ice sheet. Reduced ice growth beneath such oil-filled depressions would then exaggerate the differences in ice thicknesses around the oil providing an easily detected anomaly in the overall ice configuration. Slots or holes cut in ice and creation of subsurface troughs using subsurface insulation could be prepared to intercept a potentially advancing oil mass, thereby providing early detection of its presence as well as a point for recovery of the oil (fig.4.25.).

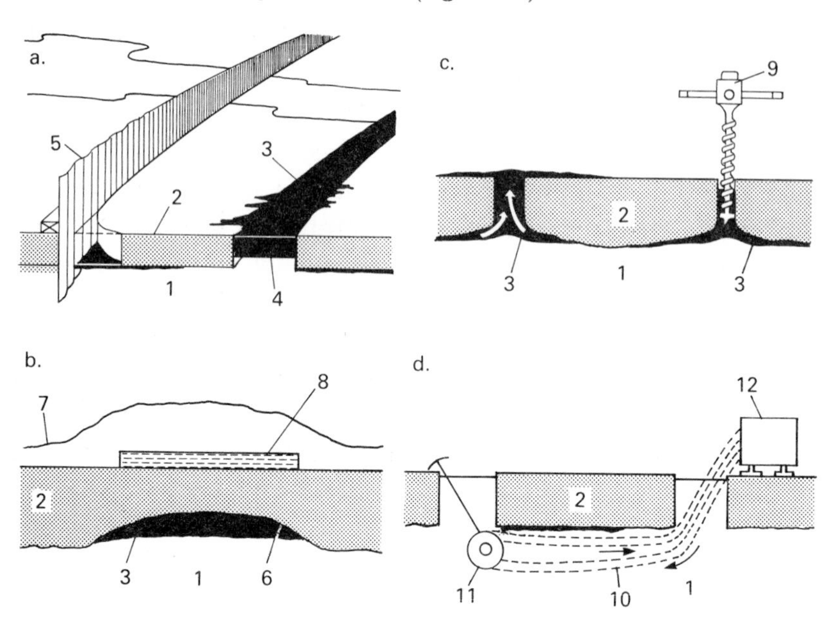

FIG.4.25 Oil behaviour and detection under ice; (a) ice slots and through-ice barriers; (b) development of subsurface trough using insulation at surface; (c) drilling through ice into subsurface pools; (d) rope mop skimmer beneath ice; (1) water; (2) ice; (3) oil; (4) slot; (5) through-ice barrier; (6) trough; (7) snow; (8) synthetic insulation; (9) ice auger; (10) rope mop; (11) pulley; (12) rope mop skimmer.

Crushed ice is capable of storing great quantities of oil in cavities and cracks. Oil contained in ice is relatively insensitive to flowing water. Spreading of oil in crushed ice is small and depends upon viscosity, block sizes, ice quality, etc. Blasting could be used to form crushed ice with a high "absorption" ability, which is the most efficient method for this purpose. Open gaps with only a few percent ice covering are made with underlying charges, where the lifting effect of the explosives is exploited. The charge size could be specified for different ice thicknesses and gap sizes. The spread and throw of ice and oil can be predicted. The size and position of charges when making gaps or fields can be determined as a function of ice thickness, charge type and its position below the water surface.

Surface charges give a downward directed blasting effect without lifting. Placing of charges in sawn grooves with or without overburden is of importance for efficiency. A few centimetres of snow packed on top of the charge can considerably amplify the blasting effect. The result of blast is largely the decomposition of the ice covering. This broken ice with stored oil can then be lifted with buckets into a container or other collection device.

In flowing water with oil spills, downstream collection gaps of sufficient size for oil collection could be prepared in advance. Line gaps could be placed perpendicular or at a certain angle to the flow direction. The edge of the ice should be initiated by making a cut with a motor saw before blasting and clearly marked after blasting. These cuts constitute an indication of fracture and serve as an obstacle for further propagation of cracks. The edge of the gap should be strengthened by reinforcement and thickening of the ice cover in order to be capable of supporting the load from heavy machines.

4.4.4 Surface Collecting Agents

The potential for effective application of collectants in cold climates due to the low temperatures of air and water is not considerably reduced. Obvious constraints exist relating to various product pour points and to the storage/pumping requirements prior to application. The collectants possess reasonable concentrating properties over a wide range of temperatures and water salinities. They could be used even in the presence of ice to drive thin and inaccessible oil layers on water into thick layers that are recoverable or combustible.

4.4.5 Use of Dispersants

In cold climatic conditions, mechanical collection of oil is sometimes very difficult, if not impossible, and chemical dispersants may be used to decrease the overall impact of the spill. Although this eliminates the hazards to surface dwellers (sea birds, mammals) and to shorelines, it increases the threat to the water column and benthic resources.

There are two Arctic phenomena which can reduce the exposure of subtidal organisms to surface-dispersed oil. Commonly, there is a barren or impoverished zone of animals between 2 and 10 m depths. The usual presence of a low salinity surface water layer would suppress the downward mixing of a surface-dispersed oil slick below this zone. The sediments containing the highest oil levels were caused by untreated oil stranding on low energy beaches. After two years of exposure the oil contents were not only an order of magnitude higher than those produced by chemically dispersed oil, but they also constituted a much longer term event and therefore had a greater potential for impact.

4.5 Disposal of Oil and Oily Debris

4.5.1 Purpose of Disposal

The task of oil spill clean-up can not be completed until the recovered material is safely and permanently disposed of. Insufficient attention is given to such problems as means of dealing with oily waste by recycling, destruction and dumping. Generally the responsibility for the ultimate disposal remains with local waste disposal authorities.

As a general rule, spills of persistent oils such as crude, heavier grades of fuel oil and some lubricating oils are likely to require disposal since clean-up of non-persistent oils is not usually necessary. Oil collected from the water will probably be free of solid debris, but it is likely to contain large amounts of water present as an emulsion. But oil stranded on beaches will contain considerable amounts of solids, which are generally difficult to separate, such as sand, wood, plastics, seaweed and solid tarballs.

Ideally as much of the collected oil as possible should be processed through a refinery or oil recycling plant. This is, however, rarely possible due to weathering of the oil and contamination with debris. Thus some other form of disposal is usually required. This includes direct dumping, stabilisation for use in land reclamation or construction of minor roads and destruction through natural processes or burning (fig.4.26). The final choice will depend upon the amount and type of oil and debris, the location of the spill, environmental and legal considerations and cost estimates.

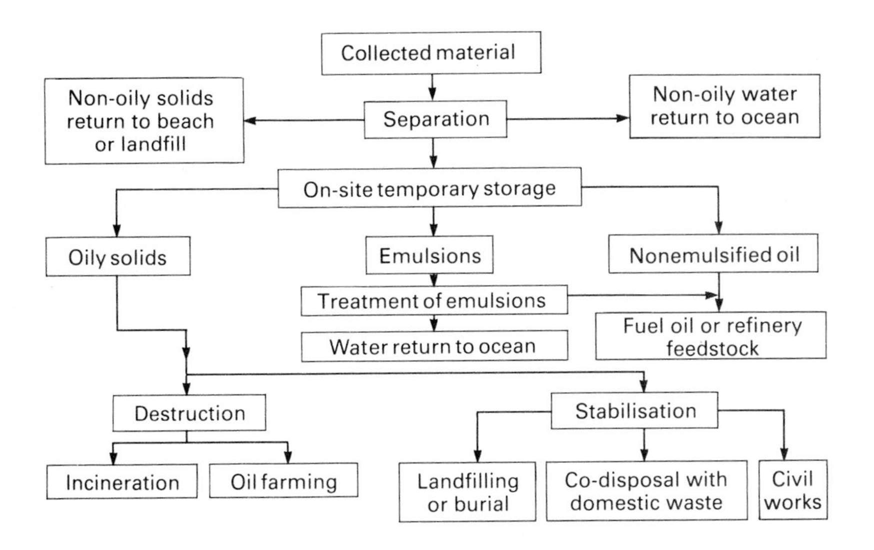

FIG.4.26 Diagram for final disposal of collected material.

Collected material must be transported to reception facilities, which consist of:

- an unloading site;
- a temporary storage site;
- a treatment plant, including incineration.

The lay-out and size of these facilities will depend on the size of an expected spill, the mode of transportation of oil residues, the spillage site and the form of the residues. The transport systems should be capable of dealing with both pumpable and nonpumpable oil. Transport should be possible by road and by sea. Individual facilities can be situated in one locality or they can be distributed over a fairly large area depending on individual possibilities. Cleaning facilities for personnel, equipment and vehicles should be established in order to ensure that the pollution does not spread to public roads and personnel accommodation. For final treatment of collected oil and oily debris, different methods are used. Thus they should be separated and temporarily stored apart in different disposal sites from the beginning of the operation.

4.5.2 Unloading Sites

The unloading site is the place where unloading, emptying, transfer and temporary deposition or storage take place. In the case of oil transported by sea, the unloading site is located at a quay approachable by barge, ship or boat. In the case of land transport, the unloading site is located in a suitable spot adjoining the road. The necessary equipment at the unloading site will consist of:

- for pumpable oil residues carried by sea in tankers or soft containers:
  - mobile receptacle or tank volumes of sufficient storage capacity. In successful combating operations quantities up to 400 m^3 may arrive in barges. In the worst case, unloading could be done temporarily to open light containers, which subsequently are emptied into other receptacles or tank trucks;
  - a pump or pumps with a total capacity of 200 to 400 m^3/h. The pumps should be fitted with a loose hose and with a suitable hose connection in the suction end, which will be connected to the tank or to the collection receptacles. If pumpable residues are delivered in open containers, these should be emptied with a suction device such as a sludge exhauster.
- for solid and pumpable oil residues carried by sea in containers, arriving at a rate of up to 100 containers per day with a volume of 5 m^3 depending on the mode of transport (1 to 2 in case of boat transport, 8 when carried on barges and 20 to 30 when carried by a vehicle ferry):

  - mobile crane;
  - place for temporary deposition of empty containers, drums etc. until they can be driven off the storage site or back to sea;
- for pumpable oil transported by road:
  - sludge exhausters for direct pumping to loose receptacles, liquid containers, railway tank cars. In the case of pumping to a tank, a pump (approx. 100 m^3/h) with a higher delivery head may be required;
  - loose receptacles or liquid containers and drums.
- for solid, land transported oil residues:
  - crane equipment for unloading;
  - site for storage until solid residues can be conveyed elsewhere.

While handling oil residues such as unloading, emptying, transferring, etc. there are generally high risks of oil being spilled out onto the ground. If the ground is not sufficiently dense, the dispersion can be reduced by spreading out sorption agents, limestone meal etc. Contamination of ground water should be avoided.

4.5.3 Temporary Storage Sites

A temporary storage site should be prepared when the size of the oil spill is largely known. At this juncture, either existing facilities are rented or new facilities should be arranged. For pumpable oil wastes, tanks should be used, but for nonpumpable oil wastes open pits should be built.

A temporary storage yard can be planned and fitted out in several different ways. The storage site can consist of an accommodation yard with a hardened surface, which should be dimensioned for traffic with 10 trucks (axle loading 10 tons, bogie loading 16 tons). The area should be fenced in, possibly with a movable fence and guarded. Decontamination agents should be available. It should be well lightened and possess offices and rest rooms as well as cleaning facilities for personnel.

Alternatively, the temporary storage facility can consist of a goods shunt yard with side tracks for railway tank cars.

A third solution is a permanent tank farm for storage of pumpable oil. This must be approved for storage of the oil product concerned. At the tank farm, a pump line must be provided from the unloading station possibly in the form of a temporary line or hose. At the permanent tank farm there should be provisions for heating, if there is a risk of the oil becoming highly viscous.

The time for which temporary sites will be required depends on the treatment capacity, but a storage period of 2 to 12 months duration should be provided for. The temporary storage site may need to be operational within one or two days after an alarm.

When considering locations for temporary storage sites, the possibility of regional cooperation between several local authorities should be borne in mind. It might be necessary in larger local self-governement districts to have several sites, on which intermediate storage sites can rapidly be built up.

During the planning stage, a search should be made for a suitable site, where the longest distance between a possible contamination site and a storage site is approximately 50 km. The pits should be located at places with suitable geohydrological prerequisites (dense ground substratum, low ground water level), insensitive surroundings and possibilities of checking the surface and ground water. Storage site should never be placed in a marsh or swamp area. Municipal garbage dumps and rich clay fields could be used for this purpose. The disposal site must also have good road connections.

Drawings should be produced describing how the disposal site is to be constructed, how much work is to be done and what machinery is to be used in order to build it in a given time. Before the pit plan becomes definitive, consultations should be held with local county councils and final approval should be reached by local authorities.

Several small pits of 600 to 700 m^3 can be constructed on one storage site. The pits are made with tight bottoms and sides using plastic foil or rubber fabric, a dense layer of soil or betonite-sealed soil. If the pit is to be sealed with plastic foil or rubber fabric, a layer of fine-grained sand should be placed beneath, so that it rests on a smooth underlying surface. The plastic foil should have fixed locking devices, which are attached to the mound of the pit so that the edges of the plastic foil are prevented from blowing away.

If the pit is sealed with betonite, a layer of soil with a high sand content and low clay content is first applied. The soil is saturated with water by irrigation. A layer of a suitable grade of betonite is spread over the moist soil. The normal dosage is 10 to 50 kg/m^2. The betonite is worked into the top soil layer to a depth of 50 to 100 mm with a cultivator, harrow or suitable implement. The soil layer is compacted and finally covered with a protective layer of ordinary soil or sand (0.25 to 0.5 m), which should be watered.

Around each pit a mound is built to divert rain water. Space is needed around the pit to enable the waste to be dumped into the pit by lifter-dumper trucks. Special movable tipping ramps are required in order to distribute the load from the vehicle, when tipping the waste into the disposal site.

The width of the pit is determined by the accessibility for an excavator, drag line bucket or wheel loader with extended arms. Thus its width should be less than 10 m, if the pit is accessible from both sides. If plastic foil is used, its width may limit the width of the pit. Its length is not limited, but it is not advisable to have the pit excessively long in order to limit its volume. The pit should reach the ground, which is moistened by the ground water, as the oil leaking through the tarpaulin or plastic sheet moves very slowly through wet sand. If the ground under the pit is dry, any oil leak will easily penetrate it and contaminate the ground water. Thus the pit depth should not exceed 2 m. Its bottom and sides should be soaked with a large quantity of water before covering it with plastic sheet, which should be protected by a layer of sand or gravel (fig.4.27.).

The full capacity of the pit should never be used because of the risk of overflow in the case of a heavy rainfall. A drain pipe with

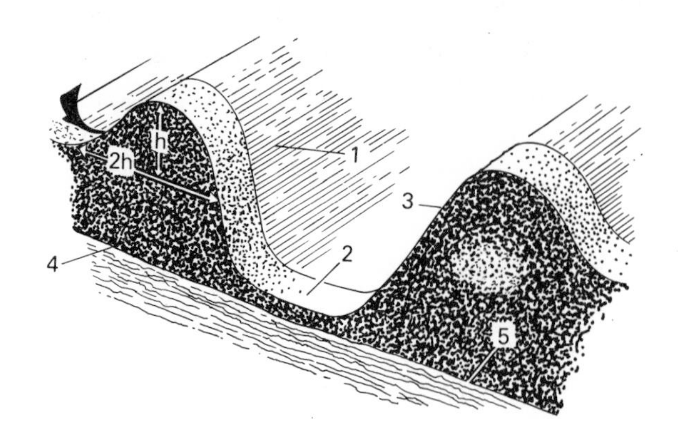

FIG.4.27 Partially buried pit for storage of liquid oil and debris; (1) tarpaulin or plastic sheething; (2) sand; (3) mound; (4) soil; (5) ground water.

a valve should be fitted at the bottom of the pit for draining rain and seawater. A tube for taking samples of ground water should be provided upstream as a reference point and downstream within the area. Similarly, a surface water drain should be arranged. During use, samples of ground water and surface water are taken at regular intervals and analysed with respect to oil.

The pit is gradually filled with solid, nonpumpable oil residues. It is of great importance to keep different grades of oil separate and place them in different pits:

- oil residues with a relatively high oil content, oil lumps;
- oily loose material;
- oil lumps and oily material in plastic bags.

Tipping into a pit is done with the aid of a mobile crane, crane truck, or lifter-dumper truck as needed. The materials are evened out in the pit with the aid of a dragline bucket or a load-tractor with an extended arm.

The oil is drained from collected material and is collected in the bottom of the pit. The solid material could be removed from the pit by means of a dragline bucket and loaded into containers or dumper trucks for conveying to a treatment facility. After digging up, the pit is cleaned in a suitable manner and any plastic material must be collected very carefully. The site must be restored to its original condition. Sampling of ground and surface water may need to be done for a certain length of time after restoration of the site.

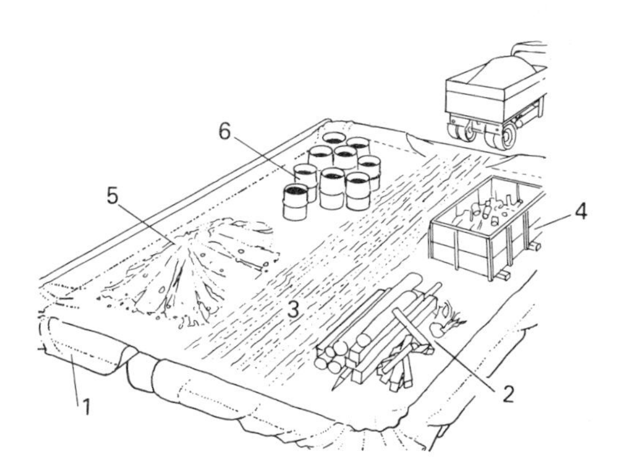

FIG.4.28 Temporary storage for oil contaminated wastes; (1) tarpaulin or plastic sheet over a bank of soil or sand; (2) wreckage; (3) sand layer to protect the coverage; (4) container for rubbish; (5) contaminated beach material; (6) drums for liquid oil from separator.

Temporary storage of pumpable oil waste and oily water could be done in tanks suitable for this purpose. Approval should be obtained during the planning period. The total volume of the tanks should be in accordance with the expected size of the spill, but not less than 2500 m^3. Tanks should be insulated with heating coils and the heat source should be secured. Reception and discharge lines to a harbour or reception site and to railway side line are of great advantage. Access to oil separators is required. The tank depot should be rented as soon as the need has been established and taken over within 24 hours.

Oil contaminated solid debris such as wreckage, shingles, pebbles, sand as well as drums, containers, etc. may be placed on a lane surface protected by a bank of soil or sand and covered with a tarpaulin or plastic sheet. The area of traffic should be protected by a layer of sand (fig.4.28).

Plastic bags should be regarded as a means of transportation of oily material rather than storage, because they tend to deteriorate in sunlight, releasing their contents. After being emptied, the bags will have to be disposed of separately.

4.5.4 Disposal Options

Some of the recovered oil might be suitable for eventual processing or blending with fuel oils. Potential recipients of such oils are refineries, contractors specialising in recycling waste oils, power stations, cement and brick works. But the recovered oil must meet narrow specifications, i.e. the oil must be pumpable, have a salt content below 0.1% for processing through a refinery or less than 0.5% for blending into oil fuel. Small pieces of debris can be removed by passing oil through wire mesh screens.

Oil collected from water usually contains a considerable amount of water, which should be separated. This separation is achieved by gravity in any collection device, the water being run-off or pumped from the bottom of the tank. The extraction of water from water-in-oil emulsions is more difficult. Unstable emulsions can be broken up by heating to a temperature of 60 to 66 °C, with a maximum not exceeding 80 °C and allowing the oil and water to separate by gravity. More stable emulsions require the use of demulsifying chemicals (0.1 to 0.5% of the bulk volume to be treated), which also tend to reduce the viscosity of most oils rendering them more pumpable. The best type of chemical agent and optimal dose rate should be experimentally established by trials on site. The treatment is best carried out during the transfer of emulsions from the collecting device to a tank or from one tank to another to ensure good mixing. After separation, the water phase will contain most of the emulsion breaker and up to 0.1% of oil. Thus care should be taken in its disposal.

Stabilisation of oily wastes by adding inorganic substances such as quicklime (calcium oxide), cement, pulverised fuel ash, etc. produces an inert product, which does not allow the oil to leak out. The stabilised material could be used for land reclamation and road construction of restricted load carrying capacity. The amount of binding agent required is dependent mainly on the water content of the waste rather than the amount of oil and is best determined experimentally on site. For quicklime the amount required is 5 to 20% by weight of the bulk material to be treated. The binding agents could be used with the oily sand immediately after its collection. This mixture is a clean, easily handled material, which can be readily transported and stored.

The disposal of tarballs and similar aggregates of weathered oil mixed with sand can cause many problems, but experiments where the sand/oil mixture was heated to the required temperature in a mixing vessel showed good results. The final mixture can be used as a foundation for tertiary roads.

On occasions it may be preferable to carry out primary mixing in pits at the site of the spill to render the oiled material more suitable for transport. The final treatment can then be undertaken at a larger reception facility using specialised equipment.

Oily wastes could be used for landfilling by agreement with local authorities. Sites should be located well away from fissured or porous strata to avoid the risk of contamination of ground water. Disused quarries or mines are best suited for this purpose. Only dry or pasty oily wastes with a relatively low content of oil mixed with binding agents or untreated solids (not exceeding 3% oil) could be used for landfilling. Spreading over a site a few metres above general ground level is preferable to filling holes or depressions. The deposit must be covered with an impermeable layer and then with a thick layer of soil (fig.4.29.). Test wells around the deposit should be established to make it possible at intervals to analyse whether oil dispersion is occurring.

The disposal of oily wastes containing up to 20% oil along with domestic wastes is often acceptable, although the degradation of oil due to lack of oxygen will be rather slow. Domestic waste easily absorbs oil with little tendency of the oil to leach out. The oily waste should be deposited on top of at least 4 m domestic waste either in 0.1 m thick subsurface strips or in 0.5 m deep slit trenches to allow free drainage of water. The oily material should be covered by a 2 m layer of domestic refuse to prevent emergence of oil to the surface. The quantity of oil should not exceed 1.5% of the total volume of the site.

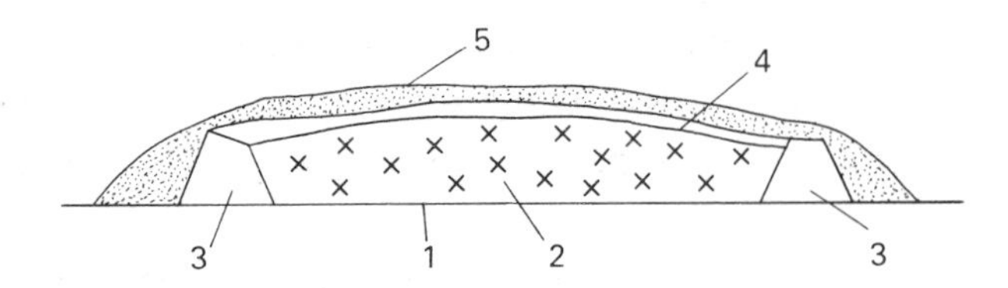

FIG.4.29 Landfilling of oily wastes treated with binding agents with less than 3% oil; 1 - ground; 2 - waste; 3 - waterproof bund wall; 4 - impervious blanket; 5 - soil blanket.

In the case of light contamination of sandy beaches by oil, oily debris or tar balls, it is permissible to bury the collected material at the back of the beach well above high water mark, provided there is no risk of damage to vegetation. A covering of at least 1.0 m should be sufficient.

If it is not possible to reclaim oil or to use it after some kind of stabilisation, it may be necessary to destroy oil and oily debris. Two methods are recommended:

- biodegradation;
- incineration.

In biodegradation, advantage is taken of the decomposition of oil by microorganisms. The rate of degradation depends upon temperature and availability of oxygen and appropriate nutrients containing nitrogen and phosphorus. Some oil components such as resins and asphaltenes are resistant to degradation and even after prolonged periods, up to 20% of the original material may be left unaffected. In order to enhance the degradation, some products containing oil degrading bacteria and other microorganisms as well as nutrients could be added.

In "oil farming", the oil and debris are distributed over land specially prepared for this purpose by loosening the topsoil by means of a harrow or ripper and encircling it by diversion channels. The soil pH should be adjusted with lime to a value higher than 6.5 to provide a suitable environment for microbial growth. Fertilizers, such as urea, ammonium phosphate, etc. may be added in quantities of 10 parts of nitrogen to 1 part of phosphorus per 100 parts of oil. The use of natural sorbents such as straw and bark during the clean-up are preferable to synthetic materials since they break down more rapidly.

The oil debris is spread out evenly over the surface in a 20 to 100 mm thick layer at a rate of about 100 tons/ha and allowed, if possible, to weather until it no longer appears wet and sticky. After the weathering period, the debris should be thoroughly mixed into the soil with a plough or a rotarator. If the area is frequented by birds, the ploughing should take place immediately. This mixing should be repeated at increasing intervals starting with 4 to 6 weeks to seasonally after 2 years. Frequent mixing increases the aeration and hence the rate of decomposition.

After the degradation of most of the oil, the soil should be capable of supporting a wide variety of plants including trees and grass. If crops are grown, they should be carefully monitored for excessive heavy metal contents.

"Composting" techniques could be employed, particularly if natural sorbents such as straw, peat and bark have been used. If the mixtures contain relatively low levels of oil, they can be stacked into heaps, which retain heat. Thus this method is particularly suitable in colder climates, where degradation through "oil farming" is slow.

Incineration through direct burning of oily debris is not recommended due to the resulting atmospheric pollution and spreading the absorption of oil into the ground. A tarry residue remains on the ground due to incomplete combustion. Thus special incinerators must be used. Some of these are described in chapter 2.

4.5.5 Disposal Techniques for Arctic Oil Spill Response

Disposing of oiled debris from Arctic spills presents problems not encountered in temperate regions. The remoteness of potential spill sites, the wide range of environmental conditions, the lack of support facilities such as roads and dump sites, ice covered water for significant periods of the year and the presence of permafrost make it impossible to use many of the standard disposal techniques used in moderate climates. Permafrost makes the southern practice of burial impossible for most Arctic spills because disturbance of the permafrost layers at a disposal site may be a more important problem than the disposal of the spill.

The hardware for the disposal of oil must conform to the following criteria:

- air-portability by helicopters;
- ease of assembly and minimum of support equipment;
- simple construction;
- durability;
- reliability and minimum of operation.

Offshore activities may cause a blow-out, releasing oil and gas under the winter ice cover until the well flow is stopped by natural bridging or relief well drilling. Conventional clean-up and disposal techniques would generally not be very useful under these circumstances. The gas and oil from a blow out stay under the ice and within a short time become encapsuled by a thin layer of new ice. They remain trapped in this manner until break-up of the ice cover the following spring. In early June, oil pools start to appear on the ice surface. This is the result of both ablation (ice above the entrapped oil melting down to expose oil) and migration (movement of the trapped oil through ice brine channels). Once most of the oil has pooled on the surface of the ice, the pools could be ignited using air-deployable igniters brought by helicopters to the site. About 30% of the oil could be disposed of by burning, about 30% is lost through evaporation and 18% is cleaned up manually. This leaves about 22% of the oil volume to disperse naturally into the water column during the break-up.

One of the problems encountered during any major spill is the disposal of recovered oil. Depending on the characteristics of the oil originally spilled, the recovered fluid can range from light fuel oil to heavily weathered residual oil ranging from one to several thousand centipoise and containing large volumes of water. Thus the burners must have:

- ability to atomise and burn fluid oil with viscosities of 400 cP and emulsions containing up to 60% water by volume;
- ability to deal with suspended solids in the fluids;

- high combustion efficiency with smokeless burn;
- selfchecking propane igniter and automatic ignition and shut-off equipment.

Air-portable incinerators serve for disposing of oil soaked debris in an environmentally accepted manner.

Burning of oil in Arctic conditions requires a fireproof boom corresponding with the following criteria:

- ability to survive long-term exposure to heat generated by burning crude oil(about 900 °C);
- ability to contain and burn oil in sea state 2 to 3 °B and current speeds up to 0.4 m/s;
- ability to withstand contact with small ice features;
- ability to survive for long periods at sea.

Such fireresistant booms are described in Chapter 3.

CHAPTER 5

Limitation of Spills and Contingency Planning

5.1 Limitation of Spills

5.1.1 Types of Emergency Situations

The types of emergency situations with which the response teams are expected to deal are as follows:

- a fixed point of discharge, with a single release of oil, i.e. a tanker with one or more of its tanks breached due to collision or grounding releasing a limited amount of oil over a short space of time. Rapid deployment offshore and close to the shore will minimise beach pollution, will prevent oil spreading over large areas and will reduce the likelihood of very viscous water-in-oil emulsions being produced;
- a fixed point source with limited oil release, but discharge continuing for days or weeks;
- a fixed point source discharging large quantities of oil over a long period of time, i.e. an oil well blow-out;
- a moving, continuous point source of discharge, i.e. a damaged tanker proceeding slowly under her own power and steadily losing oil;
- oil approaching the shore from any source.

The above emergency situations occur with different frequencies. Hydrocarbon releases have been determined for the following groups of causes:

- shipping, including transfer operations such as loading and unloading of tankers and bunkering operations;

- riverside and shore-based tank storage damage;
- pipe line damage;
- blow-outs in off-shore operations.

Observations have shown that by far the most frequent sources of release are incidents during loading and unloading operations. Generally the volume of oil released by these causes is limited. The biggest volumes of oil are released by shipping accidents and off-shore blow-outs. The consequences of oil releases are proportional to the frequency of releases, the probability that the release will reach a given target and the value of the target and its production. The overall results of risk calculations should give an overall priority index.

For an effective recovery operation there are basically two operations available:

- to attack the spill at its source. If it is possible to bring equipment to bear close to the point of discharge, there is a reasonable chance of effective containment and subsequent recovery of the oil;
- to defend sensitive areas from the effects of spillage by use of booms.

Any intermediate action can result in endless hours of chasing windrows of oil across vast areas of the sea, giving very limited recovery of oil at great expense. In considering such a diversity of potential situations, occurring under a variety of sea and weather conditions, it is obvious that a wide range of options exist with respect to types of booms, skimmers and various oil pollution response equipment for a particular case. No single method could be adopted for all spills, but each case has to be considered individually. The quantity of deployed equipment should be more than sufficient in order to get satisfactory results.

All the equipment should be on a planned maintenance schedule and frequent training exercises should ensure that the operators are fully conversant with the characteristics of each piece of equipment. A number of training courses should be held each year.

5.1.2 Prevention and Minimisation of Oil Spills

Improvement in prevention of accidental oil pollution can be achieved by introducing far reaching measures in the following areas:

- port inspection in order to enforce the international requirements for construction and equipment as well as the use of seagoing vessels;
- control over maritime navigation along the coast with the obligation to report if a ship carrying petroleum products is damaged at a distance less than 50 miles from the coast line;
- traffic separation schemes to push back ships further away from the coastline;
- high penalties for maritime traffic violation.

The necessary means for applying these preventive measures must be provided. Coordination centres should be established to coordinate sea rescue operations, insure the monitoring of maritime navigation and monitor pollution. There should be optical monitoring and radar coverage of the coastline. There must be full readiness to bring efficient help at any time to vessels in distress.

The International Convention Relating to Intervention on the High Seas in Cases of Oil Pollution Casualties, 1969 and the Protocol Relating to Intervention on the High Seas in Cases of Marine Pollution by Substances other than Oil, 1973 were prepared to give cognisance to the need to protect the interests of coastal states against the grave consequences of a maritime accident resulting in the danger of oil pollution of the sea and shorelines. It was recognised that under such circumstances measures of an exceptional character to protect such interests might be necessary on the high seas and that these measures do not affect the principle of freedom of the high seas. No such measures may be taken against warships or other ships owned and operated by a state and used only on government non-commercial service.

When an accident occurs in national waters, the Administration or a relevant Government Authority has the right to intervene, if the situation was assessed by the Authority and it was concluded

that the nature or the degree of action taken by the shipowner or his agents is not satisfactory. The Authority can either issue instructions or advice to the shipowner and his agents as to how they should proceed or take direct operational control through an intervention team sent aboard a tanker in distress.
Most important is the problem of towing the disabled vessel. Thus tugs of appropriate size and power (Table 5.1.) with sufficiently experienced crews should be provided. Towing exercises should be carried out regularly in different weather conditions.

TABLE 5.1 Resistance to tow in still weather conditions.

Angle of yaw (°)	Resistance of a tanker to a tow (tons) with the speed through water							
	2 knots	3 knots	2 knots	3 knots	2 knots	3 knots	2 knots	3 knots
	24,000 tdw		68,000 tdw		112,000 tdw		260,000 tdw	
0	2.3	4.3	4.7	8.5	6.2	11.3	7.9	14.0
10	5.5	11.4	10.8	25.5	14.8	30.5	15.2	30.0
20	8.7	18.7	17.3	37.0	23.4	50.2	39.6	85.0
30	13.0	28.3	25.8	56.0	35.0	76.5	51.8	112.7

Evaluation and intervention teams consisting of experienced personnel competent in various shipboard functions should be capable of being placed aboard a damaged vessel by helicopter. Their task is to evaluate the danger aboard a ship, to inform the appropriate Authorities and eventually to undertake any necessary immediate action on board, such as preparation for towing, short term repairs, preparation for evacuation, fire fighting, etc.
France, having had the experience of "Amoco Cadiz" and several other serious accidents, has introduced a rigorous prevention policy relying on European and international legislation, both reinforced by national measures:

- inspection procedures have been strengthened in order to enforce the rules for construction, equipment and use of seagoing vessels and to verify whether the vessel conforms to the safety legislation of international conventions, notably the MARPOL 73/78 convention;
- information available to the Authorities regarding maritime navigation along French coasts has been improved. Any tanker carrying petroleum products and entering French terri-

torial waters, or when damaged at less than 50 nautical miles from French coastline must identify herself;

- vessels carrying potential pollutants must stay at a greater distance from the French coastline. Navigating within 7 miles off the coast is forbidden, except in specially listed channels and estuaries, which give access to the ports. The traffic separation scheme off Ushant was modified by IMO in 1979 in order to push back the traffic of hydrocarbons and dangerous substances to waters further than 27 miles from the coast;
- penalties for maritime traffic violations and for oil spill pollution accidents have been strengthened.

The actions to limit the pollution caused by a damaged ship will be as follows:

- lightering the vessel in difficulties by means of heliportable pumping devices;
- confinement and recovery at sea by all possible means. The storage and transportation of the recovered products at sea could be accomplished by means of floating tanks having capacities of 100 to 200 m^3. The use of existing vessels of large storage capacity such as coastal tankers or hopper dredgers fitted with a recovery boom is strongly advised;
- treatment of oil spills by dispersants in unfavourable weather conditions and in emergency situations to a predetermined extent in strictly prescribed areas. Special spraying equipment and ships as well as aeroplanes and helicopters should be provided in emergency applications;
- use of sorbents and other products and the recovery of agglomerated slicks by trawling;
- confinement and recovery in coastal areas - of great importance in particularly ecologically sensitive areas such as coastal salt marshes;
- oil recovery on shore with modern highly specialised equipment - more productive and selective than previous methods;
- waste treatment such as sand washing on the beach to avoid transportation costs and quicklime treatment for solid residues;

- restoration of polluted zones by means of underwater polluted sand washing, beach sand washing, hot water jets, use of efficient cleaning products with low toxicity, etc.;
- storage of polluted waste;
- stockpiling of pollution products and equipment;
- personnel training and antipollution exercises.

5.1.3 *Situation Evaluation*

First information about an incident involving a discharge of hydrocarbons into the sea should be given by the master of the ship or any off-shore installation to the Coast Guard or any other shore radio station. This information should contain the following data:

- ship details: name, country of registry, tonnage, route, number of holds/tanks and their contents, heating of oil cargo;
- cargo oil including bunkers: type(s) of oil, quantities, oil characteristics;
- incident details: time, position, cause (collision, stranding, fire, explosion, etc.), crew casualties;
- location and extent of damage, preventive measures undertaken by crew;
- weather present and forecasted, temperature, sea state, wind force and direction, seawater temperature;
- rescue requirements: SAR requirements, tugs or other ships in the vicinity;
- master: name, nationality;
- shipowners: name, address, telephone, telex, telefax;
- shipowners agents: name, address, telephone, telex, telefax;
- cargo owners: name, address, telephone, telex, telefax;
- cargo owners agents: name, address, telephone, telex, telefax;
- hull insurers: name, address, telephone, telex, telefax.

The master is obliged to contact the shipowners and the Coast Guard. They must be well informed about any further development of the situation as they will have to make timely and effective decisions regarding the handling of the ship and the spill. The most important response action is to prevent further spillage.

After sounding all tanks and estimating the amount of escaped oil, the master should endeavour to transfer the cargo internally to any space available. But if the pump suction located at the bottom of the breached tank is in the water, then the transfer of cargo oil is possible only by means of portable equipment. Due regard is to be given to earthing such equipment and to avoidance of splashing as there exists the danger of a possible explosion due to static electricity.

In the case of grounding, soundings should be taken around the ship in order to establish the position of the ship in relation to the sea-bed. The possible transfer of the oil cargo from the damaged tank to other cargo tanks may not cause excessive pressures of the ships bottom in other tanks on the sea-bed, thus making the breaking-out of the stranded vessel impossible. If the sea-bed is rocky or big boulders are scattered over the sea-bottom, the master should not try to break out the stranded vessel by the ships own power before lightering the tanks. Protruding rocks or boulders, which penetrate the ships bottom, plug up the holes in the shell, thus minimising the leak. But when the ship is backed out, these rocks may rip the ships bottom open and a large spill will occur.

Attention should be given to fire hazards from the flow of oil outside the ship due to evaporation of light ends, causing flammable gases around the damaged ship. All ships in the vicinity should be notified to stay outside the flammable vapour zones. Similar information should be given to salvage/rescue ships and Authorities.

In the case of damage below the waterline, closing the pressure/vacuum valves will reduce the oil out-flow. But if the freeboard is more than 3.5 m, then the strength of the deck should be taken into consideration and the valves should not be closed. In the case of an inert gas system being connected to the venting system, the damaged tanks must be isolated from the inert gas system. Air leakages into the damaged tank should be avoided.

In case the above measures do not stop the leakage the transfer of some or all of the oil cargo to another ship or barge should be considered. This operation can not be carried out by the ship's crew alone and they must be assisted by specialised personnel using special pumps and other equipment.

Aerial surveillance should start immediately in order to provide reliable information concerning the size and location of the spill and its movement in order to choose the most appropriate response options, to note changes in the appearance and distribution of oil over time, to forecast which coastal and marine resources or areas are under threat and to observe and report the effectiveness of the response measures.

If the situation can not be coped with by the crew and if in the master's judgement the assistance of professional salvors is necessary, then the owners are obliged to arrange a contract with a salvage firm to handle the situation. Normally all such contracts are on a "No cure - No pay" basis, by which no remuneration is paid to the salvor if the ship is not brought to a place of safety. The major principles of such contracts are as follows:

- the salvor agrees to endeavour to salvage the ship and take it to a named place or a place of safety;
- in addition he agrees to endeavour to prevent the escape of any oil from the vessel whether cargo or bunkers;
- if the salvage is successful the remuneration will be on the normal "No cure - No pay" principle. The remuneration is determined after the completion of the salvage operation by agreement between the parties or, in the event of failure, to agree by arbitration;
- if the salvage is unsuccessful or only partly successful, or if the salvor is prevented from completing this contract, he will be paid reasonable expenses plus an increment of up to 15% of those expenses.

The Administration of a coastal state has the right to take powers to intervene because of a grave and imminent threat of pollution of its shores. One of the most important contributions to this is ensuring the prompt and efficient salvage of the ship and cargo. This power is given to the Maritime Administration of Coastal States by "The 1969 Intervention Convention" if they are parties to this convention. What constitutes a grave and imminent danger can only be decided at the time, taking into account the type and quantity of the pollutant, the circumstances surrounding the casualty, the existing and forecast weather, the range and strength of tides and currents, the availability of salvage equipment and expertise and such other factors as may be peculiar to that particular accident. If the interests of the shipowner, cargo owner and salvor do not coincide, then the Administration may be held responsible for incurring unreasonable and excessive costs.

Another important decision has to be taken, i.e. when the intervention should cease. The Administrations should base their judgement on the opinions of highly qualified experts, whom they should have within their country or commission from International Organisations such as IMO.

5.1.4 Salvage

Salvage is a highly technical operation, which should be placed in the hands of professional salvors. There are several salvage methods such as: lightering, air lifting, heaving, breaking-out a stranded vessel and towing. Any of these methods or a combination could be used, depending on the situation. They have to be carried out with the utmost care, because of the imminent danger of explosion and fire.

To reduce the draft of a damaged tanker the air lift method could be used. The displacement of water is replaced by air pressure, especially when tanks have been opened to the sea below the waterline. In order to minimise the risk of pollution, oil should be removed from cargo tanks and bunkers prior to pressurisation. The cargo is pumped by means of flexible hoses ashore or to a suitable vessel brought as close to the casualty as is safe and where necessary utilising portable pumps and inner gas equipment. This operation of lightering is conducted until the ships draft is reduced to a necessary value allowing the ship to be freed from its stranded position. Lightering is also employed to reduce the pollution risk or to minimise the risk of fire and explosion or to minimise the stresses and strains in damaged hull girders.

Prior to lightering, extensive fire fighting precautions must be taken. Fire fighting contractors specialising in oil fires should fit the ship with additional, supplementary fire fighting equipment. The ships structure should be examined by divers in order to establish the extent of damage. A lightering plan should be worked out showing the sequence of tanks to be off-loaded with simultaneous ballasting to maintain the ships attitude and minimise stresses in the ship's structure, especially adjacent to fractured tanks. Empty tanks should be crude oil washed. The temperature of the oil should be taken into consideration, because if the temperature of the oil drops below a certain point, pumping might become difficult and be considerably slowed down due to increased viscosity. Lightering operations might be interrupted by adverse weather conditions. All the equipment should be well fastened and protected from heavy seas.

On tidal shores, the tidal lift could be taken advantage of instead of lightering. With the incoming tide, the displacement of the stranded vessel will increase and thus the pressure on the sea-bed is decreased. In such cases, there may be a chance of liberating the stranded vessel pulling her by tugs and/or beach gear, which is heavy rigging drawn up tight at low tide and using the rising tide to exercise a pull on the stranded vessel. This procedure will probably have to be repeated several times to liberate the ship. The ground leg of the beach gear consists of an anchor of high holding capacity in a wide range of bottom conditions and a wire fastened to the ship.

In the heaving procedure, the same beach gear is used, but the force exercised on the wire is by a winch or hydraulic jack. To increase the pull on the ship, a series of blocks is fitted to the ground leg. The force should be measured by means of a dynamometer.

Breaking out a stranded vessel generally requires removal of the ground beneath the ship. This could be done by excavating machines, by high velocity water or air streams, by scouring from the propeller wash of small support ships or finally with explosives. Natural scouring by waves and currents, especially at the ends of the ship, should be monitored in order to avoid excessive hogging pressures on the ships bottom. Sometimes, natural scouring may occur amidships depending on the sea-bed profile. In such a case excessive pressures might occur which could break the ship.

Another method of reducing break out forces is the use of pontoons or barges rigged alongside to support part of the weight of the ship.

If the ship is not too hard aground or has been lightered, tugs of appropriate size could be used to supply the moving force for the stranded ship. Generally, the tugs could be most speedily available. The tug uses a wire rope with a nylon spring and/or a chafing chain.

After lightering and freeing the damaged ship, it might be necessary to move her into a new position to facilitate beach clean-up and to bring her to safety. This should be done after a thorough inspection by divers with due consideration to the strength of the ship's girder in order to avoid the ship breaking thus causing additional spillage. Any bottom protrusions would be especially important. Some of the holes may need patching with timber or concrete. In extreme cases some reinforcement of the ship's structure might be required before commencement of towing. Recently, water displacing foam and polyurethane balls and pellets have been used in salvage operations. These jobs call for highly specialised crew and appropriate equipment, which could be supplied by professional salvors.

Towage must be carried out with the utmost care using the required number of tugs, and a salvage ship should assist. The ship should remain on this new station where all the tanks should be water washed and rendered gas free. Only then can the ship be towed to a repair yard.

5.2 Aerial Surveillance

5.2.1 General Requirements

The surveillance and tracking of an oil spill is an essential component of any spill response. Air surveillance is possible by visual observation and by the use of remote sensing techniques. There are many techniques available, which are applicable under a variety of environmental conditions, which are perpetually changing. Several requirements refer to remote sensing systems. The system should:

- have the ability to detect and identify oil and to give some identification of its thickness in all weather conditions;
- have detection capability independent of oil properties;
- be light and simple so it can be installed on an aircraft of opportunity and used by observers, who could be easily trained at the site;
- have a large area coverage even against the smallest targets, detailed monitoring and close-up inspection capability;
- give high quality, high resolution photographic evidence from a camera system;
- produce the output readily interpreted in real time and data together with position, date and time, information in an image format that can be handled at the operation site;
- be selfcontained and require minimal maintenance and operational support;
- possess the possibility of transferring or receiving all the information to/from ships or ground stations.

To carry out aerial observation of oil at sea, a flight route plan should be prepared taking into consideration ship traffic routes, offshore activities, pipe line layout and special areas.

5.2.2 Visual Observation

Visual observation is only possible during the daytime and under favourable weather conditions. In order to conduct systematic observations at sea, the chart should have a grid with a distance between the grid lines of five nautical miles for a flying altitude of

300 m and a visibility of more than 8,000 m. (fig.5.1.). On each track, a search area of 5 to 10 km on both sides of the track line could be observed. The aircraft should fly on a base line between the grid lines, investigating each observed oil spill visually and recording it (time, position, dimensions, ships information, etc.).

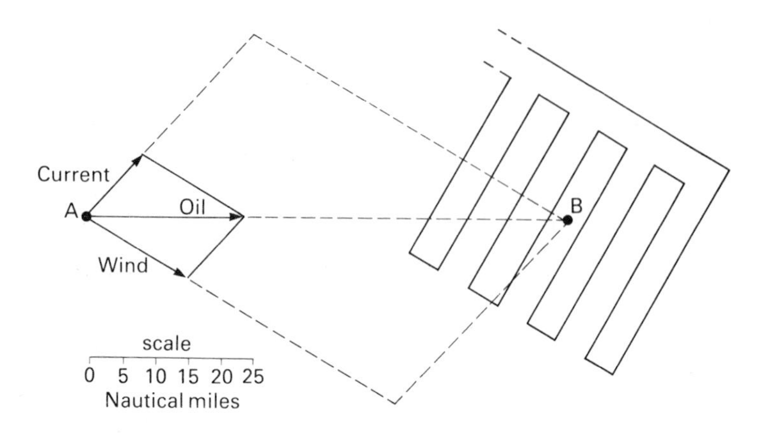

FIG.5.1 Movement of oil from position A and cross-wind ladder search pattern over B (predicted position of oil after 3 days at wind speed 25 knots and current speed 0.5 knot). Arrow length and directions represent distances applicable to movement during 24 hours.

When oil is spotted, it is advisable to arrange a ladder search across the direction of the prevailing wind, since floating oil has a tendency to become aligned in long and narrow windrows parallel to the direction of the wind. An accurate assessment of the quantity of oil is virtually impossible due to the difficulty of estimating the thickness and coverage of the floating oil. However, the approximate magnitude can be estimated by the colour of the slick, which determines the thickness of the floating oil (Table 5.2.). In order to avoid distorted views, it is necessary to look perpendicularly at the oil slick when assessing its distribution. Depending on the position of the sun, it may prove more profitable to fly a search pattern in the opposite direction to that originally planned. Sun glasses can give the observer some relief from eye strain caused by strong light, and glasses with polarising lenses will actively assist the detection of oil at sea under certain light conditions on account of the polarisation differences in light

TABLE 5.2 Slick appearance.

Oil type	Appearance	Approximate	
		Thickness (mm)	Volume (m^3/km^2)
oil sheen	silvery grey	>0.0001	0.1
oil sheen	iridescent	>0.0003	0.3
crude and fuel oil	black/dark brown	>0.1	100.0
water-in-oil emul. mousse (50-70% water content)	brown/orange	>1.0	1,000.0

reflectance off oil and off water. Visual observations are limited by difficulties in distinguishing between oil and other phenomena, which produce anomalies in the otherwise apparently uniform sea surface. Phenomena which may be mistaken for oil by inexperienced observers include cloud shadows, seaweeds, plankton blooms, differences in colour of two adjacent water masses, ripples on the sea surface, patches in shallow water, sewage discharges and subsurface sand banks. Even the wind at low speeds producing localised areas of capillary waves on an otherwise calm sea can cause the remaining flat areas to be reported as oil. Oil will be more persistent in calm conditions and so be more likely to be seen, but this capillary wave phenomenon in calm weather may be a significant contributory factor in increasing the number of erroneously reported spills.

5.2.3 Airborne Surveillance Systems

During the night or in poor weather conditions, aerial observation is only possible by the use of remote sensing equipment. Search areas on each side of the track line depend on long range remote sensing equipment. The total integrated system is built up from several sensors and subsystems in a modular design. Data from the sensors are processed to enhance features relevant to maritime applications and to obtain a suitable image format for presentation. During processing, alphanumeric annotation is added for image identification. After processing, imagery from all sensors can be recorded, documented and displayed in real time to the operator on a colour TV screen. All imagery from the aircraft can be relayed to ships and land-based units using a microwave image link. The ship and shore equipment has the same presentation and documentation facilities as the aircraft.

A minimum sensing system for oil slick detection should consist of a side looking aiborne radar (SLAR) and an infrared (IR) line scanner. SLAR is used to detect slicks at long range, while the IR line scanner is used to provide information on the slick thickness and its surface, and thus on the oil quantity. Oil detection is possible because the SLAR is specially designed to observe level variations in the sea clutter. An oil spill makes the polluted surface considerably smoother than the surrounding unpolluted water. The smooth surface gives less radar back-scatter, which makes the oil appear as a dark area on the radar display. The antenna of SLAR has no moving parts, does not require any gyro-stabilisation and therefore is relatively light.

One of the outstanding features of SLAR is automatic target positioning. The operator can use a light pen to mark selected targets on a display. Then the geographic position of the target is calculated and presented in the data block on the image.

The IR/UV (infrared/ultraviolet) scanner system is used at close range to obtain high resolution imagery. Its main applications are inspection of suspected oil discharges, surveys of accident sites and monitoring of pollution in a ship's wake. In the near infrared (IR) range (3 to 5 μm) the oil has greater emissivity than the water and in the far IR range (8 to 14 μm) the difference in radiation is greater (0.998 for water and 0.94 to 0.98 for oil). This low emissivity causes the oil to appear cooler by a temperature difference of 3 °C. Since IR imagining systems can detect 0.2 °C temperature differences, thus the oil has a high contrast for the sensor.

The IR channel can be operated day or night. It gives information on the spreading of oil and also indicates the relative thickness within the slick. Usually 90% of the oil is concentrated within less than 10% of the visual oil slick. With IR information, clean-up operations can be directed for maximum efficiency. Detection limits in terms of oil film thickness are required. This system does not work well in rainy or foggy conditions, since IR radiation is absorbed by water vapour.

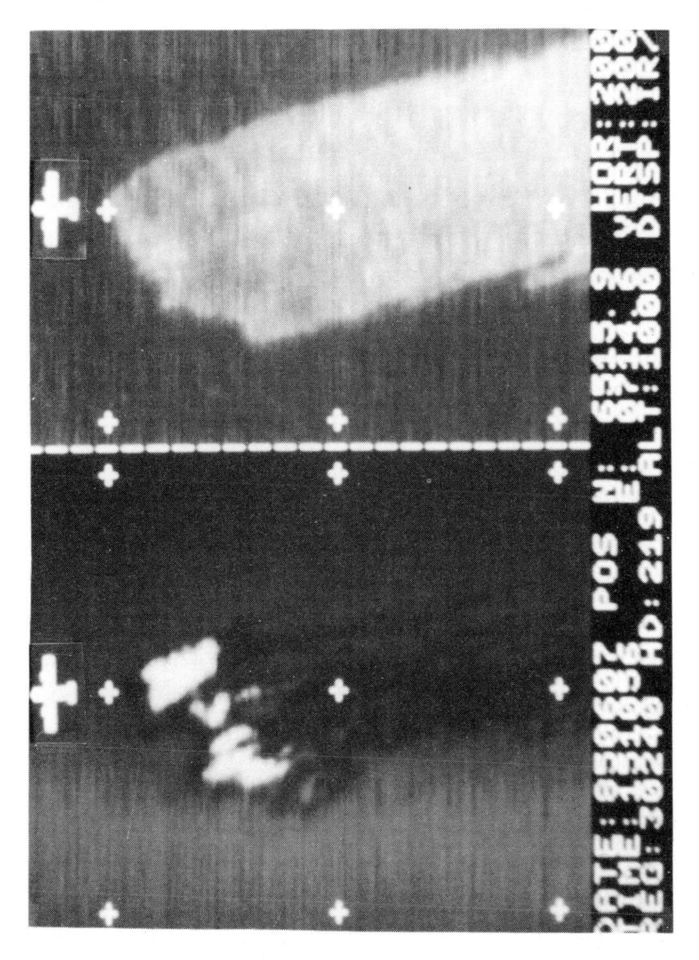

FIG.5.2 IR/UV registration of an oil slick; the thicker oil is clearly visible in the IR channel, while the UV shows the total area of the oil film.

In the near ultraviolet (UV) range (0.3 to 0.4 μm) the atmosphere has a high transmission and oil has a higher reflectance than water. At wavelengths shorter than 0.29 μm the atmosphere is opaque. Thus, this region of the spectrum is useful for detecting oil on the water surface in daylight since the UV sensor requires the slick to be illuminated by sunlight, which acts as a source of UV radiation.

The UV channel adds confidence to IR registration. In specific situations, the UV information is used to distinguish natural phenomena, such as cold upwelling water, from suspected pollution. Detection of oil in the UV band is possible because oil has a higher reflectivity than water for UV radiation, which allows slick thicknesses down to 0.01 μm to be measured (fig.5.2.).

There are two further scanning techniques under development:

- a scanning microwave radiometer has the advantage over IR systems of detecting slicks in rainy or foggy conditions and it should be used in conjunction with an IR detector rather than as a replacement for one. This radiometer gives different results for emulsified and non-emulsified oils, because they have different emissivities and different calibration is required. As the spilled oil forms water-in-oil emulsions within a few hours, operators can not be sure which calibration curves to use to convert brightness temperatures into slick thicknesses. A microwave radiometer can provide a slick thickness profile along a flight track, which could be used for volume calculations.
- a laser fluoro-sensor can provide a coarse classification of oil slicks by measuring the slick density, but these measurements until now have had to be verified with in-situ measurements carried out by survey vessels, equipped with thickness measuring and probing facilities.

A camera system is indispensable for identification of oil slicks and for evidence of violations such as illegal discharges by ships. On the photographs, relevant navigation data such as time and location have to be displayed. This system should include one vertical camera and one handheld camera for oblique pictures. Additionally, a TV system should be fitted on board an aircraft. A hand-held TV camera is connected to the image link system. It provides a highly convenient method of obtaining both overview and close-up images, which could be stored and transmitted together with other sensor data. The aircraft can send a frozen TV image of an unknown radar target to a ship or ground centre for immediate identification. The operator of the whole aerial surveillance system aboard an aircraft can observe

the indications of any chosen sensor on a TV screen. Any of the displayed images can be stored on digital tape for later playback and sent to a ground station or a shipborne receiver via an image link. All images are automatically annoted with date, time and position from the aircraft navigation system.
The above two systems (photographic and TV) have the common deficiency that no all weather identification could be guaranteed. In this field a lot of work still needs to be carried out.

5.2.4 Possible Future Developments

Current equipment can be relied upon to detect oil slicks and to indicate areas of relatively thick and thin oil. The IR technique detects only thicker oil and provides information on two thickness ranges though precise values can not be attributed to them. Thus the extent of the total area affected by an oil spill can be measured, but not the amount of oil involved, because the thickness can not be determined with sufficient accuracy. In order to provide more information on thickness, attention has been turned to microwave radiometry. This passive technique is analogous to the detection of infrared radiation, but it occurs in the microwave region of the spectrum, where it may be compared with radar, which is the active technique in this spectral region. Preliminary results on microwave are encouraging and this technique is now the subject of considerable investigation. However, this technique will measure oil thickness from 0.1 mm upwards, which is likely to confine this technique to major spills. Laser fluorometry is being investigated to facilitate the classification of an oil spill as fresh crude oil, weathered crude oil, light fuel or heavy fuel oil as well as a possible means of detecting oil below the sea surface.
Difficulties remain in identifying ships in poor light, which is a major drawback in producing evidence in cases of illegal operational discharges. Thus developments are now taking place on the use of low light level television as a means of detecting ships names during night operations. However, the results of application of low light level television by aircraft are not very hopeful.

Another prospect is the obligation to install identification markers on board ships.

Data processing on board the aircraft and data transmission to a ground based television monitor and hard copy printer is another field of extensive research work.

Consideration has been given to applying satellite mounted detectors for oil spill response management and illegal discharge monitoring. Particularly attractive is the frequency of passage over any particular point on the earth's surface by the detectors on board. As with the earlier equipment available for aircraft use, satellite equipment has not been designed or selected primarily for oil detection, but a new generation of satellites is being developed, which offers improved performance and opens new possibilities.

5.2.5 Operational Requirements

Fixed wing aircraft with good all round visibility and appropriate payload carrying capacity are best suited for aerial surveillance of open waters. Helicopters and airships, because of their flexibility, would prove of advantage for surveying coastal waters with many cliffs and islands and for assisting any clean-up operations near the coastline. The endurance of helicopters is about 2 to 4 hours, of fixed wing aircraft 5 to 6 hours and of an airship at least 20 hours. Over the open sea the speed (150 to 170 knots) and payload of fixed wing aircraft is preferable.

The aircraft should be properly fitted out with all the above described equipment, which should be properly maintained on land and serviced in flight. For each flight a detailed plan should be prepared with a surveillance grid. The frequency of surveillance flights depends on the importance of the area to be surveyed and the possibility of an oil spill. The observation method and the type of aircraft determine the flight hours requirements. A guideline is to fly over each selected area every day of the week over the whole year. Over areas with high traffic density and with large offshore activities, the flights should be carried out several times per 24 hours. On the basis of these considerations a monthly plan for aerial surveillance activities should be worked out.

Airborne detectors should be capable of surveying a wide swathe width per aircraft pass, of providing a display of images thus obtained and a means of recording and annotating them with data and time and the relevant position, altitude and heading of the aircraft.

The overall equipment package should be of a size suitable for use in light aircraft and should be compatible with the auxiliary power supply normal to small aircraft. For radar, power requirements are related to the desired detection range, which in turn is related to the operational altitude appropriate to the chosen aircraft. There is a range of design options and it is important to optimise them for a given application. Radar and infrared detection systems are generally produced for purposes other than oil detection and it is by no means certain that a given instrument will give the best possible results for an oil spill. Among the parameters which can be varied are thermal and spatial resolution in infrared, wavelength, pulse frequency, polarisation and signal treatment in radars.

It is essential that the data gathered in flight together with all navigational annotations should be available in real time on board the aircraft itself. This enables a continuous check to be made on the correct functioning of the sensors and recording equipment. It also enables the aircraft crew to make immediate use of the data in the control of spillage response operations or in pursuit of evidence for prosecution. Recorded data can be played back at the ground control centre permitting further consideration of information. Reliable records are essential, if evidence is to be presented in court.

When a ship violating the antipollution rules has been located, a photographic run (fig.5.3.) should be carried out. The photographs should be taken at an altitude of 200 to 100 m above the sea commencing when crossing the sailing course of the ship at a right angle at a distance of about 5 to 3 ship lengths in front of the vessel. Then, a run parallel to the sailing course is made

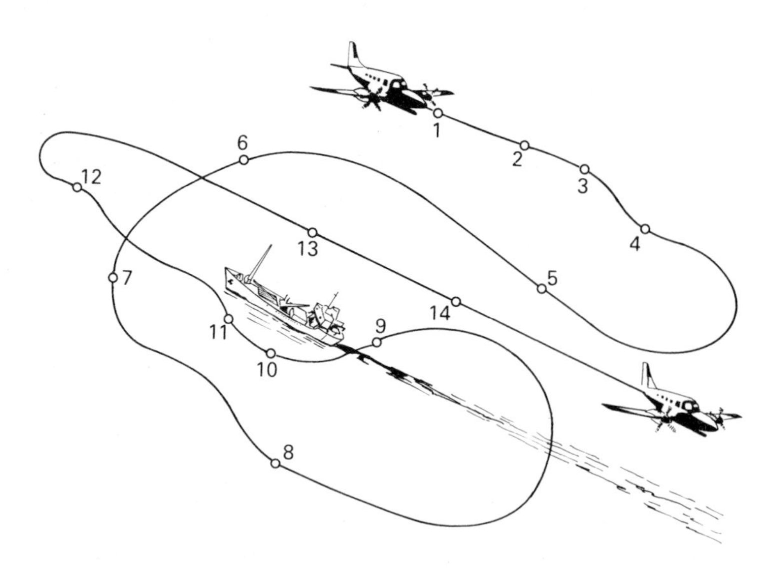

FIG.5.3 Photographic run of a fixed wing aircraft; (1) beginning of the run (altitude 300 m); (2) SLAR polaroid photograph; (3) SLAR file dump; (4) descent to 150 m; (5,6,7) photographs of ship's wake; (8) descent to 60 m; (9) name and port registry photograph; (10) ship's side photo; (11) ship's bow photograph; (12) ascent to 150 m; (13,14) IR/UV photograph.

at 100 m altitude above the sea. At the end of this run, the aircraft should turn and fly at right angles to the ships course at a distance of about 1.5 to 5 ship lengths behind the stern. After a turn, a second parallel run is made with SLAR working, the flight will be continued until the name of the vessel and the oil sheen behind the ship are clearly visible.

At night and under poor weather conditions, an alternative for the camera system may be a thermal observation system.

5.3 Oil Spill Training

5.3.1 Purpose of Training

Training is a process of acquiring valuable skills, knowledge and abilities. A student must acquire a basic understanding of the operating principles and skilful operation of the devices to their best advantage. Learning strictly by experience provides valuable lessons, but such a method proves costly, inefficient and incomplete in the long term. On the other hand, classroom and field programmes in a spill centre are readily available and they are successful in providing general knowledge to a broad audience. Thus careful planning of overall training is required. A series of short programmes, well designed and sequenced gives better results than an intensive programme. In a site-specific programme, the personnel can be trained in familiar surroundings, with their own equipment and at their own facilities under realistic local conditions. Thus the cost effectiveness is improved. The objectives of such training are as follows:

- to develop competence in personnel in the operation of equipment;
- to teach safety in the operation of spill equipment, vehicles and warehouses and in dealing with other potentially hazardous situations that might be encountered at spill sites;
- to teach operators the proper use of equipment for maximum life;
- to make equipment operators competent instructors of the use of this equipment, which will be needed on the spill site while working with a number of company personnel and local, casual workers;
- to teach environmental/habitat sensitivity, so they may learn to make decisions on the spill response priority.

To facilitate this training, imparting both cognitive and manipulative skills, visual aids and training manuals are of great importance. Special emphasis should be placed on simulation exercises in order to familiarise the participants with their roles, relations, responsibilities, interfaces and communication procedures during a spill clean-up operation. Simulations emphasise both active participation and higher level cognitive learning. For this purpose a computer should be used to enhance the activity, especially in the area of spill surveillance and tracking. Characteristic wind patterns, currents, tides and similar environmental data are used to predict spill movement, which is shown on the computer monitor. The weather input data changes as well as the effects of the response team interventions could be reflected on the spill movement, shown on the monitor, which adds valuable realism to the simulation process. Computer simulation has great value in training on-scene coordinators and managerial personnel.

5.3.2 Training Programmes

All the personnel involved in contingency planning and actual oil spill combating must be adequatly trained to handle all circumstances related to the task they are expected to perform. All manuals, equipment and deployment techniques are only as effective as the people using them. There are three levels of training, depending upon the field of responsibilities:

- operator training in the use of spill response equipment in order to familiarise each trainee with the equipment and with the conditions to which it is best suited with its limitations, most probable causes of failure and possible remedies. Operational and maintenance procedures should be theoretically and practically explained. It is most important that the operators are well acquainted with servicing all types of equipment used onshore and offshore for oil spill combating, transportation of oily waste, its storage and disposal. They must be conversant with safety precautions in order to avoid any accidents. For this purpose exercises should be held in all weather conditions ending with wash-up of equipment and debriefing to discuss any difficulties and improvements which could be made. For these exercises, modern observation, photography and video recording should be used, which enable assessment of the exercise afterwards. Some inert material such as foam or polysterene chips could be used to demonstrate the effectiveness of the exercise. It is very important that such exercises are conducted four times a year in order to have the operators trained in all seasons of the year and on all types of coastline. Occasional refresher courses should be held at time intervals of not longer than four years or less, i.e. after introduction of new equipment;
- on-scene coordinator training in managing oil spill response activities in the field. This task requires a comprehensive

knowledge of spill management functions. The training curriculum must encompass specific subjects such as combating equipment, its use and maintenance, containment and recovery, use of dispersants, shoreline clean-up, disposal techniques. A second set of subjects will give general knowledge on types of oil and their behaviour in marine environments, risks of spills, observation and quantification, organisation and contingency planning, communications, public relations. Some of these subjects should be followed by practical (in-field) and by simulation exercises as well as "question and answer" sessions. The course should be concluded with a written exercise in order to reinforce the main points of the course;

- management training to give sound knowledge to managerial staff on marine pollution combating, on methods used, on specific requirements for contingency planning jointly with operational resource and organisation needs, on damage to the environment, on sources of compensation and preparation of claims. It is advisable that the participants attend a full scale exercise to gain practical experience of the combating equipment and a real clean-up operation. It should be remembered that people in managerial positions have very limited time because of the responsibilities of their normal functions.

The training of on-scene coordinators and managers could be of a general nature or applied to specific areas, where local or regional environmental regulations are in force and where special environmental conditions prevail.

Detailed programmes should be worked out for each subject with the consideration of training level, which should be well supported by video tape, instruction booklets, etc. A mix of lectures and video tapes is found to be best. The optimum size of a group is found to be about 20 participants, but it should never exceed 40. Attention should be given to the fact that the participants have little experience in taking notes. Thus appropriate material should be issued to them, preferably before the lecture, which will be most helpful in revision and learning.

The programmes could be worked out individually for each course or in modular form for all levels of courses. In this case every course will consist of several modules depending on the subject areas in which the student is expected to be knowledgeable, and on the depth of knowledge required.

5.3.3 Programme Implementation

Training programmes can be implemented using one or two of the options given below:

- central training facility;
- travelling instructor teams;
- correspondence training;
- programmed instructions;
- computer assisted training;
- audio-visual training packages.

In choosing the most appropriate option, consideration should be given to the problem of time, group size, location, frequency, facility availability and cost. Central training facilities could be well equipped for all levels of courses with teaching aids and demonstration items. Highly qualified instructors could be employed, as not all of them would want to travel and lecture in informal and not always adequate circumstances.

In order to keep all personnel interested in response problems and to maintain team spirit, each base should conduct monthly safety meetings with different themes for each month. These meetings should consist of a detailed discussion of that theme and a "question and answer" session about its safety concerns. In addition, presentations of new video-tapes and tapes concerning other fields of interest such as first aid, fire fighting, etc. should be arranged.

5.4 Contingency Planning

5.4.1 Drafting Contingency Plans

A contingency plan should provide an orderly and formal procedure for responding to marine pollution incidents, maximising the use of local resources and providing the means for acquiring assistance from overseas when required. There should be four grades of contingency plans depending on the area of responsibilities. They should all be complementary with different degrees of detail. The following plans should be worked out for:

- ports and terminals;
- regions of coastlines, coinciding with the administrative subdivision in a country;
- national coastline in its totality;
- countries situated round a certain sea area like the Baltic, the North Sea or along the coast of a particular ocean (international plans).

Before any serious consideration can be given to producing a contingency plan, responsibilities should be clearly divided and assigned to one governmental agency (lead agency), which will be responsible for coordination during writing and for implementation of the plans of each degree. The responsibilities should be divided:

- within the port or harbour - administrative authority of that port or harbour;
- on beaches and foreshores - the relevant state or regional authority;
- in territorial seas and high seas - governmental authority.

Several agencies and organisations (support agencies) will participate in writing individual chapters on different aspects, based on the wide expertise of each organisation specialising in such fields as ship operations, chemical engineering, marine salvage, meteorology, environmental protection, fisheries, legal problems, etc. It is a very complex task, which should be undertaken with the utmost responsibility before a successful response to oil pollution could be undertaken. This lead agency, which might have the form of a National Response Organisation, must have enough authority to initiate action, to deal promptly with large spills or with an incident which may lead to a large spill. It must be sufficiently funded and free to introduce any amendments to the plans following the experience. It must control regional and local contingency plans.

The objectives of the plans are to:

- provide an appropriate system for detection and reporting of spills of oil or other toxic substances in local waters or threats of such spills;
- identify high risk areas and priority coastal areas for protection and clean-up;
- coordinate the efforts of government departments, service organisations, local authorities and private organisations to prevent, contain and, where necessary, clean-up local oil spills or marine pollution incidents in a manner which will protect public health and minimise the impact upon the environment. The response should be prompt and made by well-trained personnel with the best available equipment;
- provide a command structure and command centres appropriately equipped for the magnitude of the spill including frequently updated agency and personnel lists with telephone numbers;
- establish a reporting procedure to provide control and decision-making information for the cabinet committee and the on-scene coordinator (OSC) and to enable the public to be kept informed, when a marine pollution incident threatens or exists;
- provide adequate oil spill response equipment and well trained crews;
- ensure that complete and accurate records, including those of all expenditures, are maintained to facilitate the recovery of costs and full documentation of the incident;
- coordinate scientific support for the OSC and maintain a proper "chain of custody" for all environmental samples.

A national oil spill contingency plan is a document that endeavours to predetermine actions to be taken in the event of a major oil spill. It differs from other contingency plans in that in addition to a spill risk analysis, the oceanographic, ecological and general environmental factors are taken into account in the planning, as well as the logistics and economics to be derived from a fairly broad based, nation-wide programme.

Site-specific contingency plans are different in nature and intended only for relatively minor local spills, but they should be incorporated into the national plan to coordinate assistance in the event of a major spill.

It must be emphasised that no two spill situations are alike. However, features common to each oil spill can be of use in preplanning activities. The information contained in the plan is for guidance and not as a replacement for common sense. The information will be of assistance when decisions have to be made in connection with the actual spill, act as a reminder to the on-scene coordinator (OSC) of the order of priorities, and provide the necessary know-how to the various groups to carry out assignments.

The task of any person dealing with a major spill is difficult, requiring at any level and in any function of the spill team a talent for improvisation and, ideally, a great deal of experience from previous spills and prior knowledge of national and regional plans. It should be kept in mind that even the best plan can not replace such qualifications. The plan brings together relevant information and practical advice from past spill incidents and will ease the job of the OSC and of other people on the spill control team.

In order to facilitate the reading of the plan, it must be designed to provide for rapid interview. With this objective, a quick reference check list and outline of subsequent actions should be compiled to serve as an instant reminder of important considerations and actions at the time of the spill.

Drafts of all maps should be reviewed by various agencies and individuals in an effort to acquire specific local information that may not be available through the usual government and published data basis. Agencies that receive the maps for review are provincial/wildlife and natural resource agencies.

When a spill is reported, it is necessary to determine its course and size to decide the level of response required. The report of an oil spill may come from many sources - even from the general public or oil company personnel. The person receiving notification of an oil spill should record as much information as possible to answer the following questions:

- who is reporting the spill (name, telephone number, address, date and time of observation);
- is this a first-hand report or passing on of information supplied by others? If by others, who actually saw the spill (name, phone, etc.);
- what is the actual exact geographic position of the spill? (latitude and longitude, or position relative to coastline);
- what is the best estimate of the source, course and extent of the oil spill (name, size and type of the ship, collision or grounding, type and quantity of oil spilled);
- weather and sea conditions;
- actions already taken or intended to respond to the incident.

An alert procedure should be included in the plan. The initial and subsequent reports should be disseminated to the lead and support agencies by the fastest possible means.

Some general outlines, as for a national contingency plan, should be followed in preparing port contingency plans. Potential locations of spills and types of oil are easily identified in ports and all emergency incidents such as collisions, grounding, fires, explosions, personnel casualties should be considered. Precise organisation, methods and means of communication, stockpiles of equipment should be established. The OSC must be designated (Harbour Master or Captain of the Port, or Terminal Operator). The personnel must be well trained in the use and deployment of available equipment and fully aware of duties and responsibilities. In the contingency plan, provision should be made for response escalation to obtain additional resources, if required. Of special importance would be the problem of evacuation of employees working within the affected area, prevention of movement of the public and unauthorised vessels and vehicles at the site of the incident. Accurate cost records should be kept.

Every contingency plan should be exercised on a regular basis as a:

- paper exercise to verify communication procedures;
- operational exercise involving deployment personnel, equipment and material.

All deficiencies should be identified and any amendments to be made to improve the effectiveness of the plan.

5.4.2 Mobile Command and Communication Systems

The clean-up of a large oil spill involves a wide array of seaborne and land-based equipment and a great number of people. The management of such an operation is complicated and confused by the emergency atmosphere and adverse and changing weather conditions. A clearly identified and properly equipped command and communication centre is very important. It should be situated in close proximity to the operations, at a short driving distance. If a majority of the operations will be afloat, then the centre should be situated aboard a ship. There are many questions to be considered such as what is the total shoreline to be cleaned up, how will messages be received and passed on, will there be meetings held and how often, where will the crew sleep and eat, will the clean-up operations be restricted to daylight only, what are local roads and traffic conditions. Such questions and many others have to be considered in detail, all the pros and cons will have to be taken into account. Only then should a decision be taken.

Office buildings should never be chosen, despite obvious advantages, such as all telephone facilities, conference rooms, eating facilities, etc., because of many disadvantages such as disturbance of the normal work of that office, and visits from news media, security problems, etc. A hotel or a public building could be used, but spills generally occur where no suitable office areas are available and therefore it is advisable to have office units transportable by road or air. They could be in the form of a trailer with a complete outfit to accommodate the spill managers office, communications room, switchboard and some working spaces. A diesel driven generator should be provided for lighting inside and outside, radio operations, heating, hot water, air conditioning,

etc. A central telephone switchboard is essential.
To ensure timely information exchange and responsiveness to changing circumstances, a wide array of radio equipment must be provided. Most essential are portable radios with several channels, but they must be ignition proof and they should not require hand operation. The radio should be worn at the waist with the speaker-microphone clipped to a lapel or collar. All vehicles and aircraft should be fitted out with portable radios, which, however, have a limited range. Thus a repeater could be mounted on top of a tall building or a hill.
Base station radios are generally of two types: one type has the same frequency capabilities as the portables. A second type provides a wider array of frequencies and is used for longer range communications. When an aircraft is fitted with a remote sensing system, then the base station should be adequately equipped to receive all information from this aircraft.

5.4.3 Oil Spill Clean-up Costs

Oil spill clean-up costs will depend on several factors such as:

- location of spill - if spilled oil does not come ashore due to advantageous direction of winds and currents and stays offshore until it breaks up naturally, little or no clean-up response will be necessary;
- population density of the coastline affected - if the spill happens in uninhabited areas, it will pass virtually unnoticed, whereas if it happens in a built-up area, there will be a great demand for a thorough cleaning;
- composition and physical properties of spilled oil - light crude and refined oils evaporate and disperse naturally except when persistent water-in-oil emulsions are formed. Heavy crude oils and water-in-oil emulsions are highly persistent and viscous, causing widespread contamination and thus some of the costliest clean-up operations;
- spill quantity has some impact upon clean-up costs, but not as much as location, weather conditions and oil type. Small spills can be more expensive than large ones;

- method of clean-up:
  - at sea by mechanical recovery or by use of dispersants;
  - on the shore by manual collection supported by various mechanical equipment.

In actual clean-up operations it is very difficult to estimate exactly the quantity of oil collected, which would give an idea of clean-up effectiveness. The greater portion of any large spill is broken up by natural processes, including wave action, dispersion and biodegradation. Cost figures can be given with some precision, but the data on oil treated are not precise. It is quite easy to measure the quantity of liquids skimmed or pumped, providing the measurements are taken at the correct points, but it is more difficult to ascertain the true oil content in the mixture of oil and water collected. Any estimate of oil dispersed at sea or oil quantities on a shoreline and in collected oily waste will include errors, because estimates are based more on judgement than on measurement. Thus the interpretation of unit costs ($ per t or bbl) must be made with care.

Most offshore spills are combated both at sea and on the shores. Thus the approach in North America and the Far East is to concentrate on shoreline clean-up. Any work at sea involves the use of aircraft and vessels, which are inherently expensive to operate. There are on record some very successful offshore mechanical recovery operations, but there are some examples of very wasteful efforts since no oil was actually collected at sea due to adverse weather conditions over a certain length of time, after which time the oil had fragmented and became too viscous to pass through recovery pumps. After a few days the oil remaining at sea began to come ashore along many miles of shoreline. In such cases attempts made at sea did not materially alter the pattern of shoreline impact and no shore pollution was actually prevented.

The application of dispersants simply accelerates the natural dispersion rate. In practice, it becomes impossible to distinguish between dispersion attributable to natural water movement or to dispersant application. In the case of aircraft and vessel deployment, it is very difficult to estimate how much oil was treated by aircraft or by ships. In general, it could be said that aerial spraying is more accurate and less wasteful of dispersant, which fully compensates for higher operating costs of aircraft. The use of large aircraft only becomes cost effective when a large spraying operation is planned.

The costs of clean-up operations vary considerably with average costs between $2,000 and $4,500 per t ($300 to $700 per bbl). There are clean-up operations known where costs were as low as $71 per t ($10 per bbl) and as high as $21,000 per t ($3,270 per bbl). Chemical dispersant costs approximately $1,600 per ton. Dispersant application from aircraft costs around $4,000 per ton of dispersant, which at a dispersant to oil ratio of 20 means about $200 per ton of oil. Application from ships tends to be cheaper - costs are likely to lie in the range of $500 to $ 3.000 per ton of dispersant ($25 to $150 per ton of oil). However, aerial application offers higher spraying rates and better control over treatment of spilled oil.

Clean-up of polluted shores and disposal of recovered material range in cost from $650 to $6,500 per ton depending on the prevailing local rate for casual labour. It should be kept in mind that the efficiency of clean-up operations decreases with time as oil becomes scarcer. Complete clean-up can never be achieved in practice. A vital point in any clean-up operation is deciding when it is reasonable to stop any further action and thus expenditure.

When boats and small craft become oiled, then the easiest way to clean them is to use dispersants or some suitable cleaning liquid. But in some countries the use of chemicals for this purpose is connected with serious restrains and it is quite impossible to clean boats in the water. The cleaning costs vary considerably depending on the degree of contamination, type and size of boat. If oil is not allowed to dry on the hull surface, it could be cleaned relatively easily and cheaply ($50 to $150 per boat). If the boat has to be lifted out of the water and the hull repainted or recoated, then the cost will be doubled or even more.

Many spills give rise to additional claims for alleged damage to exploited and unexploited natural resources. These claims can amount to many tens of millions of dollars. Light oils, being highly toxic, may cause mortality of marine plants and animals, may constitute a fire hazard if spilled in confined situations and may result in a wide variety of third party claims due to temporary closure of port areas, nearby industries, beaches and amenities.

The costs of disposal of oily waste will mainly depend on:

- mode of transport;
- type of disposal site;
- type of incineration plant.

Typical transportation costs over a distance of 100 miles amount to \$4 to \$6 per t. Disposal at a simple landfill site can be as low \$3 to \$7 per t, but it may reach \$100 per t in a controlled site with impermeable membranes and leachate monitoring stations. In a simple incineration plant the costs of treatment will be \$3 per t of oily waste to \$250 per t in a fully enclosed plant with smoke emission control and other safety features.

Some additional costs will have to be contended with such as administrative support costs. In making decisions on preferred response techniques, the OSC should be guided by the above rough estimates.

5.5 International Cooperation

5.5.1 Introduction

Since the late 1960s it has become obvious that close international cooperation in marine pollution combating is essential in order to control major pollution of marine environments. Oil spreads very quickly and it may threaten shorelines of not only one, but several coastal states. Oil response is difficult and expensive. It calls for resources of equipment and personnel appropriate to the extent of the spill. The larger the spill, the greater the resources that must be allocated to provide and maintain the combating force. The size of any spill caused by a tanker accident or an offshore blow-out may exceed the capacity of any one country and the forces of neighbouring countries should be joined at very short notice. Thus regional (multinational) and bilateral agreements must be concluded, which form the legal basis for immediate intervention.

TABLE 5.3 Regional Agreements for Cooperation in Combating Pollution in Cases of Emergency in Force.

Region Date of entry into force	Legal Instrument	Organisation responsible for Secretar.Duties	Contracting Parties
1. North Sea August 1969	Agr.f.Coop.in Dealing with Poll.of the North Sea	Bonn Agreement Secretariat	Belgium, Denmark, France, F.R.G., Netherlands, Norway, Sweden,U.Kingdom
2. Nordic Area Sept.1971	Agr.for Coop.in Measures to deal with Poll.of the Sea by Oil	London Direct cooper. between govern. concerned	Denmark, Finland, Norway,Sweden (Copenhagen Agreement)
3. East Asian Seas Region 1976	ASEAN Contingency Plan	ASEAN	Indonesia,Malaysia, Philippines,Singapore, Thailand
4. Mediter. Sea Reg. February 1978	Protocol conc.coop. in Comb.Poll.of M. Sea by Oil and oth. Harm.Substances (Barcelona Emerg. Protocol 1976)	UNEP (Coordin. unit for the M.Action Plan Athens	Algeria,Cyprus,Egypt, France,Greece,Israel, Italy,Lebanon, Libyan A.J.,Malta,Monaco, Morocco,Spain,Syrian A. R.,Tunisia,Turkey,EEC
5. Kuwait Action Plan Region July 1979	Prot.conc.Reg.Coop. in Comb.Poll.by Oil and other Harm,Sub. in Cases of Emerg. (Kuwait Emer.Prot)	Reg.Org. for the Protect.of Marine Envir. Kuwait (ROPME)	Bahrain,Iran, Iraq, Kuwait,Oman,Quatar, Saudi Arabia,United Arab Emirates
6. Baltic Sea May 1980	Conv.on Protec. of the Mar.Envir. of the Baltic Sea Area (Helsinki C)	Baltic Marine Envir.Protec. Commission Helsinki	Denmark,Finland,G.D.R., F.R.G.,Poland,Sweden, U.S.S.R.
7. Wider Carribean Region March 1983	Prot.conc.Coop.in Comb.Oil Spills in the Wider Carrib.	UNEP	Antigua and Bermuda, Bahamas,Barbados,Belize, Colombia,Costa Rica, Cuba,Dominica,Domin.Rep.

After the 1967 "Torrey Canyon" disaster, multinational agreements and conventions started in 1969 with the Bonn Agreement (The Agreement for cooperation in dealing with pollution

Region Date of entry into force	Legal Instrument	Organisation responsible for Secretar.Duties	Contracting Parties
			France,Grenada,Guatemala, Guyana,Haiti,Honduras, Jamaica,Mexico,Netherlands, Nicaragua,Panama, St.Christopher & Nevis, Saint Lucia, Saint Vincent & Grenadines, Suriname,Trinidad & Tobago,U.Kingdom, USA,Venezuela,EEC
8. West and Central African Reg. August 1984	Prot.conc.Coop. in Comb.Poll. in case of Emergency	UNEP	Cote d'Ivoire,Guinea, Nigeria,Republ.of Cameroon,Senegal,Togo
9. Red Sea and Gulf of Aden Region August 1985	Prot.conc.Reg.Coop. in Comb.Poll.by Oil and oth.Harm.Subst. in Cases of Emerg.	Arab League Educat.Cult. & Scient. Org. (ALESCO)	Palestine (repr. by PLO),Saudi Arabia, Sudan,Yemen Arab Rep.
10. Eastern African Reg. June 1985	Prot.conc.Coop.in Comb.Mar.Poll.in cases of Emerg.in Eastern Afr.Reg.	UNEP	Comoros,France,Kenya, Madagascar,Mauritius, Mozambique,Seychelles, Somalia,United Rep.of Tanzania,EEC
11. South-East Pacific Region July 1986	Agreem.on Reg.Coop. in Comb.Poll.of the S-E Pacific by Hydrocarbons or oth. Harm.Subst.	Permanent Comm.of the S-E Pacific (CPPS)	Chile,Colombia,Ecuador, Panama
May 1987	Supplement.Protoc.		Chile,Colombia,Panama,
12. South-Pac. Region November 1986	Prot.conc.Coop.in Comb.Poll.Emergen. in the South Pacific Region (SPREP Protocol/Emergen)	South Pacific Commission (SPC) Noumea New Caledonia	Amer.Samoa,Australia, Cook Islands,Fed.Stat. of Micronesia,Fiji, France, French Polynesia,Guam,Kiribati, Marshall Islands,Nauru, New Caledonia,New Zealand,Niue,Northern Mariana Islands,Palau, Papua New Guinea, Pitcairn Island,Solomon Islands,Tokelau,Tonga, Tuvalu,U.Kingdom,U.S.A., Vanuatu,Wallis & Futuna, Western Samoa

of theNorth Sea by oil). Table 5.3 shows all the regional agreements for cooperation in combating pollution in cases of emergency (in force till 1987).
In addition, several bilateral agreements concluded between two neighbouring countries on cooperation in combating oil at sea have been established in support of regional agreements. And so in Europe following bilateral agreements have been concluded:

- Danish - German joint maritime contingency plan on oil combating (DENGER-plan);
- Anglo - French joint maritime contingency plan (MANCHE-plan);
- Norway - United Kingdom joint contingency plan for counter-pollution operations at sea (NORBRIT-plan).

It is expected that more bilateral agreements will be established in the near future.

5.5.2 International Convention on Oil Pollution Preparedness, Response and Cooperation, 1990

The above Convention was adopted in November 1990 at a conference attended by delegates from 90 countries. The Convention recognizes that in the event of pollution incident, prompt and effective action is essential. This in turn depends upon the establishment of oil pollution emergency plans on ships and offshore installations, and at ports and oil handling facilities, together with national and regional contingency plans as appropriate. The Convention is intended to encourage this process and at the same time to establish a frame work for international cooperation in responding to pollution emergencies which will enable maximum resources to be mobilized as quickly as possible.
The main features of the Convention are as follows:

- Oil pollution emergency plans. Ships will be required to carry detailed plans for dealing with pollution emergencies. Masters of ships, port authorities and others will be required to report pollution incidents without delay and the Convention defines the actions to be taken when a report is received. If the incident is sufficiently serious other States likely to be affected must be informed and details must also be given to IMO.

- Response systems. The Convention encourages the establishment of national and regional systems for responding pollution incidents. These systems should include features such as a national contingency plan, the pre-positionong of oil spill combating equipment, and exercises in dealing with spills.
- International cooperation. This is a key feature of the Convention. Parties to the Convention agree to cooperate and provide advisory services, technical support and equipment at the request of other Parties. The financing of the costs involved is dealt with in an Annex to the Convention. Research and development is encouraged, as is the development of standards for compatible oil combating techniques and equipment. The Convention also makes provision for technical cooperation in such matters as training and the provision of advisory services and equipment. Resolutions call upon the IMO Secretary-General to develop a comprehensive training programme in cooperation with the Governments and oil and shipping industries, and to draw up a plan on the establishment of oil spill combating equipment stockpiles.

IMO itself is designated to perform a number of functions. These include the provision of information services, the provision of education and training programmes, cooperation in research and development.

The Convention will enter into force 12 months after being ratified by 15 States. In addition to the Convention, the conference adopted several resolutions. One of them envisages expanding the scope of the Convention to hazardous and noxious substances other than oil.

5.5.3 International Marine Oil Spill Contingency Plans

The above named Conventions/Agreements/Protocols are being improved year by year and are well established throughout the world and especially in Northern European waters. Their contents are perfected by a continuous exchange of information and ideas and by international exercises and joint combating actions. In future, this pattern will be adopted in other maritime regions with multilateral agreements dealing with administrative problems, financial matters, research and development, protection of marine environments, while pure operational cooperation will be dealt with in bilateral plans.

Each government intending to participate in international contingency plans must first develop and implement a national oil spill contingency plan. Each national plan may differ, but they should contain several common basic elements such as:

- designation of the competent national authority responsible for oil spill matters;
- description of national oil spill response organisation;
- clear definitions of zones of responsibility for taking action such as surveillance, reporting, alerting and response activities;
- identification of likely sources of oil spills, vulnerable resources at risk and priorities for protection;
- existing resources and strategies for combating spills and the size of spill which can be dealt with at the national level;
- identification of logistic support of activities within the country available for response;
- identification of storage for recovered oil and disposal methods.

The above information should be supplemented by coloured maps to facilitate reading. Plans must remain simple and easy to operate. An international contingency plan should establish a framework within which two or more governments can cooperate to facilitate the operational aspects of oil spill surveillance and response. It can include, but is not limited to:

- information exchange;
- the use of vessels, aircraft and oil spill response equipment;
- arrangements for the assumption of the lead role by the State in whose waters a pollution incident occurs;
- clear definition of command structure and liaison for joint response operations;

- identification of priority coastal and sea areas;
- arrangements for vessel operation in, or overflying of, the territory of other States;
- the conduct of paper and live exercises to test the adequacy of the plan.

The State in whose zone the spill occurs should assume the lead role and be initially responsible for all actions. The transfer of responsibility from this State to another must be clearly defined in any international contingency plan. This State is obliged to inform neighbouring States immediately if it appears likely that the spill may affect their sea areas and shores. That information should include the date, time, position, type and amount of oil spilled, the prevailing and forecast weather conditions and foreseen actions. It should be continuously updated, as the situation develops. Available meteorological and hydrographic data should be used to predict a general spill movement. Visual observation of any spill is essential and all resources should be used for this purpose, such as aerial surveillance, military or commercial ships and aircraft. The results of early predictions and actual observations should be analysed and corrections introduced. The results of observations and updated predictions should be transmitted to the neighbouring States, which may be affected by the spill, until it no longer threatens any State in the area.

An international contingency plan must include agreed lists for each State, detailing the authorities or organisations assigned under national plans. These lists should identify contact points with appropriate telephone and telex numbers, which must be available on a 24 hour basis. Procedures should be developed for requesting, offering and accepting assistance in the event of a spill incident. To facilitate on-scene radio communications, it is essential that prior agreement be established on the assignment of specific frequencies and a working language for operational response to an oil spill.

An international contingency plan should contain lists of response equipment and specialist personnel available, procedures for mobilising equipment and materials and the relevant charges.

It is essential to indentify resources from outside the region, so that a reasonable response to worst-case situations can be speedily mounted. The equipment, materials and personnel should be moved to places where it is needed without undue delay. Special arrangements should be made with customs, immigration, etc. to make this movement quite free. For obtaining compensation for oil spill clean-up costs, international regimes should be observed and exact records of actions taken and equipment and other resources used should be kept. These records could later be used for subsequent evaluation of all actions taken during the spill incident to upgrade the plan, which should in any case be reviewed on a periodic basis.

5.5.4 Regional Agreements

The 1969 Bonn Agreement was at first of an informative nature without any explicit obligations on contacting parties to establish a national contingency organisation and to cooperate in combating operations. In 1976 it was decided that the contracting parties should meet on a yearly basis and a working group on Operational, Technical and Scientific questions concerning counter-pollution activities was established (BAWG/OTSOPA). This group worked out guidelines for multinational combating operations such as command structures, communication schemes, alarm procedures and exercise programmes as well as an exchange of information on equipment and material combating organisation. All this information is collected in a manual, which is updated at regular intervals. In 1983 the Agreement was revised and it came into force in 1985. It contains new elements such as coverage of other harmful substances than oil, rules for sharing the costs in joint operations, rules for contribution of each contracting party towards the annual expenditure of the Agreement, the possibility of modifying the common boundaries of respective responsibility zones.

The 1971 Copenhagen (Nordic) Agreement is based on the 1967 Agreement between Nordic Countries concerning cooperation regarding the control of compliance with the international oil discharge provisions. This new Agreement includes cooperation aspects of combating beyond the Bonn Agreement. The contracting states have established a contingency organisation for combating considerable oil slicks at sea. The Nordic Agreement deals only with oil and not with harmful substances other than oil. The level of activity has been high from the very beginning. The contracting parties have annual meetings and working groups have been established. Many joint combating operations and intensive exercise activities have created very effective cooperation.

The Convention on the Protection of the Marine Environment of the Baltic Sea Area was signed in 1974 and it came into force in 1980. The Helsinki Convention is the first international convention on the protection of the marine environment embracing pollution from all sources: land based, ship and airborne pollution as well as pollution from exploration and exploitation of sea-bed resources. The Convention includes all undertakings by seven Baltic Sea States to cooperate in combating marine pollution by oil and harmful substances. In the combating obligations this Convention goes further than the new Bonn Agreement and even further than the Copenhagen Agreement and is more specific. A recommendation has been adopted saying that in establishing national contingency plans, the contracting parties will aim at developing their ability to deal with spillage of oil so as to enable them:

- to combat oil spillages of up to 10,000 tons of oil in a period not exceeding 10 days of operation at sea;
- to maintain a state of readiness permitting the first oil combating unit to start from its base within two hours of being alerted;
- to reach within six hours any part of a spill that may occur in the response region of the country in question.

The contracting parties have also agreed to aim at the inclusion of the capability to combat spillages of harmful substances other than oil in their combating operations. A manual has been introduced containing procedures and guidelines together with information on national contingency organisations. Several communication exercises and joint operations at sea have been performed.

5.5.5 Bilateral Agreements

In addition to the above named multilateral agreements, there are a number of bilateral agreements, which are the outcome of combating experience gained in recent years by various coastal states. This experience shows that cooperation in combating pollution at sea by more than three neighbouring countries is very rare. Such close cooperation calls for more detailed planning than is possible within the framework of the larger multilateral agreements. Bilateral agreements collate the framework for close bilateral cooperation and are the backbone of plans which can give more attention to specific local factors.

In North European waters there are several bilateral plans:

- the DENGER PLAN signed in September 1982 between Denmark and the Federal Republic of Germany with the purpose of close cooperation between the two adjoining states in the case of extensive pollution in the German Bight of the North Sea and in the Western Baltic Sea. These sea areas are very sensitive to oil pollution and these two states are of the opinion that the existing international cooperation established under the Bonn and Helsinki Agreements should be extended by this plan, which contains detailed alarm procedures, command structures, communication schemes and exercise programmes. Financial obligations are not a part of the contingency plan.
- the MANCHE PLAN (the Anglo-French Joint Maritime Contingency Plan) for cooperation in cases of accidents, which might occur in the Channel and which are liable to affect British and French interests at the same time. It was agreed to act in accordance with the plan to avoid confusion and to increase the effectiveness of measures taken. The Plan determines the principles and procedures of cooperation in action to deal with pollution of any kind and in Maritime Search and

Rescue Operations (SAR). The area of implementation of the Plan lies within the Bonn Agreement area.

- the NORBRIT PLAN (The Norway-United Kingdom Plan for Counterpollution Operations at Sea) came into operation in 1983. It is mainly oriented towards spills resulting from major incidents at off-shore installations, such as blow-outs and damage to submarine pipelines. But it is not confined to such events and will apply to any spills within the Northern Part of the North Sea, which are of sufficient severity to warrant joint action. The NORBRIT region is that part of the area of the North Sea, which lies south of latitude 62 ° N (Bonn Agreement area is south of 61 ° N) and is contained within a band following the median line between Norway and the United Kingdom and extending to a distance of 50 nautical miles on either side of that line. The Plan may be also activated in the case of oil spills outside the region by agreement between the two countries. The Plan contains guidelines for implementation of the plan, determination of lead and supporting rules, command, control and liaison requirements.

5.5.6 Compensation for Pollution Damage

The intergovernmental regime for compensation for damage caused by oil spills from tankers is based on two international conventions:

- the 1969 International Convention on Civil Liability for Oil Pollution Damage (Civil Liability Convention), which lays down the principles of strict liability for the shipowner and provides a system of compulsory insurance to secure adequate compensation for the damage caused by oil pollution;
- the 1971 International Convention on the Establishment of an International Fund for Compensation for Oil Pollution Damage (Fund Convention) is supplementary to the Civil Liability Convention and provides compensation to victims, when the compensation under the Civil Liability Convention is inadequate.

The International Oil Pollution Compensation Fund (IOPC FUND) was established in 1978 to administer the regime of compensation under the Fund Convention. This Fund is financed by persons who receive crude oil and heavy fuel in States-Parties to the Convention. It provides supplementary compensation to those who can not obtain full and adequate compensation for oil pollution damage under the Civil Liability Convention. The IOPC Fund is relieved of its obligation to pay compensation if the pollution damage resulted from an act of war or was caused by a spill from a warship and if it is not proved that the damage resulted from an incident involving one or more laden tankers. Claims against the IOPC Fund can be made directly by all victims of oil pollution damage.

Two industry voluntary schemes provide compensation for oil pollution damage resulting from spills of persistent hydrocarbon mineral oil. These schemes are:

- the TOVALOP scheme (Tanker Owners Voluntary Agreement Concerning Liability for Oil Pollution) is covered by the shipowner's Third Party insurer, like the Shipowner's Liability under Civil Liability Convention;
- the CRISTAL scheme (Contract Regarding an Interim Supplement for Tanker Liability for Oil Pollution) is financed by cargo owners.

Both of the above schemes cover pure threat situations, i.e. preventive measures taken when no oil is actually spilled. TOVALOP applies also to spills from tankers under ballast. The regimes were revised in 1987, the maximum compensations payable were increased and the revised agreements have global application, i.e. regardless of whether or not the Civil Liability Convention and the Fund Convention apply to the incident.

The possibility of legal proceedings against the offending vessel require the identification of the offender. Thus samples of spilled oil should be taken, properly labelled and witnessed and submitted for analysis. Samples should also be taken from the oil cargo, bunker tanks and machinery spaces of suspected offenders for comparative analysis with the spilled oil. A custodial chain may be asked to be proved in court.

Claims for clean-up costs and damage can be brought against the owner of the ship which caused the oil spill and, if the limit of liability is exceeded, against the IOPC Fund (if the country where the pollution damage is caused is a member of the IOPC Fund) or CRISTAL (if the cargo owner is a member). Normally an Administration will coordinate the submission of various claims and it is essential that accurate detailed records are kept to support such claims. Each claim should contain:

- the name and address of the claimant or representative if any;
- the identity of the ship involved in the incident;
- the date, the place and specific details of the incident;
- the type of oil, the clean-up measures taken and the kind of pollution damage as well as the places affected;
- the amount of the claim.

The IOPC Fund provides a "Claim Manual" to claimants on request in order to facilitate the presentation of claims against the IOPC Fund.

The claims should be broken down depending on the amount and nature into different categories such as:

- costs of preventive measures and clean-up operations, giving a summary of events with a description of work carried out and working materials used, labour costs (number and categories of response personnel, regular and overtime rates of pay, days/hours worked) equipment and material costs (types of equipment used, rate of hire, consumable material quantity and cost), transport costs (number and types of vessels/aircraft/helicopters/vehicles used, number of days/hours operated, rate of hire or operating costs), costs of temporary storage and final disposal of recovered oil and oily material. Daily work sheets should be compiled by supervisory personnel. Standard work sheets should be used. To ensure that adequate control of expenditure is kept, a financial controller should be assigned to the response team;
- replacement and repair costs, giving the extent of pollution damage to the property, description of items destroyed, dama-

ged, needing replacement or repairs (boats, fishing gear, clothing, roads with their locations). Numerous claims will be made from the public and private sector such as fishermen, pleasure boat owners, marina operators, etc.;

- economic loss, giving the nature of the loss, including demonstration that the loss resulted from the incident, comparative figures for profits earned in previous periods and for a period during which such damage was suffered. These losses would be incurred by the restriction of fishing activity, closure of coastal industrial and processing installations, loss of income to operators (hotels and restaurants), increase of costs of alternate protein sources for coastal communities affected by spillage.

In case of spills of non-persistent oils, spills from unladen tankers and spills from ships other than tankers, compensation will have to be sought under the applicable national law. States are free to legislate as they consider appropriate as regards such spills. A State may provide in its national law that claims in respect of damage to harbour works, basins and waterways and aids to navigation shall have priority over other claims. However, as regards States-Parties to the Civil Liability Convention, all claims for pollution damage against the owner of the tanker from which the oil escaped must be given equal priority.

Protection and Indemnity Associations (P&I CLUBS) exist to provide shipowners with insurance for liability they may incur to third parties, i.e. anybody other than the insured shipowner. Numerous risks covered by P&I CLUBS do include oil pollution liabilities, even that coming from bunker oil leakage from ships other than tankers. P&I CLUBS will provide advice on the shipowner's rights and duties and negotiate on the shipowner's behalf with the appropriate authorities to take fast and effective action to minimise the damage and subsequent liabilities. At a later stage the Club will assist in determining eventual liability for damage and the extent of the compensation.

Glossary of Technical Terms

A

ablation of ice – melting down of ice above entrapped oil to expose the oil;
abrasive action – grinding down of a surface;
absorbent – substance capable of taking up oil through capillaries;
absorption – process by which one substance penetrates the interior of another substance so that the absorbed substance disappears physically;
accelometer – an instrument for determining the acceleration of the system with which it moves;
accommodation spaces – spaces used for public spaces, corridors, lavatories, cabins, offices, hospitals, cinemas, games and hobbies rooms, pantries containing no cooking appliances and similar spaces;
accretion – process of transportation of sand and building up of beaches;
acoustic wave – a wave due to the compressibility of water;
adhesion – tendency of a material to remain in physical contact with another material;
administration – the Government of the State whose flag the unit is entitled to fly;
adsorption – process of oil adhering to the surface of some substance without actually penetrating its internal structure generally leading to sinking;
advection – horizontal and vertical flow of sea water as a current;
aerial surveillance – tracking of an oil spill from the air;

aerosol – system in which liquid or solid particles are distributed in a finely divided state through a gas, usually air. Particles within aerosols are usually less than 1 micron (0.001 mm) in diameter and are more uniformly distributed than in a spray. The solid particles may be windblown pollutants carried for great distances, depending on wind velocity and the size of the particles;

agglomeration – process in which precipitation particles grow by collision with and assimilation of other particles. Agglomeration may result from coalescence (if two colliding particles are both liquid water), from accretion (if one particle is an ice crystal or snow flake and the other a supercooled water droplet) or from the clustering of crystals;

air lift – method in which the displacement of water is replaced by air pressure in a tank, which pushes the water out of the tank through the breached bottom;

air or water streams – method of oil containment where the force of air or water directed as a stream can be used to divert or to contain an oil slick;

alcohols – class of organic chemical compounds charaterised by the presence of a hydroxyl (OH: oxygen–hydrogen) group attached to a carbon atom. Alcohols are important solvents and are used to a certain extent in the preparation of chemical dispersants;

algae – water plants of very simple structure, characterised by possessing chlorophyll and so being capable of producing food by photosynthesis. Algae range from unicellular organisms to large, complex colonies of kelp and from the blue–green algae to yellow, green, brown and red organisms. Algae are found in almost every habitat. In the oceans, vast numbers of minute species float suspended in the upper levels of the water, while shores are covered with many and varied forms from high tide to depths of 30 feet or more;

alkanes – class of hydrocarbons (compounds of hydrogen and carbon) characterised by branched or unbranched chains of carbon atoms with attached hydrogen atoms of general formula C_nH_{2n+2};

alkenes – class of straight or branched chain hydrocarbons similar to alkanes but characterised by the presence of carbon atoms united by double bonds. Alkenes have the general formula C_nH_{2n};

alkyl – a generic name for any organic group or radical formed from a hydrocarbon such as methane, ethane, propane, etc. and so producing a univalent unit (methyl, ethyl, propyl, etc.);

ambient – local or surrounding conditions: primarily used with reference to climatic conditions at some point in time, e.g. ambient temperature;

amphibious vehicle – vehicle capable of moving on land and on water;

amplitude – in an ocean surface wave, the vertical distance from the still water level to the wave crest;

anaerobic – a term used to describe a situation or an area characterised by a lack of oxygen. A term also used with reference to organisms such as some bacteria which can survive and grow in the absence of gaseous or dissolved oxygen;

anchor dragging – anchor pulled over the bottom of the sea under excessive load;

anchoring system – a system used for keeping a ship or any floating structure fast to the sea bottom;

API gravity – a scale developed by the American Petroleum Institute to designate an oil's specific gravity, or the ratio of the weights of equal volumes of oil and pure water. API gravity is dependent on temperature and barometric pressure, and is therefore generally measured at 16 °C and one atmosphere pressure. Water with a specific gravity of 1.0 has an API gravity of 10°. A light crude oil may have an API gravity of 40°. Oils with low specific gravities have high API gravities and vice versa. API gravity can be calculated from specific gravity using the following formula:

$$\text{API}^\circ = (141.5/\text{specific gravity at } 16\ ^\circ\text{C}) - 131.5;$$

aquaculture – cultivation of plants and animals in water;

aromatics – a class of hydrocarbons characterised by rings containing 6 carbon atoms. Benzene is the simplest aromatic and most aromatics are derived from this compound. Aromatics are considered to be the most immediately toxic hydrocarbons found in oil, and are present in virtually all crude oils and petroleum products. Many aromatics are soluble in water to some extent, thereby increasing their danger to aquatic organisms. Certain aromatics are considered long–term poisons and often produce carcinogenic effects;

asphalt – a black or brown hydrocarbon material that ranges in consistency from a heavy liquid to a solid. The most common source of asphalt is the residue left after the fractional distillation of crude oils. Asphalt is primarily used for surfacing roads. It may result from weathering of crude heavy oil on the beach;

atoll – a ring–shaped coral reef, often carrying low sand islands, enclosing a lagoon;

atomiser – a device for producing a fine spray;

auk – an Arctic marine bird, incapable of flying, and nesting on rocky ledges in large colonies. Auks feed on fish, crustaceans and plankton;

B

backshore – the area of the shoreline above the high tide mark. The backshore is only inundated with water during exceptionally strong storms or abnormally high tides accompanied by high winds and therefore does not support characteristic intertidal flora and fauna. Granular materials for the replacements of oil contaminated beach material excavated during shoreline clean–up programmes are frequently taken from backshore areas;

bar – a ridge of sand across the mouth of a river or the entrance to a bay or along the shoreline, deposited by currents or tides;

barrel – a unit of liquid (volumetric) measure for petroleum and petroleum products, equal to 35 Imperial gallons or approximately 160 litres;

barren – land not good enough to produce crops, or plants not producing fruit or seeds;

barrier or containment barrier – any non-floating structure which is constructed to contain or divert spilled oil. Barriers are generally improvised and, unlike booms, are usually left in place until a clean–up programme is completed. Sorbent materials may be used in barrier construction to recover spilled oil. Generally they are used in streams or ditches too shallow for conventional floating booms;

bathymetry – information related to the depth of water;

bayou – a swamplike estuarine creek, also called a slough;

beach gear – gear consisting of an anchor of high holding ability and a strong wire rope fastened to the ship. This gear is used for liberating stranded ships;

Beaufort wind scale – a system of estimating and reporting wind speeds. It was invented in the early XIX century by Admiral Beaufort of the British Navy. In its present form it equates wind scale, wind speed, a descriptive term and visible effects on the sea surface;

belt skimmer – a skimmer with a moving belt conveying oil by adhesion from the water surface;

benthic – related to the bottom of a body of water;

benthos – the plants and animals that live on the sea bottom. These include permanently attached or immobile forms (e.g. sponges, corals, oysters), creeping forms (e.g. crabs, snails) and burrowing animals (e.g. worms). Barnacles, the larger seaweeds and seasquirts are also members of this group;

benzene – colourless, odorous aromatic hydrocarbon C_6H_6 ("benzol"), insoluble in water;

berm – a relatively flat erosion surface brought to grade during a previous cycle of erosion;

bilateral agreement – agreement between two neighbouring countries for immediate intervention in case of an oil spill;

bioassay expression (96hLC$_{50}$) – concentration of oil producing a mortality of 50% of the test organisms within 96 hours;

biodegradation – the degradation of substances resulting from their use as food energy sources by certain microorganisms including bacteria fungi and yeasts. The process with respect to oil degradation is extremely slow and limited to a great extent by temperature, nutrients and oxygen availability. Although more than 100 species of microorganisms have the ability to utilize hydrocarbons as an energy source, no single species can degrade more than 2 or 3 of the many compounds normally found in oil;

biogenic – a living entity springing from previously existing living entities;

biological agent – microorganisms (primarily bacteria) added to a water column or soil to increase the rate of biodegradation of spilled oil. Alternatively, nutrients added to the water (in the form of fertilisers) to increase the growth and biodegradation capacity of microorganisms already present;

boiling point – the temperature at which a liquid begins to boil; specifically, the temperature at which the vapour pressure of a liquid is equal to the atmospheric or external pressure. The boiling point of crude oils and petroleum products may vary from 30 to 550 °C;

bivalve – molluscs with a hinged, double shell: e.g. oyster, mussel, clam;

blower – apparatus for forcing air into or through something;

blow-out – sudden, uncontrolled escape of oil and gas from a drilling well;

blubber – fat of whales and other sea animals, from which oil is obtained;

boom or containment boom – a floating mechanical structure which extends above and below the water surface and is designed to stop or divert the spread or movement of an oil slick. Booms are an integral part of virtually all clean-up programmes following an oil spill on water;

boom draft – distance from the normal calm water line to the bottom of the skirt;

boom failure – failure of a commercial or improvised boom to contain oil due to excessive winds, waves or currents, or improper deployment. Boom failure may be manifested in oil underflow, oil splashover or structural breakage;

boom height – distance from top to bottom of boom;

boom skirt – that portion of the boom from the lowest point to the bottom of the flotation;

bore – a rapid rise in water height that propagates into estuaries;

boulder – large piece of rock or large stone rounded by water;

bow – front or forward end of a ship;

breadth of a ship – the extreme width from outside of frame to outside of frame at or below the deepest subdivision load line;

breaker – large wave breaking into foam as it advances towards the shore;

brine channels – small passages formed in the lower surface of first–year sea ice formed by the exclusion of saline water or salts during rapid freezing. The oil under first year ice will migrate through the brine channels once the spring melting process begins;

bubble barrier – a containment barrier which takes advantage of the two–way currents produced when a rising curtain of bubbles reaches the water surface. The system has been used with some success in relatively calm harbours, but requires considerable maintenance when submerged perforated pipes which are used to produce the bubble curtain become covered with redistributed bottom silt;

bulk carrier – an ocean–going vessel specifically designed to transport large quantities of a single product such as grain or coal;

bulkhead deck – uppermost deck up to which the transverse watertight bulkheads are carried;

bunker "B" – a relatively viscous fuel oil (No.5 fuel) used primarily as a fuel for marine and industrial boilers;

bunker "C" – a very viscous fuel oil (No.6 fuel) used as a fuel for marine and industrial boilers;

bunkering – accepting fuel oil into a ship's bunker tanks;

buoy tender – ship tending navigational markers;

burning agent – compounds or materials such as gasoline which are used to ignite and sustain combustion of spilled oil which otherwise will not burn. Burning agents are generally required to burn weathered oils since volatile, low flash point hydrocarbons are rapidly lost through evaporation;

burrow – hole made in the ground by benthic organisms;

C

cabbeling – sinking due to mixing of waters of equal density but differing temperatures and salinities;

capacitance – ratio of electric charge to the related change in potential;

capillary – tube with hair–like diameter;

capillary action – the process whereby the force of attraction between a solid and a liquid causes the liquid to be drawn into the porous internal structure of the solid;

capillary wave – a wave in which surface tension forces dominate. If the wind velocity is sufficient to overcome surface tension the ripples will occur on the water surface. The surface tension will restore the smooth surface;

carbon number – the number of carbon atoms present in a single molecule of a given hydrocarbon. The physical and chemical properties of hydrocarbons tend to vary with the number of carbon atoms and these properties are frequently described in terms of carbon number ranges for specific classes of hydrocarbons; alkanes with carbon numbers 1 to 4 are gaseous at ordinary temperatures and pressures;

carcinogenic – causing cancer;

catalyst – a substance added to a reacting system, which alters the rate of the reaction without itself being consumed, e.g. silica or alumina are used during the refining of petroleum to increase the rate at which large hydrocarbon molecules are split into smaller ones (catalytic cracking);

catamaran – a ship with two parallel hulls;

celerity – wave speed;

centrifugal skimmer – a mechanical skimmer design which operates by the creation of a water vortex or whirlpool which draws oil into an area within the device where it can subsequently be pumped off;

chemical barrier – a barrier formed by chemicals which act as surface modifiers to inhibit the spread of an oil slick on water. When placed on the water surface next to an oil film, these

chemicals repel the oil as a result of their surface tension;
chemical dispersion – with respect to oil spills, this term refers to the creation of oil–in–water emulsions by the use of chemicals made for this purpose. With regard to shoreline clean–up, chemical dispersion is the process of spraying chemical dispersants to remove stranded oil from rocky shoreline areas which are not considered as biologically sensitive;
chevron shape – shaped in bent lines;
chlorophyll – green colouring matter in leaves of plants playing a basic role in photosynthesis;
chocolate mousse – the name given to a water–in–oil emulsion containing 50 to 80% water. These emulsions are very stable, have a butter–like consistency and are only formed with a relatively viscous oil in the presence of considerable wave action;
clapotis – a standing wave formed by reflection from a seawall;
clay – soil or sediment particles which are less than 0.004 mm in maximum dimension. Most clays are produced as a result of the weathering of coarser rock materials. Clay particles are smaller than either sand or silt;
cliff – a steep face of rock at the edge of the sea;
coalescence – the process of several liquids coming together and uniting into one substance;
coastal state – the Government of the state exercising administrative control over drilling operations of the unit;
coastline – the line that forms the boundary between the coast and the shore;
cobble – a stone worn round and smooth by water;
cobble beach – a beach composed primarily of gravel having a size range from 64 to 256 mm. This type of beach is also referred to as a shingle beach. By comparison, boulder substrates are greater than 256 mm, while pebble substrates range in size from 4 to 64 mm;
collection efficiency – volume of oil recovered relative to the volume of liquid collected;
columnar ice – ice consisting of tall, upright small diameter pillars arranged one under the other;

column stabilised unit – a unit with the main deck connected to the underwater hulls or footings by columns or caissons;

composting – decomposition of organic matter by bacteria for agricultural use;

containment – the process of preventing the spread of oil beyond the area where it has been spilled in order to minimise pollution and facilitate recovery;

contamination – the process of making dirty, impure or diseased;

continental shelf wave – a low–frequency wave trapped along the continental shelf that can only propagate with the coast on its left in the southern hemisphere and the coast on its right in the northern hemisphere. Their period must exceed the inertial period;

contingency plan – a plan for action prepared in anticipation of an oil spill. It consists of guidelines developed for a specific industrial facility or an entire region to increase the effectiveness and speed of clean–up operations in the event of an oil spill and simultaneously protect areas of biological, social and economic importance;

contract "No cure – No pay" – a contract by which no remuneration is paid to the salvor, if the ship is not brought to a place of safety;

control stations – spaces in which the unit's radio or main navigating equipment or the emergency source of power is located or where the fire recording or fire control equipment or the dynamic positioning control system is centralised or where a fire–extinguishing system serving various locations is situated;

convection – transfer of heat from one part of a liquid or gas to another by the movement of heated substances or a process of vertical circulation being driven by vertical density differences;

convergence – coming towards each other;

copepods – the most numerous crustaceans, found in fresh and marine waters free living or living on fish (parasitic);

Coriolis effect – any object moving above the earth with constant velocity is deflected relative to the surface of the rotating

earth. The deflection is found to be to the right in northern hemisphere and to the left in the southern hemisphere;
countermeasure – action to prevent or control pollution by oil spills;
crab – a ten–legged shellfish with a broad flat body;
crest – top of a wave;
critical tow speed – speed at which a boom fails to maintain a mechanical barrier on the water surface by planing, submarining or mechanical failure;
critical velocity – the lowest water current velocity which causes loss of oil under the skirt of a containment boom;
crude oil – any oil occurring naturally in the earth, whether treated or untreated to render it suitable for transportation;
crustacean – any member of a numerous class of animals mostly living in water with hard shell (crabs, lobsters, etc.) – popularly called shellfish;
cusp – low mounds of beach material separated by crescent–shaped troughs;

D

dead ship condition – a condition under which the main propulsion plant, boilers and auxiliaries are not in operation due to the absence of power;
deadweight – difference in tonnes between the displacement of a ship in water of a specific gravity of 1.025 at the load waterline corresponding to the assigned summer freeboard and the light weight of the ship;
debris – scattered wreckage of no practical value;
decontamination – removal of impurities or poisons;
defoliation – destruction of leaves in plants;
degradation – gradual decomposition which occurs in stages with well–marked intermediate products;
density – relation of weight to volume;
Department of Environment – Government Department responsible for the protection of the environment of the countryside and the sea coast;
deponium – site of depositing debris;

deposition – the process of laying or putting down solid matter by water;

desalination – removal of salt and other dissolved minerals from seawater;

detrital material – matter produced by wearing away from rock etc., such as sand, silt, gravel;

detritus – loose material that results from rock disintegration or abrasion. Also, suspended material in the water column including fragments of decomposing flora and fauna and faecal pellets produced by zooplankton and associated bacterial communities;

diffusion – spreading in every direction;

discharge – in relation to harmful substances or effluents containing such substances means any release from a ship and includes escape, disposal, spilling, leaking, pumping, emitting or emptying. It does not include dumping under the London 1972 Dumping Convention, release of harmful substances directly arising from the exploration, exploitation and associated off–shore processing of sea–bed mineral resources or release of harmful substances for the purposes of legitimate scientific research into pollution abatement or control;

dispersants – chemicals which reduce the surface tension between oil and water and thereby facilitate the break–up and dispersal of the slick throughout the water column;

dispersion – the distribution of spilled oil into the upper layers of the water column by natural wave action or application of chemical dispersants;

dissipation – persistent loss of mechanical energy due to the presence of friction–like resistance to motion, which eventually exhausts the total energy of the system and causes it to come to rest;

dissolution – the act or process of dissolving one substance in another. Specifically, a process contributing to the weathering of spilled oil whereby certain "slightly" soluble hydrocarbons and various mineral salts present in oil are dissolved in the surrounding water;

distillation – making pure;
diurnal – once every 24 hours;
divergence – the process of moving further from a point or the movement of objects from each other as they progress;
diversionary boom – a boom changing the direction of oil spill movement;
diverter barrier – a barrier turning the oil spill movement in another direction;
diving system – plant and equipment necessary for safe conduct of diving operations;
dose – the amount of something given at any time;
draft (draught) – vertical distance from the moulded base line amidships to the subdivision load line in question;
drift – a movement due to currents, winds, tides etc.;
droplet breakaway – a type of boom failure resulting from excessive current velocity. The head wave formed upstream of the oil mass contained within a boom becomes unstable and oil droplets are torn off and become entrained in the water flow beneath the boom;
dry sand blasting – cleaning by means of dry sand accelerated in a stream of compressed air;
dumping – placing of rubbish where it may be unloaded and deposited. It means any deliberate disposal at sea of wastes or other matter from vessels, aircraft, platforms or other man–made structures as well as any deliberate disposal at sea of vessels, air–craft, platforms or other man–made structures. It does not include the disposal at sea of wastes or other matter incidental to or derived from the normal operations of the above named objects;
dune – mound of loose, dry sand formed by the wind near the seashore;

E

earth's crust – the outer portion of the earth;
ebb – associated with a falling tide;
ecosystem – a community functioning as a dynamic system within the environment;

edge wave – gravity waves of period less than the inertial period, which are trapped along a beach or along a coastline and can only propagate parallel to it;
emissivity – capability of sending out or giving off (of light, heat, smell, etc.);
emulsification – the process whereby one liquid is dispersed into another liquid in the form of small droplets. In the case of oil, the emulsion can be either oil–in–water or water–in–oil. Both types of emulsions are formed as a result of wave action, although water–in–oil emulsions are more stable and create special clean–up problems;
enclosed spaces – spaces delineated by floors, bulkheads and/or decks, which may have doors and/or windows;
encounter rate – frequency of meeting a wave;
encrustation – covering with a crust;
environmental sensitivity – the susceptibility of a local environment or area to any disturbance which might decrease its stability or result in either short or long–term adverse impacts. Environmental sensitivity generally includes physical, biological and socio–economic parameters;
erosion – any process of wearing away sea shores, causing reduction of land surface towards sea level by various agencies of weathering, stream action, glacial action, wind action etc;
estuary – a partly enclosed coastal body of water in which freshwater (usually originating from a river) is mixed and diluted by sea water. Estuaries are generally considered more biologically productive than either adjacent marine or freshwater environments;
eutrophication – an oversupply of nutrients to a water body;
evaporation – conversion process of a liquid to a vapour. The rate of evaporation depends on the volatility of various hydrocarbon constituents, temperature, wind and water turbulence and the spreading rate of the slick. Evaporation is the most important process in the weathering of most oils;

F

faecal pellets – solid or semi–solid excretion products (faeces) which are enclosed within a thin membrane, such as are produced by zooplankton and some other invertebrates. If organisms ingest oil which is dispersed in the water column, these faecal pellets may contain oil globules;

fathom – an imperial unit of measure equal to 1.83 m;

fauna – animals in general, or animal life as distinguished from plant life (flora). Usually used with reference to all the animal life characteristic of or inhabiting a particular region or locality;

fertiliser – plant chemical food or artificial manure;

fireproof boom – a boom resistant to flames and heat which does not allow fire to spread beyond it;

first–gross–loss tow speed – tow speed at which the oil preload is first lost from the boom by drainage;

first–loss tow speed – tow speed at which the contained oil is first shed from the boom by entrainment;

first–year ice – in the Arctic, ice formed during the winter of a given year and not containing residual or polar pack ice from previous years. It is characterised by the presence of brine channels and is relatively porous in comparison with pack ice;

fjord – a narrow, deep, steep–walled inlet of the sea usually associated with a deep glacial trough;

flash point – the temperature at which vapours produced by a crude oil or petroleum product will ignite when exposed to an ignition source such as a naked flame;

flexible panel – a section of a fireproof boom, which can easily bend without breaking;

floater – a body used to support a net or boom end floating in water;

flocculation – formation of small, loosely aggregated bits of material (floccules), usually clay, forming through electrochemical reactions in salt water;

flood – associated with a rising tide;

flora – plants in general or plant life as distinguished from fauna (animal life). Usually used in reference to all the plant life inhabiting or characteristic of a particular region or locality;

flotation – being supported by air, gas or on the surface of a liquid; ability to move with moving air or liquid;

fluorescence – the process of emission of electromagnetic radiation by a substance as a consequence of the absorption of energy from radiation; the ability of substances to take in radiations and emit them in the form of light;

flushing – rapid movement of water washing away particles of oil;

foreshore – part of the shore between sea and land that is cultivated, built on etc;

fractional distillation – separation of a mixture of liquids such as crude oil into components having different boiling points. Fractional distillation is the primary process in the refining of crude oils;

freeboard – the part of a floating boom designed to prevent waves from washing oil over the top. The term is also used to describe the distance from the water surface to the top of the boom. Freeboard is generally also applied to the distance from the deck of a vessel to the water line;

freezing – process of turning water into ice;

fronts – regions of separation between fluid masses of different physical properties;

fuel oils – refined petroleum products having specific gravities in the range from 0.85 to 0.98 and flash points greater than 55 °C. This group of products includes furnace, auto diesel and stove fuels (No.2 fuel oils); plant or industrial heating fuels (No.4 fuel oils) and various bunker fuels (No.5 & 6 fuel oils);

fumigation – an air pollution event characterised by a plume spreading downwards only;

fungus (fungi) – plant (plants) without leaves, flowers or green colouring matter, growing on other plants or on a decaying matter, e.g. old wood;

G

garbage – all kinds of victual, domestic and operational waste excluding fresh fish and parts thereof, generated during the normal operation of the ship and liable to be disposed of continuously or periodically, except those substances which are defined or listed in other regulations;

gasoline – a mixture of volatile, flammable liquid hydrocarbons used primarily for internal combustion engines and characterised by a flash point of approximately –40 °C and a specific gravity from 0.65 to 0.75;

gastight doors – a solid, close–fitting door, designed to resist the passage of gas under normal atmospheric conditions;

gastropods – a large group of marine and freshwater species such as snails, slugs and allied forms constituting a class of the phylum Mollusca;

gelling agents – chemicals which increase the viscosity of oil and can be applied to an oil slick to reduce its rate of spread over the water surface. However, gelling agents are rarely used due to their expense, the large volume required and their slow action;

globulation – forming of tiny drops from an oil slick under the influence of breaking waves;

glycols – any of the class of organic compounds belonging to the alcohol family, but having two hydroxyl groups (OH) attached to different carbon atoms. The simplest glycol is ethylene glycol, a compound used extensively for automobile antifreeze and contained in many modern chemical dispersants;

granular ice – ice consisting of small crystals;

gravel – small stones with coarse sand;

gravity – force attracting objects towards the centre of the earth;

gravity wave – any wave in which gravity is a dominant restoring force;

grooming – cleaning and arranging the fur or feathers;

groundfish – fish species normally found close to the sea bottom throughout the adult phase of their life history. Groundfish feed extensively on bottom fauna and include species such as cod, halibut and turbot;

groundwater – water present below the soil surface and occupying voids in the porous subsoil. The upper surface of the groundwater is referred to as the water table. Contamination of groundwater supplies is a major concern when oil is spilled on land since groundwater supplies springs and wells and passes into surface water supplies in many areas;

GRP – glass reinforced plastic;

H

habitat – the space occupied by an individual or a group of individuals including everthing in that space both living and nonliving;

harmful substance – any hazardous, noxious or other substance which, if introduced into the sea, is liable to cause pollution;

hazardous area – all those areas, where, due to the possible presence of a flammable atmosphere arising from drilling operations, the use of machinery or electrical equipment without proper consideration may lead to fire or explosion;

head wave – an area of oil concentration which occurs behind and at some distance from a containment boom. This area of oil thickening is important to the positioning of mechanical recovery devices (i.e. skimmers) and is the region where the droplet breakaway phenomenon is initiated, where current flow underneath the boom exceeds critical velocity;

HLB – hydrophilic/lipophilic balance;

hose – flexible tubing;

hydraulic dispersion – one of various shoreline clean-up techniques which utilises a water stream at either low or high pressure to remove stranded oil. These techniques are most suited to removal of oil from coarse sediments, rocks and man–made structures, although care must be taken to avoid damage to intertidal flora and fauna;

hydrocarbons – organic chemical compounds composed only of carbon and hydrogen. Hydrocarbons are the principal constituents of crude oils, natural gas and refined petroleum products;

hydrophobic agent – a chemical or material which has the ability to resist wetting by water. Hydrophobic agents are occasionally used in the treatment of synthetic sorbents to decrease the amount of water absorbed and hence increase the volume of oil they can absorb before becoming saturated;

hydrography – the description and study of seas, lakes, rivers and other waters and especially the configuration of underwater surfaces;

I

icebreaker – ship with strong, curved hull used for breaking a passage through ice;

igniter – a mechanism for igniting solids, liquids or gases;

impedance – the complex ratio of a force–like quantity (force, pressure, voltage, temperature or electric field strength) to a related velocity–like quantity (velocity, volume velocity, current, heat flow, or magnetic field strength);

improvised boom – a boom constructed from readily available materials such as railroad ties, logs and telephone poles. Improvised booms may be used as temporary containment structures until more suitable commercial booms arrive at spill site. They may also be used in conjunction with commercial floating booms to divert oil into areas where the commercial booms are positioned;

incident – an event involving the actual or probable discharge into the sea of a harmful substance or effluents containing such a substance;

incinerator – furnace or enclosed fireplace for burning debris;

inert gas – gas without active chemical properties;

ingestion – taking in or swallowing;

inshore – close to the shore;

interface – surface or area common to two or more systems;

intertidal zone – the portion of the shoreline between the low tide mark and high tide mark which is covered by water at some time during daily tidal cycles. The size of the intertidal zone varies with the tidal characteristics of a given region, as well as the steepness of the shoreline;

invertebrate – not having a backbone or spinal column, e.g. insects, worms, molluscs;

iridescent – changing colours as light falls from different directions, showing rainbow–like colours;

IR/UV scanner system – infrared/ultraviolet scanner system, used in aerial surveillance at close range to obtain high resolution imagery;

isothermal –having the same mean temperature;

isotropic – denotes a medium whose properties are the same, in whatever direction they are measured;

J

jet – a narrow strong current;

jet fuel – a kerosine or kerosine–based fuel used to power jet aircraft combustion engines;

joystick – control lever moving in all directions;

K

kelp – large, brown algae or brown seaweeds;

kerosine – a flammable oil characterised by a relatively low viscosity, a specific gravity of approximately 0.8 and a flash point near 55 °C. Kerosine lies between the gasolines and fuel oils in terms of major physical properties and is separated from these products during fractional distillation of crude oils. Uses for kerosine include fuels for wick lamps, domestic heaters and furnaces, fuel or fuel components for jet aircraft engines and thinner in paints and insectide emulsions;

kinetic energy – energy of a moving body because of its motion;

knot – measure of speed for ships, 1 knot = 1 nautical mile per hour $\approx$ 0.5 m/s;

L

lagoon – a shallow body of water separated from open sea by a reef barrier usually connected to the sea;

landbased pollution – pollution of the sea caused by discharges from land reaching the sea water borne, air borne or directly from the coast, including outfalls from pipelines;

landfilling – filling a groundhole with oil spill debris;

lanugo – dense layer of fur on newly born seals;

laser – a device for generating, amplifying and concentrating light waves into an intense beam in one specific direction;

layering – forming one horizontal division;

lead agency – the authority within a national government designated under the Contingency Plan as having overall responsibility for response to a marine emergency;

lee – the region toward which the wind blows. The lee of an object is sheltered;

leeway – wind-induced drift of an oil spill; sideways drift of a ship or oil slick in the direction towards which the wind is blowing;

levee – a dike or embankment to protect land from inundation;

levitation device – a device causing oil to rise and float in the air in defiance of gravity;

light ends – a term used to describe the low molecular weight, volatile hydrocarbons in crude oil and in petroleum products. The light ends are the first compounds recovered from crude oil during fractional distillation process and are also the first fractions of spilled oil to be lost through evaporation;

lightering – unloading a ship at sea in order to decrease its draft in case of an emergency;

light-weight – displacement of a ship in tonnes without cargo, fuel, lubricating oil, ballast water, freshwater and feed water in tanks, consumable stores, passengers and crew and their effects;

litigation – going to law;

littoral zone – a zone affected by the action of both tides and breaking waves;

lobster – shellfish with stocked eyes and chitinous, segmented exoskeletons, covering a three–part body. They have eight legs and two claws. Lobsters are bluish–black before and scarlet after being boiled;

longshore transport – transport of sand along the shores by longshore currents;

lubricating oils – oils used to reduce friction and wear between solid surfaces such as moving machine parts and internal combustion engine components. They are refined from crude oil through a variety of processes including vacuum distillation, extraction of specific products with solvents, removal of waxes with solvents and treatment with hydrogen in the presence of a catalyst. The viscosity of a lubricating oil is its most important characteristic since it determines the amount of friction and whether a thick enough film can be built up to avoid wear from solid–to–solid contact;

M

machinery spaces – all machinery spaces containing propelling machinery, boilers, and other fired processes, oil fuel units, steam and internal combustion engines, generators and major electric machinery, oil filling stations, refrigerating, stabilizing, ventilation and air conditioning machinery and similar spaces and trunks to such spaces;

maintenance – all activities to support the life of equipment;

mammals – any of the class of animals which feed their young ones with milk from the breast;

mangrove – a tropical or semitropical tree found in brackish waters that has large prop root systems that are important in coastal land building;

manual recovery – a term used to describe the recovery of oil from contaminated areas by the clean–up work force using buckets, shovels and similar equipment. Manual recovery is extremely labour intensive, but plays a role in many oil spill clean–up programmes;

marine emergency – any casualty, incident, occurrence or situation, however caused, resulting in substantial pollution or imminent threat of substantial pollution to a marine environment by oil;

marsh – a brackish-water area adjoining salt water;

maximum ahead service speed – greatest speed which a ship is designed to maintain in service at sea at the deepest seagoing draft;

maximum astern speed – the speed which a ship can attain at the designed maximum astern power at the deepest seagoing draft;

metabolism – the interchange of materials between living organisms and the environment, by which the body is built up and energy for its vital processes is secured;

methane – odourless, colourless, inflammable gas (CH_4) occurring as natural gas or marsh gas in marshy areas;

metric ton (tonne) – a unit of mass and weight equal to 1,000 kilograms or 2,205 pounds avoirdupois. The metric ton is the most widely used measure of oil quantity by weight. There are roughly 7 to 9 barrels (245 to 315 Imperial gallons) of oil per ton, depending on the specific gravity of the crude oil or petroleum product;

microbial degradation – degradation caused by microorganisms;

microorganisms – plant or animal life of microscopic or ultramicroscopic size (i.e. not visible to the human eye without the aid of a microscope). Microorganisms are found in air, water and soil, and generally include bacteria, yeasts and fungi. Some microorganisms are capable of metabolizing hydrocarbons and play a role in the natural degradation of spilled oil;

migration of oil – movement of trapped oil through ice brine channels;

mineral–based sorbent – any of a number of inorganic, mineral–based substances used to recover oil because of their adsorptive or absorptive capacities. Mineral–based sorbents include materials such as vermiculite, perlite or volcanic ash and recover from 4 to 8 times their weight in oil;

mineral spirits – flammable petroleum distillates that boil at temperatures lower than kerosine and are used as solvents and thinners, especially in paints and varnishes. Mineral spirits are the common term for some naphthas;

mobile offshore drilling unit (MODU) – a vessel capable of engaging in drilling operations for the exploration and for the exploitation of resources beneath the sea–bed such as liquid or gaseous hydrocarbons, sulphur or salt;

mobile oil dike (MOD) – a nonselfpropelled, oil skimming catamaran comprising two floats, a box transverse girder and a sloping ramp, built by Nobiskrug (FRG);

molecular weight – the total mass of any group of atoms which are bound together to act as a single unit or molecule;

molluscs – animals with soft bodies and often hard shells, e.g. oysters, mussels, cuttlefish, snails, slugs;

momentum – quantity of motion of a moving body (the product of its mass and velocity);

morphology – a branch of biology dealing with the form and structure of animals and plants;

mud – sediment of rock fragments or organic detritus;

mussel – mollusc with a black shell in two parts;

N

naphtha – any of various volatile and often flammable liquid hydrocarbon mixtures used chiefly as solvents and thinners. Naphtha consists mainly of hydrocarbons which have a higher boiling point than gasolines and a lower boiling point than kerosine;

naphthenes – a class of hydrocarbons with similar physical and chemical properties to alkanes, but characterised by the presence of simple closed rings. Like alkanes, naphthenes are also saturated (i.e. they contain no carbon–carbon double bonds) and have the general formula C_nH_{2n}. This class of hydrocarbons is insoluble in water and generally boils at 10 to 20 °C higher than the corresponding carbon number alkanes;

national response organisation – governmental agency responsible for national planning and oil spill response in a country;

natural organic sorbent – natural materials such as peat moss, straw and saw dust which can be used to recover spilled oil. Natural sorbents generally absorb 3 to 6 times their weight in oil by virtue of the criss–cross arrangement of fibres within the material. However, all natural sorbents will absorb water as well as oil and virtually all sink when saturated with water, adding to problems of oil spill clean–up;

node – point of rest in a vibrating body;

normal – a mathematical term synonymous with perpendicular;

nozzle – a fitting through which a stream of liquid or gas is directed;

nutrient – a compound serving as or providing nourishment. It is needed for the growth of marine plants, consisting of nitrites, nitrates, phosphates, ammonia and silicates;

O

offshore – in a direction away from the shore or land;

oil – petroleum in any form including crude oil, fuel oil, sludge, oil refuse and refined products;

oil–in–water emulsion – an emulsion of oil droplets in surrounding water, formed as a result of wave action or by a chemical dispersant. Oil–in–water emulsions show a tendency to coalesce and reform an oil slick when the water becomes calm, although the presence of surface active agents in the oil or artificially added in the form of chemical dispersants increases the persistence of this type of emulsion;

oil slug – the name given to a downward moving oil mass which often results when oil is spilled on a relatively porous soil;

oil spill contractor – a private firm which has been formed to provide oil spill clean–up services. These independent contractors often have their own containment and recovery equipment and may be identified in local and regional contingency plans;

oil spill cooperative – organisation formed by oil companies operating in a given area for the purpose of pooling equipment and training personnel to combat oil spills;

oleophilic agent – a material or chemical which has the tendency to attract oil. Chemicals of this type may be used to treat sorbent materials in order to increase their oil recovery;

on–scene commander (OSC) – the overall coordinator of an oil spill response team. The OSC is responsible for on–site strategical decisions and actions throughout each phase of a clean–up operation and maintains close liaison with the appropriate government agencies to obtain support and provide progress reports on each phase of the emergency response;

operating conditions – conditions wherein a unit is on location for the purpose of conducting drilling operations, and combined environmental and operational loadings are within the appropriate design limits established for such operations;

ORE – oil recovery efficiency – the percentage of oil in a recovered fluid;

ORR – oil recovery rate – the volume of oil recovered per unit time;

orthogonal – a line drawn perpendicular to water wave crests;

osmosis – the passage of liquids or gases from solution through semipermeable membranes;

otter – a semiaquatic mammal of the weasel family, feeding on fish and molluscs. It has thick, red–brown to dark–brown lustrous hair;

outfall – a structure extending into a body of water for the purpose of discharging waste water or sewage;

oxidation – the chemical combination of compounds such as hydrocarbons with oxygen. Oxidation is a process which contributes to the weathering of oil. However, in comparison to other weathering processes, oxidation is slow since the reaction occurs primarily at the surface and only a limited amount of oxygen is capable of penetrating the slick or surface oil. Sunlight will aid oxidation. This is called photo–oxidation, producing other compounds which may dissolve in water or disperse more quickly;

P

PAR – photosynthetically active radiation – electromagnetic radiation between 350 mm and 700 mm within which band plant photosynthesis occurs;

paraffin – a waxy substance obtained from the distillation of crude oils. Paraffin is a complex mixture of higher carbon number alkanes that is resistant to water and water vapour and is chemically inert;

particle – smallest possible quantity, elementary particle – one of the constituents of an atom, not yet known to be composed of simpler particles;

partitioning – process of dividing into parts or sections;

pebble – small stone made smooth and round by the action of water on the seashore;

pebble beach – a beach substrate composed primarily of gravel having a size range from 6 to 64 mm. Pebble substrates are finer than cobble and coarser than sand, and can allow stranded oil to penetrate to a considerable depth;

penalty – punishment for law–breaking in the form of imprisonment or payment of a fine;

penguin – a flightless, erect bird, adapted to swimming in cold seas and found mainly in the southern hemisphere;

permafrost – permanently frozen subsoil in the polar regions;

permeability of a space – percentage of a space which can be occupied by water;

pH – logarithm (base 10) of the reciprocal of the hydrogen ion concentration;

photosynthesis – process by which the energy of sunlight is used by a green plant to keep it growing;

phytoplankton – floating plant life of natural waters;

piezzoelectric effect – the interaction of mechanical and electrical stress–strain variables in a medium;

pit – a hole in the ground from which material is dug out;

planing – exposure of the bottom of the boom skirt due to excessive tow speed;

plenum – pressure slightly above atmospheric, occurring in air or gas systems as the result of action of fans or blowers;

plunging breaker – wave that breaks suddenly and with force;

pocket beach – a small beach between two barriers;

pollution – introduction by man directly or indirectly of substances or energy into marine environments, including estuaries, resulting in such deleterious effects as hazards to human health, harm to living resources and marine life, hindrance to legitimate uses of the sea, including fishing, impairment of the quality for use of seawater, and reduction of amenities;

polyethylene – a polymer (substance composed of very large molecules that are multiples of simpler chemical units) of the alkene, ethylene, which takes the form of a light weight thermoplastic. Polyethylene has high resistance to chemicals, low water absorption and good insulating properties and can be manufactured in a number of forms. Polyethylene also has high oleophilic properties and has been used with considerable success as a sorbent for oil spill clean-up;

polymorphic ice – ice having many stages of growth;

polyurethane – any of the class of synthetic resinous, fibrous or elastomeric compounds belonging to the family of organic polymers, consisting of large molecules formed by chemical combination of successive smaller molecules into chains or networks. The best known polyurethanes are the flexible foams used as upholstery material and mattresses, and the rigid foams used as lightweight structural elements including cores for aeroplane wings. Polyurethane is also the most effective sorbent that can be used for oil spill clean–up and unlike most synthetic sorbents it effectively recovers a wide range of different viscosity oils;

porosity – a measure of the space in a rock or soil that is not occupied by mineral matter. Porosity is defined as the percentage of total pore spaces, including all voids, whether or not they are interconnected, in the total volume of the rock. The porosity of the rock or soil is determined by the mode of deposition, packing of grains, compaction, grain shape and grain size. Porosity can also be used to refer to the voids in other materials such as in sorbents;

porpoise – sea animal rather like a dolphin or a small whale;

potential energy – energy waiting to be released;

pour point – the lowest temperature at which a substance such as oil will flow under specified conditions. The pour point of crude oils generally varies from -57 to +32°C; lighter oils with low viscosities have lower pour points. The pour point of an oil is important in terms of impact to the shoreline and subsequent clean–up since free–flowing oils rapidly penetrate most beach substrates, whereas semi–solid oils tend to be deposited on the surface and will only penetrate if the beach material is coarse or the ambient temperature is high;

precipitation – condensation of vapour into drops of liquid;

prism – solid figure with equal and parallel ends and with sides which are parallelograms;

proliferation – growth or reproduction by rapid reproduction

of cells or new parts;
prop roots – roots supporting mangroves;
propane – colourless gas (C_3H_8) in natural gas and petroleum;
protoplasm – colourless, jelly–like substance which is the material basis of life for animals and plants;
PTO shafts – power–take–off shafts at the front of diesel engines;
public spaces – portions of accommodation which are used for halls, dining rooms, lounges and similar permanently enclosed spaces;
puffin – a diving sea bird related to the auk;

Q

quicklime – unslaked lime;

R

range finder – instrument finding the distance of a target;
ray – the path along which the wave energy travels. For water waves it is the same as an orthogonal;
recovery – in an oil spill clean–up, the entire process or any operation contributing to the physical removal of spilled oil from land, water or shoreline environments. General methods of recovery of oil from water are the use of mechanical skimmers, sorbents and manual recovery by the clean–up work force. The main method of recovery of oil spilled on land or shorelines is the excavation of contaminated materials;
reef – an offshore navigation hazard of rock, coral or sand whose depth is less than 10 m. A sand reef is called a bar;
reflectance – ability to send back an image;
refraction – changing the direction of wave travel with decreasing depth of water with the tendency to parallel the depth contour; a cause of a ray of light to bend aside where it obliquely enters water or glass;
regional agreement – agreement between countries within a sea or ocean area to combine their combating forces in a large spill response;
remote sensing – the aerial sensing of oil on the water surface.

The primary applications of remote sensing are the location of an oil spill prior to its detection by any other means and the monitoring of the movement of an oil slick under adverse climatic conditions and during the night;

rescue boat – easily manoeuvered power boat capable of rapid launching and adequate for quick recovery of a man overboard and towing a life raft away from immediate danger;

residual oil – the oil remaining after fractional distillation during petroleum refining; generally includes bunker oils;

residuum – part of oil left after distillation;

respiration – single act of breathing, i.e. breathing–in and breathing–out;

response – any action undertaken to prevent, reduce, monitor or combat oil pollution;

ring – an eddy formed by a strong current but detached from it;

rip – a narrow strong current flowing outward from the shore, caused by waves;

runnel – brook, open gutter for rainwater at a roadside;

S

salinity – a measure of the quantity of dissolved salts in seawater;

sanitary landfill site – an approved disposal area where materials including garbage, oil–contaminated debris and highly weathered oil are spread in layers and covered with soil to a depth which will prevent disturbance or leaching of contaminants towards the surface. Sanitary landfill sites should always be located in areas where there is no potential for contamination of groundwater supplies. This method of disposal takes advantage of the ability of some bacteria and other microorganisms to biodegrade garbage and debris resulting from oil spill clean–up operations;

scanning microwave radiometer – a device used in aerial surveillance for detecting oil slicks in rainy or foggy conditions;

scattering – the effect that causes light to deviate from a straight line;

scavenging – removal of unwanted gases from a system i.e. of products of combustion from an internal combustion engine;
scoop – a deep, ladle–like tool for taking up or dipping out quantities of spilled oil;
scouring – making dirty surface clean by friction;
sea clutter – untidy sea surface due to confused surface waves;
sea urchin – a small sea animal with a shell covered with sharp points;
sedimentation – a process of settling the matter suspended in the liquid to the bottom;
sediments – a general term used to describe or refer to material in suspension in air or water; the total dissolved and suspended material transported by a stream or river; the unconsolidated sand and gravel deposits of river valleys and coastlines; materials deposited on the floor of lakes and oceans;
self–elevating unit – unit with movable legs capable of raising its hull or footings above the surface of the sea;
semi–enclosed locations – locations where natural conditions of ventilation are notably different from those on open decks due to the presence of structures such as roofs, windbreaks and bulkheads and which are so arranged that dispersion of gas may not occur;
sensitivity – quality of easily receiving impressions;
sensitivity maps – maps used by the On–Scene Commander and oil spill response team which designate areas of biological, social and economic importance in a given region. These maps often prioritise sensitive areas so that in the event of an extensive spill these areas can be protected or cleaned–up first. Sensitivity maps usually contain other information useful to the response team, such as the location of shoreline access areas, landing strips, roads, communities, and the composition and steepness of shoreline areas. Maps of this type often form an integral part of local or regional contingency plans;
service spaces – spaces used for galleys, pantries containing cooking appliances, lockers and store–rooms, workshops other than those forming part of machinery spaces, and similar spaces

and trunks for such spaces;
severe storm conditions – conditions wherein a unit may be subjected to the most severe environmental loadings for which the unit is designed. Drilling operations are assumed to have been discontinued due to the severity of environmental loadings;
sewage – drainage and other wastes from any form of toilets, urinals and WC scuppers, from medical premises, from spaces containing living animals and other waste waters when mixed with the drainages defined above;
shallow water wave – surface waves in shallow water ($\frac{H}{\lambda} < 0.05$) which are non–dispersive with phase and group speeds of $(gH)^{\frac{1}{2}}$;
sheet breakaway – a type of current–induced boom failure resulting from the fact that a boom placed in moving water tends to act like a dam. The surface water being held back by the boom is diverted downwards and accelerates in an attempt to keep up with the water flowing directly under the boom skirt and in so doing simultaneously draws oil from the surface under the boom. As a general rule, sheet breakaway will occur when current velocity exceeds 0.36 m/s, although skirt depth, oil viscosity, specific gravity, slick thickness and angle or placement of the boom relative to the current direction have a bearing on this form of boom failure;
shellfish – all kinds of molluscs and crustaceans having shells;
shingle – small, rounded pebbles on the seashore;
shoal – shallow place in the sea where there are some sandbanks;
shoreline – the line where water meets the shore;
shoreline sensitivity – the susceptibility of a shoreline environment to any disturbance which might decrease its stability or result in short or long term adverse impacts;
shoreline type – the average slope or steepness and predominant substrate composition of the intertidal zone of a shoreline area. In any given region, shoreline type may be used to assess the type and abundance of intertidal flora and fauna, protection priorities and most suitable oil spill clean–up strategies;
silt – sand, mud, etc. carried by moving water and left at the

mouth of a river, in a harbour, etc. Sediment particles range in size from 4 to 64 microns. Silt particles are larger than clays (4 microns) but smaller than sand (64 microns to 2 mm);

siltation – process of depositing sand and mud carried by water at the mouth of river or in a harbour;

sinking agent – a material which is spread over the surface of an oil slick to adsorb oil and cause it to sink. Common sinking agents include treated sand, fly ash and special types of clay. However, these materials are rarely used because they provide a purely cosmetic approach to oil spill cleanup and may cause considerable damage to bottom–dwelling organisms;

sinusoidal wave – any wave whose form is that of a sinusoid. Most linear waves are assumed to be sinusoidal;

skirt – the portion of a floating boom which lies below the water surface and provides the basic barrier to the spread of an oil slick or the loss of oil beneath the boom;

SLAR – Side Looking Airborne Radar, used in aerial surveillance;

slick – the common term used to describe a film of oil (usually less than 2 microns thick) on the water surface;

slop tank – tank containing dirty waste water;

solubility – the amount of a substance (solute) that will dissolve in a given amount of another substance (solvent). The solubility of oil in water is extremly low;

solute – substance that can be dissolved;

solvent – a liquid able to dissolve another substance (solute);

sooty residue – black substance left by smoke on surfaces;

sorbent – a substance that either adsorbs or absorbs another substance. Specifically, a natural organic, mineral–based or synthetic organic material used to recover small amounts of oil which have been spilled on land or stranded on shorelines;

sorbent barrier – a barrier which is constructed of or includes sorbent materials to simultaneously recover spilled oil during the containment process. Sorbent booms and barriers are only used when the oil slick is relatively thin since their recovery efficiency rapidly decreases once the sorbent is saturated with oil;

sorbent surface skimmer – skimmer with a surface of a different form to which oil adheres while the surface moves through the oil;

specific gravity – relation between the weight of a substance and that of the same volume of a standard substance (water for liquids and solids, air for gases);

spreading – extending the surface;

spreading coefficient – difference between the air/surface tension;

stern – rear end of the ship;

strainer – sieve or other device, by means of which solid matter is separated from liquid;

stratification – arrangement in strata;

submarining – complete submergence of the boom due to excessive tow speed;

submersion skimmer – a skimmer with a moving belt inclined at an angle to the surface forcing the oil beneath the water surface. Subsequently the oil rises due to its buoyancy into a collection well;

substrate – material or materials which form the base of something. In biology, the base on which an organism lives. Substrate materials include water, soils and rocks as well as other plants and animals;

substratum – level lying below another;

subtropics – climatic regions bordering on the tropics;

suction skimmer – skimmer drawing oil surface layer through an orifice into a vacuum tank;

sullage – wastewater from storms, baths etc., that does not require sewage treatment;

support agency – one of the agencies participating in drafting individual chapters of a contingency plan and assigned to specific tasks in support of an oil spill response;

surface acting agents – chemicals which when added to a fluid alter the forces of surface tension between two adjacent molecules. Surface active agents generally decrease the surface tension of a fluid such as an oil and are used to facilitate its dispersion throughout the water column or prevent it from adhering to solid surfaces such as pilings, wharfs and rocks on the shore;

surface tension – the force of attraction between the surface molecules of a liquid. Surface tension affects the rate at which spilled oil will spread over a land or water surface or into the ground. Oils with low specific gravities are often characterised by low surface tension and therefore faster spreading rates;

surface tension coefficient – difference between air/water surface tensions;

surface unit – a unit with a ship or barge-type displacement hull of single or multiple hull construction intended for operation in floating condition;

surface wave – a wave on the surface of the water;

surf zone – zone of the seashore where the waves are breaking (in white foam);

surging breaker – breaker moving forward;

survival craft – craft capable of removing persons from a unit to be abandoned and capable of sustaining persons until retrieval is completed;

swash – the rush of water onto the beach following the breaking of a wave;

swathe – space left clear after one passage;

swell waves – surface gravity waves with a period of about 10 seconds generated by distant storms;

synthetic organic sorbent – one of several polymers, generally in the form of plastic foams or plastic fibres, used to recover spilled oil. Synthetic organic sorbents have higher recovery capacities than either natural organic sorbents or mineral-based sorbents and many of these materials can be reused after the oil is squeezed out;

T

tar – a black or brown hydrocarbon material that ranges in consistency from a heavy liquid to a solid. The most common source of tar is the residue left after fractional distillation of crude oil;

tar balls – compact semi–solid or solid masses of highly weathered oil formed through the aggregation of viscous, high carbon number hydrocarbons with debris present in the water column. Tar balls generally sink to the sea bottom but may be deposited on shorelines where they tend to resist further weathering;

TE – throughput efficiency – the ratio of oil recovery rate to oil distribution rate, as a percentage;

tension member – the part of a floating containment boom which carries the load placed on the barrier by wind, wave and current forces. Tension members are commonly constructed from wire cable due to its strength and stretch resistance;

tidal prism – the total volume of water that flows into a harbour or estuary with the movement of the tide (excluding any freshwater flow);

tidal wave – an obsolete name for tsunami;

tide pools – permanent depressions in the substrate of intertidal zones which always contain water but are periodically flushed with successive incoming tides. Tide pools are most frequently located near the high tide mark and often contain abundant flora and fauna which can be adversely affected when spilled oil becomes stranded in these areas;

topography – description of the features of a place or of a district;

toxicity – the capability of a poisonous compound (toxin) to produce deleterious effects in organisms such as alteration of behavioural patterns or biological productivity (sublethal toxicity) or, in some cases, death (lethal or acute toxicity). The toxic capability of a compound is frequently measured by its "acute LC50" with a standard test organism such as rainbow trout. The acute LC50 of a toxic chemical is the concentration which will result in death in 50% of the test organisms over a given time period, usually 96 hours. The most immediately toxic compounds in crude oils or refined petroleum products are the aromatics such as benzene;

transit conditions – conditions wherein a unit is moving from one geographical location to another;

trawl net – large, wide–mouthed net to be dragged along or beside a ship;
tsunami – a long wavelength surface gravity wave generated by earthquakes. The breaking of a tsunami upon a coast generally causes great damage;
tubular ice – ice consisting of crystals having the shape of tubes;
turbidity – the cloudiness in a liquid caused by the presence of finely divided, suspended material;
turbulent flow – disorderly flow;

U

ultraviolet radiation – the portion of the electromagnetic spectrum emitted by the sun which is adjacent to the violet end of the visible light range. Often called "black light", ultraviolet light is invisible to the human eye, but when it falls on certain surfaces, it causes them to fluoresce or emit visible light. Ultraviolet light is responsible for the photo–oxidation of certain compounds, including hydrocarbons, although this process is limited to a large extent by low penetration ability (in water, air or soil) of this short wavelength form of energy;

V

vaporizer – apparatus for obtaining vapour from a liquid;
viscosity – the property of a fluid (gas or liquid) by which it resists a change in shape or movement;
volatile compounds – compounds that easily change from liquid to gas or vapour;
volatility – the tendency of a solid or liquid substance to pass into the vapour state. Many low carbon number hydrocarbons are extremely volatile and readily pass into a vapour state when spilled;
vortex skimmer – skimmer working on the principle of creating a water vortex, in which oil is separated from water;
vorticity – rotational motion formed either by fluid spin or shear;

W

water–in–oil emulsion – a type of emulsion where droplets of water are dispersed throughout oil, formed when water is mixed with a relatively viscous oil by wave action. In contrast to oil–in–water emulsions, this type of emulsion is extremely stable and may persist for months or years after a spill, particularly when deposited in shoreline areas. Water–in–oil emulsions containing 50 to 80% water are most common, have a grease–like consistency and are generally referred to as "chocolate mousse";

water table – the fluctuating upper level of the water-saturated zone (groundwater) located below the soil surface;

wax – any of a class of pliable substances of plant, animal, mineral or synthetic origin. Waxes generally consist of long–chain organic compounds. Waxes are included in the residue formed following the refining of crude oil;

weathering – the alteration of the physical and chemical properties of spilled oil through a series of natural processes which begin when the spill occurs and continue indefinitely, while the oil remains in the environment. Major processes which contribute to weathering include evaporation, dissolution, oxidation, emulsification and microbial degradation;

watertight – capable of preventing the passage of water through a structure in any direction under a head of water for which the surrounding structure is designed;

weathertight – means that in any sea conditions water will not penetrate into the ship;

weir skimmer – a type of skimmer which employs the force of gravity to drain oil from the water surface. Basic components include a weir or dam, a holding tank and an external pump. As oil on the water surface falls over the weir or is forced over by currents into a holding tank, it is continuously removed by the pump;

wicking agent – woven material by which oil is drawn up to burn; it may be a substance such as straw, wood chips, glass beads and treated silica which are used to increase oxygen availability and provide insulation between oil and water during the disposal of spilled oil by burning;

wind herding – air flowing into the flames of burning oil induces

a surface water current, which opposes spreading of the slick;
windrows – rows of visible matter on the surface of choppy water aligned parallel to the wind;
wind waves – any wave generated by the wind. The term is generally used for short period (about 1 s) surface gravity waves directly generated by wind;
WSF – water soluble fraction;

Z

zooplankton – minute animal life found passively floating or weakly swimming inside natural waters;

Selected Bibliography, Further Reading List

The reader should note that the O.S.C. and the O.T.C. references are selected from the specific meetings as being representative of the two conference series.

O.S.C. – Proceedings of Oil Spill Conference, held biennially in Baltimore, Maryland, U.S.A.

O.T.C. – Proceedings of Offshore Technology Conference, held annually in Houston, Texas, U.S.A.

1. *Action plan for marine pollution.* National Swedish Environmental Protection Board. Stockholm 1987
2. Alejandro A.C., J.L.Buri, 1987. *M/V ALVENIUS: An anatomy of a major spill.* O.S.C. p.27
3. Alexander S.K., J.W.Webb, Jr., 1987. *Relationship of Spartina Alterniflora growth to sediment oil content following an oil spill.* O.S.C. p.445
4. Allen T.E., 1985. *New concepts in spraying dispersants from boats.* O.S.C. p.3
5. Baca B.J., T.E.Lankford, E.R.Gundlach, 1987. *Recovery of Brittany Coast marshes in the eight years following the AMOCO CADIZ incident.* O.S.C. p.459
6. Backlund Lars, Holmström Lars. 1983. *Second generation oil spill and marine surveillance systems now operational in Sweden.* O.S.C. p.349
7. Baltic Marine Environment Protection Commission - Helsinki Commission. Selection of various meeting documents.

8. Beach R.L., W.T.Lewis. 1983. *Testing of a prototype waste oil flaring system.* O.S.C. p.29
9. Bender K., R.Taylor.1987. *Oil spill contingency planning in Thailand.* O.S.C. p.161
10. Beraud A., Sainlos J.C., 1985. *Accidental marine pollution. French policy and response.* O.S.C. p.97
11. Biggers R.J. 1987. *Oil spill contingency planning at navy shore activities.* O.S.C. p.193
12. Blanchi A. 1981. *The economics of oil spill clean up.* Marine Technology Vol.18 No.4 p.335
13. Bocard C., J.J.Quinquis. C.Such. 1987. *A mobile plant prototype for restoration of polluted beaches by washing oily sand.* O.S.C. p.61
14. Borst M., H.W.Lichte. 1985. *Standardizing boom test procedures.* O.S.C. p.41
15. Brodie D. 1987. *Oil pollution response arrangements in Australia: the government view (including an update on dispersant testing).* O.S.C. p.181
16. Buist I.A. 1987. *A preliminary feasibility study of in-situ burning of spreading oil slicks.* O.S.C. p.359
17. Buist I.A., S.L.Ross. 1987. *Emulsion inhibitors: a new concept in oil spill treatment.* O.S.C. p.217
18. Buist I.A. et al. 1983. *The development and testing of a fireproof boom.* O.S.C. p.43
19. Canevari G.P. 1987. *Basic study reveals how different crude oils influence dispersant perfomance.* O.S.C. p.293
20. Charton B. 1988. *Dictionary of marine science.* Facts on File Publications. New York
21. Christodoulou M.S. 1987. *Experimental study and improvement of the rotating disc skimmer.* O.S.C. p.101
22. Clauss G. 1986. *Hydrodynamic problems of offshore skimming systems.* Marine Technology No.2. p.69
23. Clay J.S. 1985. *Trends in stranded tanker salvage.* O.S.C. p.177
24. *Coastal and marine management in Sweden.* The Swedish Environmental Protection Board

25. Cohen S., S. Dalton. 1983. *Self-contained oil recovery system for use in protected waters.* O.S.C. p.73

26. Concave. 1981. *A field guide to coastal oil spill control and clean up techniques.* Report No.9/81

27. Concave. 1983. *A field guide to inland oil spill clean-up techniques.* Report No.10/83

28. Concave. 1983. *Characteristics of petroleum and its behaviour at sea.* Report No.8/83

29. *Convention on the protection of the marine environment of the Baltic Sea Area, 1974.* (Helsinki Convention). Edition 1987

30. Cormack D. et al. 1987. *Remote sensing techniques for detecting oil slicks at sea - a review of work carried out in the United Kingdom.* O.S.C. p.95

31. Crocker M.J. 1985. *Design considerations for a large sweep width skimming system.* O.S.C. p.47

32. Cubit J.D. et al. 1987. *An oil spill affecting coral reefs and mangroves on the Caribbean Coast of Panama.* O.S.C. p.401

33. Dalton W.J., Heikamp A.J. 1985. *Early experiences with a single-vessel offshore spill clean-up system.* O.S.C. p.57

34. Delikat D.S., T.P.Mackey. 1984. *Marine industry standards: A general with particular reference to pollution abatement equipment.* Marine Technology. Vol.21. No.3

35. Delvigne Gerard A.L. 1987. *Netting of viscous oil.* O.S.C. p.115

36. Dietzel K. *Öl Barriere aus Superplastischem Polyurethanweichschaum.* IMT 82-312/01 p.558

37. Doherty Chad B. et al. *Open ocean pollution response - the Coast Guard system.* O.T.C. 3703 . p.245

38. Edwards D.T. 1985. *International and regional arrangements on cooperation in combating marine pollution.* O.S.C. p.201

39. Environment Canada. 1983. *A catalogue of oil skimmers.*

40. Evans C.W. 1985. *The effects and implications of oil pollution in mangrove forests.* O.S.C. p.367

41. Fast O. 1985. *Monitoring an oil spill experiment with the Swedish maritime surveillance system.* O.S.C. p. 597
42. Fast O. 1987. *Swedish Coast Guard starts using third generation maritime surveillance system.* O.S.C. p. 137
43. Fingas M.F., W.S.Duval, G.B.Stevenson. 1979. *The basics of oil spill clean-up.* Environmental Emergency Branch. Environment Canada
44. Fischel M. 1987. *Preliminary assessment of the effect of an oil spill on a Louisiana Marsh.* O.S.C. p.489
45. Fraser J.P. 1985. *Advance planning for dispersant use/non-use.* O.S.C.
46. Fraser J.P., L.M.C. Clark. *SOCK skimmer - performance and field tests.* O.T.C. 4417
47. Fries O., H-G.Hattendorf, B.Arndt, A.Westram. 1982. *Doppelrumpf-Ölfangschiff "Thor".* Hansa 18/1982 p.1152
48. Frink L. 1987. *An overview: rehabilitation of oil contaminated birds.* O.S.C. p.479
49. Gates D.C., K.M.Corradino. 1985. *OHMSETT tests of the Toscon weir skimmer and gravity differential separator.* O.S.C. p.35
50. Goodman R.H., J.W.Morrison. 1985. *A simple remote sensing system for the detection of oil.* O.S.C. p. 51
51. Gundlach E.R., T.W.Kana. 1985. *Modeling spilled oil partitioning in nearshore and surf zone areas.* O.S.C. p. 379
52. Haegh T., L.I.Rossemyr. *A comparison of weathering processes from the Bravo and the IXTOC blow-outs.* O.T.C. 3702 p.237
53. Helsinki Convention 1974-1984 - *Goals and achievements.* Helsinki 1984
54. Hillman S.O., R.V.Shafer. 1983. *Absorb: a three year update in Arctic spill response.* O.S.C. p.219
55. Hornby A.S. 1980. *Oxford advanced learner's dictionary of current english.* Oxford University Press
56. Howard S., D.I.Little. 1985. *Biological effects of low-pressure seawater flushing of oiled sediments.* O.S.C. p.439

57. Howard S., D.I.Little. 1985. *Effects of infaunal burrow structure on oil penetration into sediments.* O.S.C. p. 427

58. Huse E. *Tensile strength and elasticity of offshore oil booms.* IMT 80-317/01 p. 641

59. Hühnerfuss M. 1981. *On the problem of sea surface film sampling.* Marine Technology 12 No.5

60. IMO - *Manual on oil pollution.* Section IV - Practical information on means of dealing with oil spillages. London. February 1980

61. IMO - Selection of various meeting documents

62. IMO/UNEP - *Guidelines on oil spill dispersant application and environmental considerations.* London 1982

63. IMO/UNEP - Meeting on regional arrangements for cooperation in combating major incidents of marine pollution. 29.04.-3.05.1985

64. Jackson L.F., S.R.Lipschitz. 1985. *Sensitivity mapping: an aid to contingency planning on Southern African shores.* O.S.C. p.223

65. Kaiser T., et al. 1987. *Canadian - U.S. spill response cooperation along the Great Lakes.* O.S.C. p.177

66. Kastner S. 1985. *Current status on vessel and equipment for cleaning of marine oil spill pollution with emphasis on design and hydrodynamic problems.* Journal of advanced technology transfer. p.6

67. Kerley G.I.H. 1985. *The management of oiled penguins.* O.S.C. p.465

68. Kleij A.N., J.M.Gubbens. 1985. *Case history of a South Holland oil spill - organisation and cooperation.* O.S.C. p.293

69. Klimek R. 1981. *Ölskimmer in Kombination mit einem Hoppersaugbagger.* Meerestechnik 12 Nr.4

70. Klimek R., G.Claus. 1983. *Entwicklung eines Ölaufnahme- und -trennsysstems zum Einsatz auf See.* Marine Technology 14. No.1 p.9

71. Koblanski J.N. 1983. *An acoustical method of burning and collecting oil spills on cold open water surfaces.* O.S.C.

p.25

72. Kretschmer D., J.Odgers. 1985. *Combustability and incineration of Beaufort crude-seawater emulsions.* O.S.C. p.19
73. Kruk K.F. 1983. *Air curtain incinerator tests.* O.S.C. p.33
74. Kruth D.J. 1987. *Protecting an island's drinking water and desalination plant.* O.S.C. p.49
75. Lanau G.A., et al. *Arctic hydrocarbon transportation systems in the twenty-first century.* MTS Journal Vol.18.No.1
76. Lane P.A. et al. 1987. *Impact of experimentally dispersed crude oil on vegetation in a Northwestern Atlantic salt marsh - preliminary observations.* O.S.C. p.509
77. Le Guen Y.J.F., M.Brussieux, R.Burkhalter. 1987. *Real-time processing of oil spill remote sensing data.* O.S.C. p.71
78. Leigh J.T., W.C.Park III. 1983. *Mobile command and communications systems.* O.S.C. p.111
79. Lindblom G.P., B.S.Cashion. 1983. *Operational considerations for optimum deposition efficiency in aerial application of dispersants.* O.S.C. p.53
80. Lindstedt-Siwa J. *Oil spill response and ecological impacts - 15 years beyond Santa Barbara.* MTS Journal Vol.18. No.3 p.43
81. Little D.I., D.L.Scales. 1987. *Effectiveness of a type III dispersant on low-energy shorelines.* O.S.C. p.263
82. Little D.I., D.L.Scales. 1987. *The persistence of oil stranded on sediment shorelines.* O.S.C. p.433
83. Lord D.A. et al. 1987. *The "KAPODISTRIAS" and oil spill Cape Recife, South Africa.* O.S.C. p.33
84. Lühring C. *Doppelrumpf-Mehrzweckschiff für den Einsatz auf See als Ölfangschiff.* IMT 80-315/01 p.622
85. Lytle J.S., T.F.Lytle. 1987. *The role of JUNCUS ROEMERIANUS in cleanup of oil-polluted sediments.* O.S.C. p.495
86. MacGregor Navire *Oil Recovery Catalogue*

87. MacKay D., A.Chan, Yiu C.Poon. 1986. *A study of the mechanism of chemical dispersion of oil spills.* Environmental Service. Canada
88. Mackay D., P.G.Wells. 1983. *Effectiveness, behaviour and toxicity of dispersants.* O.S.C. p.65
89. MacNeill et all. 1985. *Dispersant tests in a wave basin.* O.S.C. p.463
90. Marchard G. et al. 1987. *French know-how in the prevention and fight against accidental oil spills.* O.S.C. p.15
91. McAuliffe C.D. 1987. *Organism exposure to volatile/ soluble hydrocarbons from crude oil spills - a field and laboratory comparison.* O.S.C. p.275
92. Meikle K.M. 1983. *An effective low-cost fireproof boom.* O.S.C. p.39
93. Meikle K.M., H.Whittaker, F.Laperrire. 1985. *An experimental high pressure waterjet barrier.* O.S.C. p.7
94. Menagie H.M. 1980. *Bestrijdingvan olieverontreininging op zee.* Shep en Werf 23/1980 p.379
95. Meyers R.J., J.L.Payne, C.B.Strong. 1987. *Oil spill training : the need continues.* O.S.C. p.565
96. Miller J.A. 1987. *Beach agitation for crude oil removal from intertidal beach sediments.* O.S.C. p.85
97. Moller T.H., H.D.Parker, J.A.Nichols. 1987. *Comparative costs of oil spill clean-up techniques.* O.S.C. p.123
98. Morris P.R. 1985. *Recovery of viscous emulsions from a firm sandy beach.* O.S.C. p.193
99. National Swedish Environmental Protection Board. *Action plan for marine pollution.*
100. Neralla V.R., S.Venkatesh. 1985. *Real time application of an oil spill motion prediction system.* O.S.C. p.235
101. Oebius H.U. 1985. *Physical requirements of oil combating equipment.* Marine Technology vol.18 No.1
102. *Oil-Crab - a new approach to oil spill removal.* Holland Shipbuilding. July 1984
103. Owens E.H. 1985. *Factors affecting the persistence of stranded oil on low energy coasts.* O.S.C. p.359

104. *Ölunfallbekämpfung im See- und Küstengebiet der Bundesrepublik Deutschland.* Hansa 6/1985
105. Pasquet R., J.Denis. 1983. *New Developments in beach clean-up techniques.* O.S.C. p.279
106. Payne J.R., G.D.McNabb Jr. *Weathering of petroleum in the marine environment.* MTS Journal Vol 18 Nr.3 p.24
107. Peigne G. 1985. *Ecumoire II : evaluation of three oil recovery devices.* O.S.C. p.13
108. Pistruzak W.M. 1981. *Dome petroleum's oil spill research and development program for the Arctic.* O.S.C. p.173
109. Pope A. et al. 1985. *Assessment of three surface collecting agents during temperate and Arctic conditions.* O.S.C. p.199
110. Potter S.G., S.L.Ross, L.C.Oddy. 1987. *The development of a Canadian oil spill training program.* O.S.C. p.577
111. Pu Bao-Kang. 1987. *Analysis of significant oil spill incidents from ships, 1976-1985.* O.S.C. p.43
112. Rapsch H-J., H-H.Münte, O.Fries. *Doppelrumpf-Ölfangschiff "Bottsand".* Hansa 4/1985 p.334
113. Rapsch H-J., H-H Münte, O.Fries. *Doppelrumpf-Ölfangschiff "Bottsand".* Schiff & Hafen 2/1985 p.46
114. Ross S.L. 1986. *In situ burning of uncontained oilslicks.* Ottawa July
115. Sasamura Yoshio. 1985. *Implementation of MARPOL 73/78.* O.S.C. p.121
116. Scholten M., J.Kuiper. 1987. *The effects of oil and chemically dispersed oil on natural phytoplankton communities.* O.S.C. p.255
117. Schriel R.C. 1987.*Operational air surveillance and experiences in the Netherlands.* O.S.C. p.129
118. Schulze R., I.Lissauer. 1985. *An overview of a field guide for Arctic spill behaviour.* O.S.C. p.399
119. Shafer R.V. 1987. *Alaska clean seas: meeting response needs for 1987 and beyond.* O.S.C. p.145
120. Shum J.S., M.Borst. 1985. *OHMSETT tests of a ropemop skimmer in ice-infested waters.* O.S.C. p.31

121. Sleeter T.D., A.H.Knap, I.W.Hughes. 1983. *Oil spill contingency planning and scientific support coordination in Bermuda: a successful model.* O.S.C. p.149

122. Smith J.B.H., C.McLellan, L.R.Pintler. 1987. *Development of an oil skimming system to meet Navy specifications.* O.S.C. p.91

123. Smith N.K., A.Diaz. 1985. *In-place burning of Prudhoe Bay oil in broken ice.* O.S.C. p.405

124. Solsberg L.B. 1983. *A catalogue of oil skimmers.* Economic and technical review report EPS 3-EP-83-1. Environmental Protection Programs Directorate.

125. Spitzer J.D. 1985. *Developing marine pollution response capability in the Wider Carribbean Region.* O.S.C. p.127

126. Stacey M.L. 1985. *Marine pollution contingency planning - recent changes in United Kingdom organisation.* O.S.C. p.89

127. Stacey M.L. 1987. *United Kingdom marine pollution contingency planning - a review of the last two years.* O.S.C. p.167

128. Stamp P.S. 1984. *Present state on regional cooperation on oil spill control: Northern Europe.* International Symposium on Regional Cooperation on Oil Spill Prevention and Combating. Copenhagen Sept.1984

129. Swiss J.J., D.J.Smrke, W.M.Pistruzak. 1985. *Unique disposal techniques for Arctic oil spill response.* O.S.C. p.395

130. Suzuki Isao, Y.Tsukino, M. Yanagisawa. 1985. *Simulation tests of portable oil booms in broken ice.* O.S.C. p.25

131. Suzuki Isao, K.Miki. 1987. *Research and development of oil spill control devices for use in cold climates in Japan.* O.S.C. p.349

132. Svenska Kommunförbundel. *Beredskap för oljeutsläpp till sjöss.*

133. Taylor J. et al. *Environmental protection in the Beatrice Field development.* O.T.C. 4419 p.481

134. Teasdale G. 1987. *Royal Navy oil spill response posture - development since the mid 1970s.* O.S.C. p. 171

135. The International Tanker Owners Pollution Federation Ltd. *Aerial Application of oil spill dispersants.* Technical Information Paper No.3
136. The International Tanker Owners Pollution Federation Ltd. *Response to marine oil spills.* London 1986
137. The International Tanker Owners Pollution Federation Ltd. *Use of booms in combating oil pollution.* Techn. Information Paper No.2-1981
138. The International Tanker Owners Pollution Federation Ltd. *Use of skimmers in combating oil pollution.* Techn. Information Paper No.5-1983
139. The International Tanker Owners Pollution Federation Ltd. *Recognition of oil on shorelines.* Techn. Information Paper No.6-1983
140. The International Tanker Owners Pollution Federation Ltd. *Disposal of oil and debris.* Technical Information Paper No.8-1984
141. The International Tanker Owners Pollution Federation Ltd. *Effects of marine oil spills.* Technical Information Paper No.10-1985
142. TOBOS 85 - The Swedish Oil Combat Program - selection of documents
143. Trudel B.K., S.L.Ross. 1987. *Method for making dispersant-use decisions based on environmental impact considerations.* O.S.C. p.211
144. U.S.Coast Guard. *National oil and hazardous substances pollution contingency plan.* 40 CFR Part 300 - 1985
145. U.S.Coast Guard. *Policy guidance for intervention in ship related marine pollution incidents on the high seas and on navigable waters of the U.S.* Comdtinst. 16451.5 -1983
146. U.S.Coast Guard. *Policy guidance for response to hazardous chemical releases.* Comdtinst.M.16465.30 - 1984
147. Walker J.D. *Chemical fate of toxic substances: biodegradation of petroleum.* MTS Journal Vol.18 No.3 p.73
148. Waterworth M.D. 1987. *The laser ignition device and its application to oil spills.* O.S.C. p.369

149. Webb C.L.F. 1985. *Offshore oil production in the Baltic Sea: a coastal sensitivity study.* O.S.C. p.99

150. Whittaker H. 1987. *Laser ignition of oil spills.* O.S.C. p.389

151. Vaca-Torelli M., A.L.Geraci, A.Risitano. 1987. *Dispersant application by hydrofoil: high speed control and clean-up of large oil spills.* O.S.C. p.75

152. *Van Nostrand's Scientific Encyclopedia.* D.Van Nostrand Company Inc. Princeton. 1968

153. Veenstra F.A. 1981. *Design requirements and the design of an oil-recovery vessel for the North Sea.* Ship en Werf 8/1981

154. Vejledning fra miljstyrelsen - Strandvensning II - 1982

155. Yazaki Atsuo. 1983. *Research and development in the Institute of Ocean Environmental Technology.* O.S.C. p.95

List of Tables

Index